THE TOTAL CONNECTION

THE TOTAL CONNECTION

C. Merrill Johnson

Edited by Edna Kimery Johnson

Royal Books
15865-B Gale Avenue, Suite 823
Hacienda Heights, California 91745

Published by:
Royal Books
15865-B Gale Avenue, Suite 823
Hacienda Heights, California 91745
(818) 330-8114

Manufactured in the United States of America by Lowell Moss Publishing Services.

Library of Congress Cataloging in Publication Data

Johnson, C. Merrill (Clifford Merrill), 1914 -
The total connection.

Includes bibliographical references and index.
1. Religion and science—1946— . 2. Eschatology.
3. Bible—Prophecies. 4. Cosmogony. I. Johnson, Edna Kimery. II. Title.
BL240.2.J545 1984 261.5'5 84-17712
ISBN 0-916821-00-5
ISBN 0-916821-01-3 (pbk.)

Dedicated in memory of my Mother

HILDUR MARY JOHNSON

Special thanks to

BERNICE STOREY

For her contribution to this project

King James Version is used except where otherwise indicated.

ACKNOWLEDGEMENTS

Our appreciation to the following for permission to use materials from their publications:

MACMILLIAN PUBLISHING COMPANY, publisher of the Phillip's translation.

HARPER AND ROE PUBLISHERS, publisher of the Moffatt translation.

THE HEARST CORPORATION, publisher of the SCIENCE magazine.

MARÇIA F. BARTUSICH, Science Writer.

ASTRONOMY magazine.

Science without religion is lame; religion without science is blind.
—Albert Einstein

Theologians and scientists fulfill two profound needs in man—the need to know and the need to not know, but to believe.
—The Author

Let me be weighed in an even balance . . .—Job 31:6

"That they may see, and know, and consider, and understand TOGETHER, that the hand of the Lord hath done this...
"Produce your cause, saith the Lord; shew us what shall happen: let them shew the former things, what they be, that we may consider them, and know the latter end of them; or declare us things for to come" (Isaiah 41:20-22).

"God...appointed...the things that are coming, and shall come..."
(Isaiah 44:6:7).

Table of Contents

Foreword xxix

Song and Dance xxx, A Total Connection xxxi, Let the Bible Speak xxxi

Preface xxxv

Self-Appointed Prophets xxxvi, Fry or Freeze xxxvi, The Lineup xxxvii, The Mushroom Cloud xxxvii

1 A Book of Hope 3

Escaping the Inferno 4

2 Social Justice 7

Equality for Unequals 9, Prejudication 9, Can God Control Evil? 11, When Deity Sits on the Bench 11

3 The Eternal Footstool of God 13

Everlasting Real Estate 14, Everlasting Mountains and Hills 14, The Eternal City of Jerusalem 15, The Indestructible Jew 15, Israel Shall Exist Forever 15, Land Promised Israel is Eternal 15, Endless Monarchy 16, Eternal Levitical Priesthood 16, Eternal Festivals 16, Permanency of Hell 17, Millennial Kingdom Extends Forever 18, Millennial Renewal Lasts Forever 18, The Eternal King of Kings 19, Eternal Aristocracy 20, Gigantic Waterhole 20

4 Before the World Began 23

The Creator's Coattails 23, Was God Sitting On His Hands? 25, Angels' Islands 25, The King's Palace 27, Palaces Within the Eternal City 27

5 The Three Heavens 29

Genesis II 29, God's Stars 30, Jesus Christ, the Enterprenuer 31, No

Mansions in the Sky 32, A Place—Not a House 33, When Death is Buried 34

6 Generations of the Heavens 35

Wingless Space Fleets 35, Giant Oysters and Strange Gold 36, A Summarized History of the Universe 36, Denizens of the Atomic Deep 37, Factories in Space 38, Brightness of Pure Energy 39, A Dateless Universe 39, The Cosmic Incubator 40

7 God's Cyclotron 41

Wheels Within Wheels 41, Celestial Weavers 42, Big Bangs 42

8 Graveyards in the Cosmos 45

Pinholes in Space 45, A Handful 47, Why the Earth Must Be Redeemed 47, Only the Originals Are Eternal 48

9 Wombs in the Universe 49

Celestial Gardens 51, Cosmic Chamber 51, Celestial Nurseries 51, Celestial Time Bombs 52, God's Atom Smasher 53, A Crisis Generation 53, Each Atom an Armed Camp 54

10 Flying Mountains 57

Intergalactic Plazas 58

11 Vanguard of the Godhead 61

Oasis in Eternity 61, Cradle Throne 62, Spiritual Rebirths 63, Spiritual Development 63, Witnesses for Christ 63, The True Church—Where? 64, Earth Gets a Facelift 65

12 The Angel of the Lord 67

The Unforgiving Angel 68, A Secret Name 68, Theophany of the Holy Spirit 69

13 What is Eternity 71

Curvature of the Universe 72, Ouasars 72, Unlimited Riches 73, Heavenly Cosmetics 74, The Bible Says What It Means 75

14 Light in Its Natural Environment 77

Why Light Bends 78, Source of Life Within Particles 79, A Nonconsumable Creator 79, Gaze and Blaze 80, Sonlight 80, Therapeutic Light 81, Faith is No Copout 81, Seven Lamps of God 82, Bodies of Ultra-Matter 82, New Dimensional Senses 83, Hostages of Mortality 84, The Mortality of Time 85, The Mortality of Life 86, Diamonds Are Not Forever 87, Removing the Stinger, Death 87

15 Doctrine of Unified Creation 89

The Original Creation 90, The Precelestial Earth 90, A Dispensational Day 90, Creation of the First Living Creatures 91, The Glowing Sky 91, Walking Ghosts 93, Generations of the Earth 94, Animal Pets 95, Mankind in Both the Image and Likeness of God 96, Final Curtain on Ancient History 96, The Big Refrigerator 98, Angelic Beings 98, Listening to Bones 98, The Mythical Unicorn 99, An Audience in Hell 99, God's Footstool: A Tombstone 100

16 Sons of the Morning 103

Pavement With Stones of Fire 104, The Prince's Wardrobe 104, Orchestra Hall 105, Earth's Original Cities 105, A Proud and Jealous God 106, Poisons of Pride and Jealousy 107

17 Sons of God 109

God's Only Begotten Son 109, Original Sons of God 110, Redeemed Sons of God 110, Mystic of Angels 111, Why Be a Sure Loser? 112, The Canon is Complete 112, Millennium and Prophets 113

18 Lucifer Blew It 115

Eternal Spirit Beings 116, Propaganda Minister 116, Lords of the Deep 117, Why Not More Visions? 117, Why Angels Cannot Be Forgiven 119

19 The Genesis War 121

Hell on Earth 121, No Honor Among Thieves 122, When the Earth Literally Died 123, The Missing Link is Found 123, Heavens' First

Blackout 124, First Black Hole in Space 125, What is Death 126, Time for Redemption 126

20 God's Olive Branch 127

Scenario of Events 127, Temptation and Sin 129, Rattling of Sabres 129, Survival of Paradise 130, A Powerful Kingdom Falls 130, Confrontation of Angels 130

21 The Wrath of God 133

Evolution in Reverse Gear 133, Frankenstein's Laboratory 134, Globe Trotting Ghosts 135

22 Resurrection of the Earth 137

Stones of Darkness 137, Earth in Suspension 138, Incorrigibles 139, God's Terrestrial Child 139, Spotlight from Heaven 139

23 The Book of Life 141

Handful of Shadowy Figures 141, God's Telescopic Vision 142, Man's Self-Destructive Faculty 142, God Holds an Election 142, Election of Grace 143, Chosen Ones 143, The Calling of God 144, Predestination 145, Hope of the Redeemed 145, A Mature Bride 146, Have Faith in God 146

24 God Ties His Own Hands 147

Umbrella Over the Gentiles 148

25 God's Arsenal 151

Earth's Canopy 151, Millions of Artesian Wells 152, God's Future Arsenal 153

26 The Craftsman at Work 155

Resurrection in the Animal Kingdom 155, Atoms Put to Rest 156, In the Likeness of God 157, Craftsman at Work 157, God the Physiological Botanist 158, Delivery From the Curse 158

27 Name Calling 161

Clonings 162

28 Principalities of Angels 163

Gates of Hell 163, Celestial Deserts 164, Prince of Princes 164

29 Celestial Hatcheries 167

Stars in Your Eyes 168, In the Throes of Death 168, Casting Shadows 169, Sounds of Silence 170, Heavenly Medleys 172

30 Princes in Heavenly Principalities 173

God's Envoys 173, The Great Prince 174, Living Luminaries 175, Star of Bethlehem 175, An Angelic Choir 175, Morning Stars 176, Christ Outshines Them All 176, Cosmic War 177

31 Original Paradise 179

In Full View of God's Throne 180, A Sanctuary 180, First Human Arrivals 181

32 Paradise on the Earth 183

All Flesh is Not the Same 183, Paradisiacal Body 183, Corporeal Body 184, Death in Life 185, Life in Death 185, Resurrection of the Body 186, A Reserve Body 187, All Must Be Changed 188, Garments of Salvation 188, Judged in the Flesh 189, Tree of Knowledge 189

33 Paradise Within the Earth 191

Soul Survival 192, Prince of Life 194

34 Chariots Are Coming 197

Mystery of the Living Dead 198, Pathway Through the Stars 198, Efficacious Atonement 199, Baptizing and Prayers For the Dead? 200

35 Heavenly Paradise 201

Madame Butterfly 201, Ivory (Wardrobe) Palaces 202, Wardrobes Within Paradise 203, Christ's Soliloquy and Prayer 204, God's Love For Gardens 205, No Islands 206, A Glowing Planet 207, Language in Paradise 207,

No Apartheid Protectorate 208, Paradise: A Place of Opportunity 208, Menu in Paradise 209, The River of Life 209, Vanishing Childish Thoughts 210, The Sexes But Not Sexuality 210, Equality of the Sexes 211, No Nights Between Days 211, "They Shall Walk With Me" 212, A World of Peace, Health, and Progress 212, Planet Paradise 213, Knowledge Shall Vanish Away 213, Hidden Beauty 214

36 Perverted Justice 217

Registrar of the Unborn 218, A Whole Person 218, Spirit Being 218, Divine Rights of the Unborn 219, Before I Was Born 220, Comedy of Errors 220, My Little Children 221, Living Too Long or Dying Too Soon 221

37 Illegitimate Parents 223

The Hidden Connection 223, Shangri-La of the Human Race 224, Demolishing the Holy Spirit's Temple 225, Real Little People 225, When Two Lives Hang in the Balance 226, An Adamic Inheritance 226

38 Kinds of Abortions 227

Infanticide 228, Divine Foresight 228

39 When Innocence Ends 231

Knowledge of the Truth 232, Little Children's Escorts 233

40 The Doctrine of Infinities 235

Mortal Mathematical Infinity 236, Mathematical Infinity of the Soul 236, Immortal Mathematical Infinity 237, Mathematical Infinity of Spirit Beings 238, Absolute Eternal Infinity 238, God in the Atom 239, The Inner Connection 239, Hypostasis of the "Spirit" 240, Properties of the Soul 241, The Amortality of the Soul 242, Chemistry of the "Body" 242

41 The Everlasting Kingdom 245

Mystery of the Kingdom 245, Pearl of Great Price 247, Launching of the Church Age 247, Importance of Works 248, God's Daughter-in-Law 249

42 Prologue to Bible Prophecy 253

43 Importance of Biblical Prophecy 257

Book of Remembrance 257, Ghosts and the Guilty 258, At the Crossroads 259, Predestination and the Individual 260, Predestination and Immortality 261

44 Foundation for Biblical Prophecies 263

Designed to Avoid Tragedies 264, When Little is Big 264, Red Letter Day 265, Shadows From the Cross 265

45 Escaping the Holocaust 267

Timber! 268, The Golden Ark of God 268, Israel's Second Flight From Egypt 269

46 The Cradle Throne 271

A Palatial Home 272, For Whom the Bells Toll 272, Our Generation in Review 273, The Guiding Light 273, What You Believe Makes A Difference 274

47 Cosmic Bridges 275

Release of Mortal Hostages 276, Believer's Wardrobes 276, New Laws of Physics 277, The Incorruptible Body 278, The Quantum Leap 278, New Biological Matter 280, Hyperspace 281, Cosmic Super-Expressways 281, Expanded Opportunities 282, God's Family Completed 283, Nephilim Giants 283, When Death is the Quarry 284

48 The Cosmic Families 285

Time-Dilation Effect 286

49 Tribulation Syndrome 289

Indissolvable Marriage 290, Development of Doctrinal Errors 291, A Throneless Temple 291, Inseparable Twins 292, Acts, Chapter 29 293, The Tribulation Temple 293

50 Harvest Time 295

Second Salvation 295, Firstborn From the Dead 296, Firstfruits of the Harvest 296, Spiritual Rains 297, Ethnic Harvest 297, Your Honor, the Judge 299

51 Space Invaders 301

Millions of Miracles 302, Rescue Operation 302, Just in Time 304, Fall in Rank 305, March Away 305

52 First Harvest Day Rapture 307

The Day of Christ 308, God's Stars on Stage 309, Kings Without Crowns 309, Honor Greater Than Crowns 310, Royal Christening 311, Midnight Riders 311, A Million Years of Ministry 312, Priesthood Apprenticeship 312, The Innocent and the Rapture 313, Who Never Heard the Gospel 314

53 Catalyst of the Tribulation 317

Gods of Confusion 318, How the Rapture Will Alter World Events 318, Syrian Confederacy 319, Strange Famine 319, Unsung Heroes 320, Why These Frightful Apocalyptic Events 320

54 Mysterious Weapon of Antichrist 321

Dust Bombs 321, Radioactive Dust 322, A New Radiological Weapon 323, Romance of Terrorism 323

55 Deadly Shadows 325

Cascading of Events 326, Soviet Union Defeated in One Day 327, Tribulation Midnight 328, Black Smoke From the Antarctic 330, Cosmic Killers on the Loose 331, Purpose of the Final Tribulation 332

56 Coronation Day of Antichrist 335

Austria: The Continental Bridge 335, Assassination of a King 336, Ram With Two Horns 336, The Little Horn 337, Jehovah Displaced 338, Morality According to Cain 338, Blinded With Rage 339, Red Horse Rider of the Apocalypse 340, From the "King" to the "Beast" 340, Healing of the Revived Roman Empire 341, Two Beastly Kingdoms Confederate 341, The International Pac-Man 342, A Radioactive City 343

57 A King's Nightmare 345

Fall of the Ancient Roman Empire 346, America and Europe at the Crossroads 346, The Treaty of Rome 348, Heartland Countries of Europe 348, Another Little Horn 349, Last Day Babylon 350, The European Triangle 350

58 Antichrist Profiled 353

Similarities 354, Contrasts 354, God Hooks the "Jaws" of the Soviets 354, God's Cannonade 355, Kingdom of Seven Principalities 355, Cosmic Monster Invades the Earth 356

59 Covenant With Death 359

The Escape Clause 360

60 Middle East Aflame 363

Legal Owner of the Holy Land 364, Israel and the Promised Land 364, Islam and Bible Prophecies 366, Bahais 366, Israel's Seven Years of Peace 366, The Apostate Church 367, Fire Over Lebanon 368, Israel's Greatest Military Victory 370, A Prophetic Shadow Over Europe 370

61 Man of Mystery 373

Man of Magic 374, Changing Signals 374, Raping of the Vatican 375, Jacob's Trouble 375, Juhad! Moslem's Holy War 376, Antichrist in the Western Hemisphere 377

62 The Lord's Royal Court 379

Universal Witnessing 380, Virtuosos 381

63 Child of Destiny 383

An Amalgamated Being 384, Assault on the Manchild 384, Last Trumpet Sound 384, Stormclouds Over America 386

64 Last Harvest Day Rapture 387

Cosmic All-Stars 388, End of the Church Age 389, Two-Act Drama 390, Revelation of Antichrist 390, A Successful Fire Escape 391, Under Fire 392, Righteous Nations 392, Heavenly Vacationists 393, Marriage of the Lamb of God 393, Seven Seals -Seven Trumpets - Seven Vials 395

65 Post-Tribulation Christians 397

Flashing Sword in the Sky 398, Terror in the Night 399

66 The Judgment Seat of Christ 401

Remitting and Retaining Sins 402, Uncivil Servants 403, Spiritual Bankruptcy 404, Tears in Heaven 404, Interlocutory Judgment 405, Turning Tears Into Laughter 406, Beyond the Gates of Hell 406, Twilight Zone 407, Blind Men With Lanterns 407

67 The Lord's Olivet Discourse 409

Walls Come Tumbling Down 409, The Church's Limited Mission 410, The Missing Dimension in Knowledge 410, Final Tribulation Period 411, Rebuilding the Tribulation Temple 411, Wise Admonitions 413, Results of Five Major Wars in the Last Days 414

68 The Last Two Prophets on Earth 415

From the Altar to the Image 416

69 United States in Bible Prophecy 419

"Isles" of the Gentiles 419, Beyond "The Sea" 420, European Holocaust 420, Mystery Nation 421, Worldwide Sanctuaries 421, Beyond the Old World 422, The Vast Terra Incognita 422, The Unknown African Continent 423, Bible and Antiquity 423, Holy Scriptures: Indisputably Accurate 424

70 Land of Outstretched Wings 425

Salute to the New Nation 425, National Emblem 426, Air Power 426, Merchant Fleets 427, Beyond the Horizon 427

71 Early American History 429

Watersheds 429, Redskins 430, The Melting Pot 430

72 Independence Day 431

Metes and Bounds Act 431, Blessed of God 432

73 The Macedonian Call 433

A Missionary Nation 434, Israel's Emergence 435, United States and Israel 436, When God Whistles 436

74 Future of the United States 439

Continentalism 439, Nuclear Arms Proliferation 440, God's Intervention 440, Berlin Wall Comes Down 441, Seven Years of Peace For Israel 442

75 World War III 445

A Graveyard City 445, The Temple Mount 446, Two "Great Tribulations" 448, European Conqueror 449, Covenant of Peace 449, Second Flander's Field 450, Prophetic Names and Places 451, Trade With Antichrist 452

76 United States in the Final Tribulation 453

Countries Under God's Direct Wrath 454, Orphans of Babylon 454, Kings of the East 455, War Against the Babylonian Coalition 457, Road to Armageddon 458, The Broken Arrow 459, Wealth Flows From America to Christ 459, All Sea Lanes Lead to Jerusalem 460

77 Sword in the Sky 461

Battle for Possession of the Earth 462, Armageddon 462, The Eagles Have Landed 463, Satan Taken Hostage 465, The Vanishing Duo 465, Heaven Declares War 466, The Seven Thunders 466, Scenario of the Little Book 467, How Christ Solves the Palestinian Problem 468, Sunrise Strip 469, Survival of Egypt 470, World War V -The Last Battle 471, Survivalists 472, Destroy the Roots 474

78 A World of Kings 477

David the Prince 478, Drug Control 478, Silent Rage 479, Gun Control 480, Wine 481, Vineyards on the Mountains 482, Sabbatical Years 482, Earth's Water Canopy 484, Arsenal of Ice and Volcanic Power 486, God Turns Up the Thermostat 486, A Water Crisis Ends 487, Millennial

Arbor Days 488, Benefits From the Heat Plagues 489, God's Suicide Infantry 490, Benefits From the Earthquakes 491, The Wings of the Wind 492, Judgments: Long-Range Effects 492, The Wings of a Dove 493, Taming Mother Nature 494, Cancer, The Genetic Time Bomb 495

79 Restoration of the Earth 497

Redemption of the Earth 498, Roses Without Thorns 500, Poking God in the Eye 500, Judgment of the Nations 501, Sanitation Department 501, Delegation of Authority 502, A Righteous Government 503, Weep No More 504, The Racial Issue 504, Millennial Officialdom 505

80 Millennial Blueprint 507

Millions of Missionaries 507, Restoration of Jerusalem 508, The Millennial Temple 509, Millennial Holidays 511, Pilgrimages to Jerusalem 511, A Royal Reception 512, City Traffic 512, Safe Cities 513, No Congestion 514, Time Machines 514, The Dove Transit Lines 515, Old Gives Way to the New 517, Land Grants 517, Iron Hand of Righteous Laws 518, Social Life 518

81 Great Leap Forward 521

From Adam to Atom 522, The Miracle Microchips 523, Redesigned Ecosystems 524, Microbiology 525

82 The Ages To Come 527

A Millennium of Reconstruction 528, The Millennial Transformation 529, A Million Years From Now 530, Super-Civilization 530

83 The Flight of Spacecraft Earth 533

The Scarlet Sky 534, Earth's Spacesuit 534, The Cosmic Corridor 535, Journey From Darkness 535, Edge of Doom 536, Bottomless Pit 537, Below Outer Darkness 537, Terminal Island 538, Backdoor of Heaven 538

84 Return to Cosmos 541

The Lost World 542, Astroplanes? 543, Satan's Channel Into the Soul 544, New Senses 545

85 Out of This World 547

Alien World 548, Cosmonova 548, Dance of Death 549, Cosmic Requiem 550, View From the Edge of Darkness 551, Cosmic Vacancy Refurbished 551, Religious Violence 552, Bright Light Ahead 552, Odyssey Ends 552

86 Oasis in the New Heaven 555

New Frontiers 557, Royal Aristocracy 557, Two Vacant Thrones on Mount Sion 558, Royal Introductions 559

87 Redemption of the Earth 561

The Holy City's Facade 562, The Royal Redeemed's Wedding Ring 562, A Huge Beacon Light 563, Distant Universes 564, Chorales 564, Museum of Redemption 565, Tour of the King's Palace 566, Physical Transformations 566, Cloned Colonies 567, Royal Transit Shuttle Lines 567, Cherubic Innkeepers 568, Prima Donnas in the New Heaven 568, From "Time" to "Eternity" 568, Enormous Salt Basins 569, Then Cometh the End 570, Too Big to Bag 570

88 The Best Kept Secret 573

The Origin of Landforms 574, The Four Dispensational Ages 575, The Last is Best 576

89 The Highest Royal Tribunal 577

The Court's Irrevocable Decisions 578, Two Narrow Roads 578, Royal Jurors on Thrones 579, Oh My God, What Have I Done! 580, Is This All There Is to Life? 581, Instant Recall 582, Degrees of Punishment 582, Justice For All 582, The Last Walk 584, Extraditions 585, Mistakes One Cannot Erase 585, The Emptying of Paradise 586, Reincarnation 587, Future Lost 587, Summation 588, Death's Dominions 588, Physical Features of Lost Souls 589

90 Genesis III - The Heavenly Father's Manifested Kingdom 591

Timetable For Eternity 592, Wonderlands in Eternity 593, Distorting the Facts 593, World Without End 594, Spacecraft Earth's Third Flight 595,

Frontiers 597, A Kingdom of Kings 598, Festivals For the Masses 598, A Partial Connection 599, Shadows or Substance 599, The Final Connection 600

Illustrations and Charts 603

Index 635

List of Illustrations

I New Heavens 603

II Interstellar Plantings 604

III Light of Eternity 605

IV How Increased Velocity Reduces Time 606

V Celestial Womb(s) 607

VI Dimensions of Infinities 608

VII Visible and Invisible Entities 609

VIII Displacement of the Universe 610

IX Flight of the Soul After One is Deceased 611

X Destinies at Death 612

XI The Sword of the Lord 613

XII Generations of the Earth 614

XIII Royal Millennial Highway System 615

XIV First Conquests of Antichrist and Middle East Confederacy ... 616

XV Babylon's Last Days: When Two Kingdoms Unite at Rome ... 618

XVI Treaty of Rome and the Ancient Roman Empire ... 620

XVII Road to Armageddon ... 622

XVIII Future-Present-Past ... 624

List of Charts

A Defeat of the Soviet Union's Forces 625

B Israel's Seven Years of Peace 625

C Rise of the European Conqueror 625

D Tribulation of the Saints 626

E Dedication Of Israel's Tribulation Temple 626

F The Reign of Antichrist as the Beast 626

G The Two Great Tribulations 627

H When the Seven Year Peace Covenant is Broken 627

I Seven Seals - Seven Trumpets - Seven Vials 627

J Rise and Fall of Antichrist and the False Prophet 628

K How the Judgments of the Lord Increase 628

L The 144,000 Sealed Israeli Christians 628

M Last Two Witnesses (Prophets) 629

N The Day of the Lord 629

O The Three Woes 629

P The Seven Thunders 630

Q The Abomination of Desolation Setup 630

R The Judgment Seat of Christ—Marriage of the Lamb 630

S The Battle of Armageddon 631

T The Millennium 631

U Period of the Outer Darkness 631

V Christ's Manifested Kingdom on Earth 632

W The White Throne Judgment 632

X Manifested Kingdom of the Heavenly Father 632

Y Verdicts From the Highest Royal Supreme Court 633

Z The Two Restraining Forces of 2 Thessalonians 2:6.7 633

AA The Unification of Two Great Kingdoms 634

Foreword

Error repeated frequently by enough people is often accepted as truth.

The author has had numerous requests to prepare this repertory of Biblical and scientific material for publication. This outgrowth represents nearly five years of diligent research and in-depth study. Albert Einstein once stated: "The death of an individual matters little if his ideas live on."

As is obvious with the following intellectual nonconformists in the scientific field, inconsistency is the product of a strong mind. Consider Galileo's speculation concerning the motions of the planets; Alfred Wagner's postulation in 1912 on the drifting continents, Lavoister's hypothesis regarding combustion; Dalton's atomic theory; Planck's belief about the quantum; Rutherford's nuclear atom observation and Einstein's theory of relativity. All were initially scorned and rejected until proven right. And in 1836, there was the new rationale of the French chemist, Laurent, which pertained to molecular structure and which conflicted with that of the aged and revered chemist, Berzelius, whose criticism ruined Laurent's career. Christianity has equally borne its share of reformists and revivalists who were vilified, such as Martin Luther and John Wesley, not to mention Jesus of Nazareth who was crucified.

It is of utmost importance that scientists and theologians realize that resistance to change in one's thinking can be crippling to a person's intellectual and spiritual progress. Albert Einstein, one of the world's most brilliant minds, did his greatest work between 1905 and the early 1920s. From that point on he so adamantly rejected the quantum theory (the unpredictable behavior of the electron particle as advanced by the physicist Niels Bohr[1]) that nothing more of significance emerged from his work. How much is lost by not maintaining an open mind! The Bible nowhere suggests that empirical evidence alone (Hebrews 11:3) is the only source to understand how things came into existence. Knowledge of the Creator is also gained through science and theory. "For the *invisible things* . . . are clearly seen being understood by the things that are made . . . ." (Romans 1:20).

SONG AND DANCE

Prejudice is a serious barrier to understanding (John 9:41; Matthew 11:16-19) as evidenced by the Pharisees who were fundamentalists that believed in the resurrection, the angels, and God's power; however, they obstinately refused Jesus Christ as their Messiah, in spite of His miracles, which they attributed to the Devil. The Sadducees were modernists who denied the views held by the Pharisees and who were programmed against Jesus in His miraculous ministry. Both parties, therefore, were guilty of prejudice and rejection of Jesus as their Messiah: "And some of the Pharisees . . . said unto him, Are we blind also? Jesus said unto them, if ye were blind, ye should have no sin; but ye say, *We see, therefore your sin remaineth*" (John 9:40,41). "But whereunto shall I liken this generation? It is like unto children sitting in the markets, and calling unto their fellows, And saying, We have piped unto you, and ye have not lamented" (Matthew 11:16,17). The Pharisees were offended that Jesus would not unequivocally accept their song-and-dance theology; likewise the Sadducees were vexed because He rejected their mournful liturgy. To the agnostics we say: if there really is a Creator, is it not logical that He would leave us something in writing, that He might receive credit for what He has done?

A TOTAL CONNECTION

The history of the earth has been a long and fascinating one, yet there remains much live drama to be played upon its stage. God's plans and purposes for the earth and its inhabitants are not enigmatic, but clear. We simply need to gather the seemingly disoriented jigsaw pieces into their proper places. Once this is done, there will emerge a beautiful tableau of God's architectural design conceived in the pre-dawn of creation. This will constitute the total connection in our quest for totality.

There is an aura of excitement as one becomes aware of God's objectives and the way in which He assembles His cast, directs the various scenes, and brings them all to a flawless climactic finale. To see how the creator has implemented the many details confirms His perfect knowledge and wisdom and enhances our appreciation of Him and His work.

LET THE BIBLE SPEAK

When that which blurs our view begins to fall away, like dust gathered upon a picture, there is exposed to the observer that glorious plan of God which is so tremendously reassuring to one's faith and which provides veracity in these days of so many conflicting voices with uncertain sounds. A spiritual warfare is waging for souls. Truth is being challenged, and error repeated frequently by enough people is often embraced as truth.

The acceptance of Christ as one's personal Savior may not be sufficient to keep one from error or instability in his faith. A taproot needs supporting roots to withstand the adverse winds when they begin to blow. The Living Word, plus the Written Word, enables a person to withstand ". . . All the wiles of the devil" (Ephesians 6:11. See also, John 14:6, 17:17).

Only to the degree that a person has enlightenment can he acurately project his views; therefore, when one injects personal opinions contrary to the total harmony of The Book, confusion occurs (Matthew 22:20; 23:13; John 9:41).

Paul, in writing to the Romans (10:6-8) shows that the Scriptures are not equal, but supreme in the content of truth. Visions, dreams, or any other sources claiming divine information, instruction, or direction are subordinate to the Written Word of God: "But the righteousness which is of faith speaketh on this wise, Say *not* in thine heart, Who shall ascend into heaven? (that is, to bring Christ down from above: Or, Who shall descend into the deep? that is, to bring up Christ again from the dead) But what saith it? *The word is nigh thee,* even in thy mouth, and in thy heart: that is, *the word of faith, which we preach.*"

The author's prime objective in writing this book is to provoke serious thinking and examination of the Scriptures in search for a united truth.[2] Paraphrasing the works of others serves little purpose. Carefully examining facts in obtaining the truth is more tedious and time consuming, but it is definitely more profitable. If this book cannot offer added insight into God's Word, there is no purpose in having it written and published.

Gnosticism, during the Apostolic period, divorced faith from knowledge declaring that man's knowledge was superior and adequate to understanding and solving mankind's problems. Faith was suited only for the masses. Paul, the Apostle, called this the philosophical approach, ". . . (a) science falsely so called" (I Timothy 6:20. Compare Colossians 2:8).

Science knows only about physical laws. Bibliology remains one of its gray areas. If inaccurate interpretations in theology and errors in science are eliminated, there will be unity instead of the wide gap of light years that separates the two views. Scientists claim that science is self-correcting; therefore, it should give the biblical texts a closer evaluation. Perhaps this book can lessen the distance separating the two positions. Christianity must cast aside "medieval dogmas" which denounce science as adverse. Both Christianity and science must align themselves more closely to the *truths* of Scripture and to scientific *facts*. Evolution, which some embrace as a science, is actually an intractible unsupportable theory in its refutation of the Creator. Why cannot science include the doctrine of creation since the origin, function, and future of the universe are the objectives of both scientists and theologians?

Why did experimental science arise in the Western World and not in Central Asia or in the Far East? It is because science is the outgrowth of the Western belief in a Creator. Hindus and Buddhists, for example, regard the physical world as gross and as an illusion; the Taoists, Confucianists, and other Eastern religionists conceive the material world as unimportant; but the Western Christian considers it to be the personal handiwork of God. Islam calls the Christian nations the Barbaric West because of their development of the sciences.

Theological discoveries are made through Scriptures and doctrine; scientific discoveries are made through observation and theory. Einstein developed the theory of relativity by means of highly involved mathematical calculations and equations. The theory of quasars, on the other hand, was developed through a long process of observation.

In semantics, which is the science of meaning, it is a basic rule that each word we read must be understood in the light of the words that are in relationship to it. We cannot make sense out of anything we read without this rule. Similarly, any presentation of a theological viewpoint must be supported by the biblical science of hermeneutics. There are two Greek words used to describe the Word of God, "Logos" and "Rhema." "Logos" has to do with the whole or totality of God's Word in the Greek New Testament and is used 318 times. "Rhema" refers to a single promise from the Lord and it appears only 70 times. These statistics show the value of bringing together the entire Word of God.[3]

The purpose of the compilation of years of prayerful and studious research is to more thoroughly examine and understand God's Word so as to evoke that certain sound of symphonic harmony from the Incomparable Maestro Himself—God! "For if the trumpet give an uncertain sound, who shall prepare himself to battle?" (I Corinthians 14:8).

[1] See page 278.

[2] It is the right of every person to ask a question and to seek its answer.

[3] "Logos" (Greek for "Word" in John, chapter one) also embodies the concept of God as invisible personality and power as well as the inaudible voice. When the "Word" (Logos) entered the flesh of Jesus Christ, God was from then on personalized through his Son Jesus Christ. Einstein failed to recognize God as the Logos in the *Person of Jesus Christ.*

Preface

God is more interested in redemption than He is in judgment.

God is more interested in redemption than He is in judgment. Einstein said that the Lord is subtle but "malicious He is not." Destruction becomes inevitable only when the candidates for redemption are so firmly entrenched in defending their own wills that restoration is made impossible. It becomes a contest of will—God's will versus that of another bent on a dangerous course which ultimately become destructive.

God, by His very nature, cannot allow infectious sin to continue unabated. If He were to do so, He would be guilty of gross callousness similar to that of an artist or sculptor passively watching his masterpiece being abused and desecrated. Instead, such a situation would bring revulsion to the artistic genius of its creator.

We shall present in this book the great extent to which God pursues redemption, an imperative in the total connection. The incorrigibles are like pieces of clay that will not conform to the intent of their creators. They must be destroyed because they have no place in the gallery of God's creation:

"But now, O Lord, thou art our Father; we are the clay, and thou art out potter, and we are all the work of thy hand!" (Isaiah 64:8).

"Surely your turning of things upside down shall be esteemed as the potter's clay: for shall the work say to him that made it, He made me not? or shall the thing framed say to him that framed it, he had no understanding?" (Isaiah 29:16).

Some may look with askance at the discards and rejects in the potter's field where lies a Judas Iscariot (Matthew 27:7,10), but they fail to observe in the great house of creation ". . . Vessels of gold and of silver (unto) honour, sanctified, and meet for the master's use, and prepared unto every good work" (2 Timothy 2:20,21).

SELF-APPOINTED PROPHETS

Periodically self-appointed prophets appear on the horizon announcing the termination of the world. Some claim spiritual revelations; some cite Scriptures out of context; others set dates which fail to materialize. We must remember man sets "dates"; God sets "times" (Acts 17:26; Ephesians 1:10). They assert the earth is coming to an ignominious end with unprecedented earthquakes bringing vast destruction and plagues of such devastating proportions as to exterminate the entire human race.

FRY OR FREEZE

Some prophets propound the annihilation of earth's residents by predicting falling meteorites that will shower destruction from the skies upon an ungodly world. Others predict that the sun will advance about one percent closer to the earth so that it will sear its inhabitants, burn everything to a cinder, and boil the oceans; or conversely, that the sun will recede approximately five percent and cause everything to freeze.

Some proclaim that the moon will vary its course and create unprecedented tidal waves which will inundate metropolitan coastal areas around the world. Others suggest that famine will scourge the earth and erase all vestige of life.

THE LINE-UP

Another prediction that failed to materialize stated that several planets would align in conjunction with the earth and sun to produce a pull of such magnitude upon the earth as to simultaneously create earthquakes, tidal waves, and volcanoes of unbelievable consequences.

There was such a conjunction of the planets on January 16, 1901, covering only a 25° arc; another on March 10, 1982, covering a 90° arc. No dire disasters occurred at those times.

THE MUSHROOM CLOUD

Many portend the end of the world will occur in a nuclear holocaust, culminating in the battle of Armageddon, and leaving no trace of life on earth.

The problem with these predictions is that if all life on earth is terminated, what happens to the many remaining unfulfilled prophecies in God's Word? Are they to go begging for fulfillment?

While it is true that the Bible prophesies great calamities, there is no biblical foundation to the supposition that the entire population on earth will be decimated. Quite to the contrary, evidence shows there will be a millennial reign of Christ on earth. This prediction would go unfulfilled if there were no survivors on earth, so how can we explain Matthew 25:31-46 where Christ will set up some nations on His left and others on His right?

The prophets of God never based their predictions upon changing current events. They uttered forth God's Word even though all contemporary events made the prophet's statements seem irrelevant, impossible, or even ridiculous.

Many prognosticators interpret Bible prophecies in the light of vacillating events of the time, only to see their predictions wiped out. Eschatology (doctrine of final things) must be based on God's Word regardless of current trends: "For God hath declared the end from the beginning, and from ancient times the things that are not yet done, saying, My counsel shall stand, and I will do all my pleasure" (Isaiah 46:10).

When trying to assimilate the complexities and responsibilities of the Creator's creations, one cannot but wonder: it must be something to be the One and Only Almighty God!

SECTION 1
THE BOOK OF HOPE

1

A Book of Hope

The Bible is primarily a collection of books of tremendous hope and assurance.

The Bible is primarily a collection of books of tremendous hope and assurance. It does, however, contain some judgmental predictions resulting in tragedies as a result of divine displeasure. It forecasts catastrophies, but it does not contain a single prediction declaring the annihilation of *all* life on earth.

During the Great Flood in Noah's time, life was preserved in the ark: ". . . eight souls were saved . . ." (I Peter 3:20). When Sodom was destroyed by divine conflagration, the righteous were escorted out of the city previous to the divine inferno: "And delivered just Lot . . ." (2 Peter 2:7).

"Though (if) these three men, Noah, Daniel, and Job were in it (judgment), they should deliver but their own souls by their righteousness, saith the Lord . . . they shall deliver neither sons nor daughters; they only shall be delivered . . ." (Ezekiel 14:13-20). "The soul that sinneth it shall die. The son shall not bear the iniquity of the father, neither shall the father bear the iniquity of the son: the righteousness of the righteous shall be upon him and the wickedness of the wicked shall be upon him" (Ezekiel 18:20).

What about those Christians who are persecuted and become martyrs at the hands of their persecutors? They are not victims at the hands of God but at the hands of *men*. God will compensate them with generous rewards at the Judgment Seat of Christ (Matthew 5:1-12; 2 Timothy 2:12).

When Jesus predicted the destruction of Jerusalem which occurred in 70 A.D. under the Roman Emperor, Titus, and his armies, He assured His disciples that they could escape the debacle if they would heed the specific signs of His warning. He said, "But there shall not an hair of your head perish . . . And when ye see Jerusalem compassed with armies . . . flee to the mountains . . . and let not them that are in the countries enter thereinto. For these be the days of vengeance . . . " (Luke 21:18-22). Josephus, the noted Jewish historian, recorded that the Christians successfully made a safe exodus from Jerusalem before the city was conquered.

Divine judgments are designed to punish the ungodly, not the righteous. Even at the close of the Millennium (1,000 years of universal peace under Christ), those who survive the deception of Satan and the judgment fire of God (Revelation 20:7-9) will escape the celestial inferno and appear on the New Earth and in the New Heaven (Revelation 21:24).

ESCAPING THE INFERNO

The earth also will survive that celestial furnace: "But the day of the Lord will come as a thief, in which the *heavens will pass away* (and) the earth and all its works will *vanish*" (2 Peter 3:10, *Weymouth*). *James Moffatt* translates it: "The heavens (set) ablaze and melt, the earth and its works will *disappear*." In Hebrews 1:10-12, the word "earth" in the original is not included in the word "they:" " . . . the heavens . . . they shall disappear . . . they shall perish . . . they shall be changed." In *Wuest's Word Studies:* "The word 'they' refers back, not to the earth, but to the heavens. The Greek text makes this clear. The pronoun is intensive, 'they themselves.'"

The accomplishments on the earth during the millennial reign of Christ will not be obliterated but will survive with the planet to become the New Earth.

The words "vanish" and "disappear" do not mean disintegration but escape. The earth will vanish or disappear from its place in the universe after which the planetary heaven will be destroyed by fire. The natural people still inhabiting the earth after the Millenium will pass unharmed beyond the cosmos before the planetary heavens are dissolved. The vacant place left by the destroyed cosmos will become the site for the White Throne Judgment (Revelation 20:11).

The New Earth will have upon its surface the white ashes of the ungodly destroyed at the end of the Millennium: "For, behold; the day cometh, that shall burn as an oven; and all the proud, yea, and all that do wickedly, shall be stubble (and) shall burn them up. (But) unto you that fear my name (ye) shall tread down the wicked; for they shall be *ashes* under the soles of your feet" (Malachi 4:1-3).

"Whose voice then shook the earth: but now he hath promised, saying, Yet once more I shake not the earth only, but also heaven.

"And this word, Yet once more, signifieth the removing of those things that are shaken, as of things that are made, that those things which cannot be shaken may remain.

"Wherefore we receiving a kingdom which cannot be moved, let us have grace, whereby we may serve God acceptably with reverence and godly fear: For our God is a consuming fire" (Hebrews 12:26-29).

2

Social Justice

God does not gamble with the souls of mankind.

"One generation passeth away, and another generation cometh, but the earth abideth for ever" (Ecclesiastes 1:4). The earth will not cease to exist as do generations.

The rhetorical question is not whether the earth can survive, but why should it? Perhaps no subject is less comprehended than God's reasoning in proceeding with creation. It would be out of character for God to create beings who were predetermined to be expendables, so redemption must provide all created beings with an equal right and opportunity under the proper circumstances to make their decisions for Christ.

"Happy are those who long for justice, for they shall surely have it. . ." (Matthew 5:6, *Living Bible*).

Man with his sovereign will chose unrighteousness, injustice, and mortality; God with His Sovereign Will chose righteousness, justice, and redemption. Sin is the misuse of one's God-given will and the abuse of freedom.

Injustice is not the design of God. It is Satan who warps the plan of Deity and places eternal souls as if they were chips on a casino table.

Einstein has stated that God does not play dice with the Universe. Neither does He gamble with the eternal souls of mankind; nor with the angel spirit beings He has created.

God has beautifully designed redemption, which involves the preservation of the present earth. Every semblance of inequality must and will have the scales balanced in the New Heaven. This is imperative in God's eternal plan. Anything less would be deliberate arrogance from God's position. If there is a polarization of the will against the Almighty, it must be the sole decision of the individual made under his own volition, not under duress.

The value of a soul in the eyes of God is priceless because He has declared that one soul is worth more than all of the world (Matthew 16:23). If this is true, then God cannot view lightly the creation of even a single being since each is produced consistent with His character. Apart from redemption, the whole plan of God in relationship with creation sinks into futility.

We are not advocating that God is remiss in exacting punishment upon the deserving; however, in the light of the severity of His judgment, it is imperative that all avenues of Justice be explored and pursued relentlessly: "declaring the end from the beginning, and from ancient times the things that are not done, saying, My counsel shall stand, and I will do all my pleasure . . ." (Isaiah 46:10).

Is God to blame for the existence of the Devil, and for all the evil that is pervading the world? In the light of God's foreknowledge (to the surprise of many) we must say in all candor that God is responsible. How then can we justify that He is an all wise and good God? This is only possible when we view His Word from the beginning to the end.

God could have refrained from creating Lucifer since He knew that he, as an archangel, would fall to become the Devil. Or God could have permitted an early death for human tyrants while they were yet in their innocence. Then why didn't God do so? When viewed in the perspective of the overall plan of God, there is justification for their existence. This only can be understood in the light of two important truths—the sovereign will, which is implanted in such creatures, and the plan of redemption. Not a single human being, even one prematurely cut off from life, is ever going to be denied the opportunity to exercise his will under fair and equitable circumstances. Justice will be administered. The guilty will be punished and the innocent victim will be reinstated to life with the opportunity to make his own decisions under equitable circumstances in the New Heaven.

The White Throne Judgment will be a time of revelation and judgment. If, in Christ's fair opinion, they deserve the chance for redemption that has been denied them, Scriptures reveal that they will receive that opportunity in the New Heaven. That is one of the prime purposes for its existence.

EQUALITY FOR UNEQUALS

"Let me be weighed in an even balance" (Job 31:6).

Is it a crime to have been born if one does not hear the Gospel of Jesus Christ, or does not have the mental capacity to understand and believe, or is prematurely terminated from life? People are not doomed eternally simply because they are born inexpediently into this world. The advantaged and the disadvantaged will be equalized because, as we shall see, the major purpose of the New Heaven is to bring into balance all injustices and inequalities.

Will Justice be the victim instead of the victor at the bar of divine judgment? As long as there are any redeeming factors remaining, God will not ignore them. What is the component of guilt that justifies no redemption? Is it not the fixation of the individual's own will or is it the circumstances? Are God's mercies compatible with His justice, or is the voice of mercy silenced? If so, then why a Judgment at all if the same indictment is rendered for everyone? It makes mockery of divine justice. There is too much unjustified harshness in the insufficiently researched doctrinal positions which pertain to this matter.

PREJUDICATION

Circumstances must not be allowed to devalue a single soul by prejudication. Each deserves the right and opportunity to exercise his sovereign will in determining his own eternal destiny. We must credit

God with justice. Is He love or is this merely a cliche? Does the Scripture renege on its own statements at the time of divine judgment?

The works (deeds) and conscience of the unredeemed will be an important factor at the White Throne Judgment. At that time the Magistrate will decree whether or not redemption remains in the offing for the defendant in the New Heaven: ". . . conscience also bearing witness . . . accusing or excusing *(in) the day* when God shall judge the secrets of men by Jesus Christ . . ." (Romans 2:15,16). ". . . Judged out of those things which were written in the books, according to their works" (Revelation 20:12). That which a person *avoids* (such ad having his name stricken from the book of life) is equally important with what a person *does* (such as accepting God's plan of redemption).

There is fairness with God. The majority who have ever lived on earth will have experienced death and resurrection prior to hearing and understanding the plan of redemption. If their names remain in the book of life, they will go into the New Heaven. Subsequently, they will have their opportunity to accept Jesus Christ under equitable circumstances through the priestly ministries of the Royal Redeemed (Revelation 1:6).[1] If their names are erased because God cannot condone their sadistic behavior and evil ways, they can expect no mercy since they themselves have demonstrated none (James 2:13).

What about the young victims of criminals denied the privilege of a normal life with all its accruements, including raising a family, as well as whatever would have been their vicissitudes of life? Are they to be denied these benefits forever? If they are unredeemed and their names are still retained in the book of life, they will possess natural bodies in the New Heaven and will recover the losses denied them under a more perfect environment. This must not be confused with the Heavenly Paradise, a temporary holding place. (See "Heavenly Paradise"). The transfer to the New Heaven in natural bodies will be a simple matter since they already have appeared that way at the final tribunal as a result of the resurrection.

CAN GOD CONTROL EVIL?

A new doctrine has emerged called "Process Theology," which teaches that God, though all-loving, is not all-controlling. Moreover, because He Himself is evolving and is not completely developed, He is unable to control evil. Conversely, Scriptures teach that God is the Absolute Sovereign and that evil is the sole result of rebellion against a righteous God. God, therefore, has provided through redemption a means whereby those who accept His plan finally will enter a world of euphoria in which the Almighty has totally eradicated evil, together with its influences and effects, by imputing His own righteousness within the redeemed to make them eternally conformable to God's righteousness. The prophetic Scriptures alone prove that God is in complete control.

WHEN DEITY SITS ON THE BENCH

As severe as the judgment may be, we can be grateful that Jesus Christ is the Judge, and not man, with his multiple prejudices and biases (John 5:22, 23; Romans 8:34).

God does include in His judgment all of the elements of justice; man in his private judgments does not. Job (8:3) said, "Doth the Almighty pervert justice?" Justice is so vital a part of God's constitutional Being and policy that He has injected a compassionate and companionable word, "justification," meaning "to set right."

When redemption has been completed and the unredeemable have been justly committed to their eternal destiny, then there will be the eternal renaissance which God envisioned before His first act of creation, thus completing the plan He formulated at the beginning:

"O the depth of the riches both of the wisdom and knowledge of God! How unsearchable are his Judgments, and his ways past finding out!

"For who hath known the mind of the Lord? or who hath been his counsellor?

"Or who hath first given to him, and it shall be recompensed unto him again?

"For of him, and through him and to him, are all things: to whom be glory for ever. Amen" (Romans 11:33-36).

"Justice and judgment are the habitation of thy throne: mercy and truth shall go before thy face" (Psalm 89:14).

[1] The Royal Redeemed are those who will be in the "first resurrection" and joined to Christ as His Royal Wife.

3

The Eternal Footstool of God

In order for the mountains and hills to be everlasting the earth must be everlasting also.

"Thus saith the Lord, The heaven is my throne, and the earth is my footstool . . . " (Isaiah 66:1). The earth not only is the footstool of God, but also it is the holding place of the enemies of God: "The Lord said unto my Lord, Sit thou at my right hand, until I make thine enemies thy footstool" (Psalm 110:1).

It will take considerable time before *all* of the enemies of Christ are gathered into hell and later released for the White Throne Judgment: "Then cometh the end, when he (Christ) shall have delivered up the kingdom *to* God, even the Father; when he shall have put down all rule and all authority and power. For he must reign, till he hath put down *all enemies* under his feet" (I Corinthians 15:24,25).

This final judgment will end when Christ delivers up His Kingdom to the Heavenly Father and when *all* of the redeemed in the New Heaven receive bodies of immortality. This latter event marks the destruction of death, but not that of the earth. Therefore, the White Throne Judgment will extend throughout the period of the New Heaven until all of the unredeemed who will accept Christ have been redeemed (Revelation 21:1; 22:11,14,15). Then when all of the enemies of Christ have been gathered within the footstool of God (earth), the final judgment will terminate. The word "enemies" indicates they have by their own volition rejected Christ, and are not merely innocent ones without either the knowledge or opportunity to accept Him. Their decision has been premeditated.

EVERLASTING REAL ESTATE

"The people... shall inherit the *land for ever*... the work of my hands, that I may be glorified" (Isaiah 60:21). "And I will give unto thee, and to thy seed after thee... the land of Canaan, for an *everlasting possession*..." (Genesis 17:8).

Nothing could be plainer. God made an eternal covenant with Abraham that the righteous shall inherit the land of promise forever, indicating that the earth will never cease to exist.

EVERLASTING MOUNTAINS AND HILLS

The topography of the earth includes its panorama of mountains and hills: "He stood, and measured the earth: he beheld (and) the *everlasting mountains* were scattered (not eliminated), the *perpetual hills* did bow: his ways are everlasting" (Habakkuk 3:6). "The everlasting hills..." (Genesis 49:26). "The *ancient mountains* (which existed when Lucifer once reigned upon the earth long before our first parents were created)... (and the) *everlasting hills*" (Deuteronomy 33:15).

In order for the mountains and hills to be everlasting, the earth must be everlasting also. Some theologians maintain that the words "everlasting" and "eternal" have two different meanings but no evidence is presented. Moreover, in assuming this point of view, how can one know which meaning is intended when the Scriptures use these terms? The implication that God may not be eternal or everlasting only makes for confusion and breathes disorganization into the structure of the Holy Scriptures. God is not given to such discrepancies and meaningless jargon.

THE ETERNAL CITY OF JERUSALEM

"As we have heard, so have we seen in the *city of the Lord of hosts,* in the city of our God: *God will establish it for ever"* (Joel 3:18,20). "I will make *(Jerusalem) an eternal excellency,* a joy of many generations" (Isaiah 60:15).

THE INDESTRUCTIBLE JEW

"And it shall come to pass, that (Judah) shall dwell *for ever"* (Joel 3:18).

"... Judah shall dwell *for ever ..."* (Joel 3:20).

ISRAEL SHALL EXIST FOREVER

"Even of the covenant which (God) made... to Israel for an *everlasting* covenant" (I Chronicles 16:16,17; Psalm 105:9,10).

"And he (Jesus) shall reign over the house of Jacob (Israel) *for ever ..."* (Ezekiel 43:7).

"Israel... and to his seed *forever"* (Luke 1:33,55).

"... (God) will dwell in the midst of the children of Israel *for ever ...* " (Ezekiel 43:7,9).

LAND PROMISED ISRAEL IS ETERNAL

"Then will I cause you to *dwell* in this place, in the *land* that I gave to your fathers, *for ever and ever"* (Jeremiah 7:7).

"And Moses sware in that day, saying, Surely the *land* whereon thy feet have trodden shall be in thine inheritance... *for ever"* (Joshua 14:9).

"Dwell in the *land* that the Lord hath given unto you and to your fathers *for ever and ever"* (Jeremiah 25:5).

Anything that God creates He owns, including the land; He can give it to whomever He chooses. Anyone else that invades or occupies the land does so illegally and is subject to eviction. Historical reasons do not stand up in the court of divine jurisprudence.

ENDLESS MONARCHY

"I have sworn unto David my servant, thy seed will I establish *for ever,* and build up thy *throne to all generations"* (Psalm 89:4).

"Shall call his name JESUS, (the) Lord God shall give *unto him the throne of his father David:* and he shall *reign* over the house of Jacob *for ever;* and of his kingdom *there shall be no end"* (Luke 1:31-33).

"The word of the Lord came unto Nathan, saying, Go and tell my servant David (I) will establish *his throne for ever,* (I) will settle him *in my kingdom for ever;* and *his throne shall be established for ever more"* (1 Chronicles 17:3,4,12,14).

How can the above promises be fulfilled if the earth comes to a fiery end?

ETERNAL LEVITICAL PRIESTHOOD

"And the Lord spake unto Aaron, Behold, I also have given thee the charge of mine heave offerings of all the hallowed things of the children of Israel; unto thee have I given them by reason of the anointing, and to thy sons, by an inheritance *for ever"* (Numbers 18:8).

"Then David said, none ought to carry the ark of God but the Levites: for them hath the Lord chosen to carry the ark of God, and to minister unto him *for ever"* (I Chronicles 15:2).

Since the original tabernacle and the Ark of the Covenant are in heaven (Hebrews 8:1-6), and the earthly ones are patterned after them, we have reason to believe that the redeemed Levites will resume their ministry and continue it not only during the Millennium, but also into the New Heaven. The eternal blood of Christ on record in heaven will replace adequately the need for more sacrifices (Hebrews 9:12; 10:10, 26). Christ will direct all priestly functions in the coming Kingdom of Christ (Hebrews 7:3; 8:2). The Levitical priesthood will serve under Christ forever.

ETERNAL FESTIVALS

"And this day shall be unto you for a memorial; and ye shall keep it a feast to the Lord throughout your generations; ye shall keep it a *feast* by ordinance *for ever*" (Exodus 12:14).

"And ye shall keep it a *feast* unto the Lord. (It) shall be *a statute for ever* in your generations" (Leviticus 23:41).

"And it shall come to pass, that every one that is left of all the nations which came against Jerusalem shall *ever* go up from year to year to worship the King, the Lord of hosts, *and to keep the feast of the tabernacles.*

"And it shall be, that whoso will not come up of all the families of the earth unto Jerusalem to worship the King, the Lord of hosts, even upon them shall be no rain.

"This shall be the punishment (of) all nations, that come not up to keep the *feast* of tabernacles" (Zechariah 14:17-19).

The feast of tabernacles looks forward to the establishment of the new Kingdom and the complete restoration of Israel when "the tabernacle of God is with men, and he will dwell with them, and they shall be his people, and God himself shall be with them, and be their God" (Revelation 21:3).

PERMANENCY OF HELL

"And I say unto you, Make to yourselves friends of the mammon of unrighteousness; that, when ye fail, they may receive you into *everlasting habitations*" (Luke 16:9).

Why is hell ("sheol" in Hebrew; "hades" in Greek) eternal? There are several reasons. First, the lost will be held in one section of these strongholds in the bowels of the earth until after the earth has been returned as the footstool of God. Then shall be fulfilled the prophecy, "For David (saith), The Lord said unto my Lord, Sit thou on my right hand, *until I make thy foes thy footstool*" (Acts 2:34).

"But to which of the angels said he at any time, Sit on my right hand, until I make *thine enemies thy footstool*" (Psalm 110:1; Matthew 22:44; Luke 20:43).

In order for those prophecies to be fulfilled, *all of the lost souls* must have had an existence on planet earth prior to their release to the White Throne Judgment: "And the sea gave up the dead *which were in it* (indicating a return of the Pangaea);[1] and death and hell delivered up the dead *which were in them;* and they were judged every man according to their works" (Revelation 20:13,14).

The above cannot take place until *after* the heavens have been dissolved with fervent heat, and the earth is no longer in its present location. Then God will set up His White Throne Judgment in the space vacated by the removal of the earth and the dissolution of the planetary heavens: "And I saw a great white throne, and One who was seated on it, from whose presence *earth and sky fled away: and no place* (that is, in the planetary heaven) *was found for them"* (Revelation 20:1, *Weymouth*).

MILLENNIAL KINGDOM EXTENDS FOREVER

"And the kingdom and dominion, and the greatness of the kingdom under the whole heaven, *shall be given* to the people of the saints of the most High, whose kingdom is an *everlasting kingdom"* (Daniel 7:27).

"And there was given him dominion, and glory, and a kingdom, that *all people, nations, and languages,* should serve him: his dominion is an *everlasting dominion,* which shall *not pass away,* and his kingdom that which *shall not be destroyed"* (Daniel 7:14).

". . . Thine eyes shall see Jerusalem a quiet habitation, a tabernacle (temple) that *shall not be taken down;* not one of the stakes thereof *shall ever be removed,* neither shall any of the cords thereof be broken" (Isaiah 33:20).

Thus we see that the conditions prevailing during the Millennium will continue onto the New Earth, which will be located in the New Heaven, and then on into the Heavenly Father's Kingdom.

MILLENNIAL RENEWAL LASTS FOREVER

The Millennium will bring many great and exciting changes to the earth and to all of its inhabitants.

The restoration and beautification of the earth will continue on into His endless kingdom: "So shall my word be that goeth forth out of my mouth:. . . it shall accomplish that which I please, and it shall prosper in the thing whereto I sent it. For ye shall go out with joy, and be led forth with peace: the mountains and the hills shall break forth before you into singing, and all the trees of the field shall clap their hands." There will be no more ghostly spires pointing their scorched and scarred branches like fingers after a devastating fire: "Instead of the thorn shall come up the fir tree; and instead of the brier shall come up the myrtle tree: and it shall be to the Lord . . . for *an everlasting sign that shall not be cut off*" (Isaiah 55:11-15).

We must believe from the foregoing that on the New Earth within the New Heaven and continuing on into the Heavenly Father's Kingdom, there will exist a great variety of plants, birds, animals, hills, and mountains gracing their surfaces. These species of plants and animals probably will be located in their suitable habitat within the New Heaven on one of its planets or asteroids.

THE ETERNAL KING OF KINGS

Christ, the King of kings, will become earth's new Monarch in conjunction with the Royal Redeemed who will reign with Him. His rule will expand into the eternal extended kingdom forever: "And the seventh angel sounded; and there were great voices in heaven, saying, The kingdoms of *this world* are become the kingdoms of our Lord, and of his Christ; and he shall reign *for ever and ever*" (Revelation 11:15).

". . . He is Lord of lords, and King of kings . . ." (Revelation 17:14).

". . . The Lord of hosts shall reign in Mount Zion, and in Jerusalem, and before his ancients (angels) gloriously" (Isaiah 24:23).

"And there was given him (Christ) dominion, and glory, and a kingdom, that all people, nations, and languages, should serve him: his dominion is an *everlasting dominion, which shall not pass away,* and his kingdom that which shall *not be destroyed*" (Daniel 7:14). In order for Christ to reign forever over the kingdoms of this world and its multilingual people, these persons must remain infinitely in the New Heaven and the Heavenly Father's "New Heavens."

ETERNAL ARISTOCRACY

"If we suffer, we shall also reign with him: if we deny him, he also will deny us" (2 Timothy 2:12).

"And hast made us unto our God kings and priests: and we shall reign on the earth" (Revelation 5:10).

"And there shall be no more curse: but the throne of God and the Lamb shall be in it: and *his servants shall serve him:* . . . and his name shall be in their foreheads . . . and they shall *reign for ever and ever"* (Revelation 22:3-5).

"Him that overcometh . . . *will write upon him the name of my God . . .* which is *New Jerusalem"* (Revelation 3:12).

The "New Jerusalem" will come down upon the New Earth when planet Earth has been redeemed and restored to its original position as God's footstool near His throne. If the "New Jerusalem," which is eternal, is the dominion of the Royal Redeemed, that means the Royal Redeemed will reign upon the New Earth as well as the Millennial Earth; therefore, the earth has to be eternal.

GIGANTIC WATERHOLE

The fact that God's Word omits mention of permanent bodies of water, such as oceans and seas, is significant because the time is coming when there will be *no more sea* (Revelation 21:1). The Pangaea (single land mass) which forms for the Millennium has one sea (singular); however, on the New Earth the Glory of God will evaporate that and replace it with salt.

There is also the possibility that some of the water will evaporate as the transition is made from the Millennial Earth to the New Earth. Christ, at that time, will be in the midst of the Millennial survivors to protect them from the cosmic furnaces just as He protected the three Hebrew children in the fiery furnace when He stood in their midst (Daniel 3:19-23). Thus, the single scar remaining on the earth will be the salt which will serve to identify *the Planet Earth as the New Earth:* " . . . shall be as Sodom and . . . Gomorrah . . . Saltpits . . . " (Zephaniah 2:9). Just as the redemption of mankind left its marks upon the immortal body of Jesus Christ (Revelation 1:7), so the proof of the redemption of the earth will be these vast saltpits.

The fertile soil, together with the waters of life flowing upon the reconstituted land mass, will combine to provide all of the sustenance necessary for the inhabitants of the Pangaea on the New Earth.

Adam and Eve lived lavishly in Eden on the fruits and vegetables which God provided for them (Genesis 1:29). When they transgressed, God's punishment was to curse the ground from which came their sustenance. When this curse is lifted the deserts will be restored to productivity and they will provide abundantly during the Millennium, on the New Earth, and in the New Heaven. The mountains, which are now unproductive, will be lowered; the swamps and rocky land will all become fruitful: "*Every* valley shall be filled (with good soil burying the rocks), and *every* mountain and hill shall be brought low (with melted snow to make them accessible and warm); and the crooked shall be made straight (with roads built to reach inaccessible places), and the rough ways shall be made smooth (with beautiful freeways)" (Luke 3:5).

As we shall discover later in this book, the exploding population will inherit productive places extending far into the vast reaches of the New Heaven as the New Earth obviously cannot accommodate them all.

[1] One land, one continent.

4

Before the World Began

What makes anyone think God was sitting on His hands doing nothing before He entered into His work of creation as we know it?

Long before the heavens and the earth were created, God existed. But where? "For thus saith the high and lofty One that *inhabited* (past tense) *eternity,* whose name is Holy; I dwell (present tense) in the high and holy *place*. . . " (Isaiah 57:15).

The first of His creation occurred within the habitation of eternity, which was a place where God set up His throne—the central hub from which all new creations would come into existence. It was not heaven. Heaven was created and later identified in Scripture as the "Heaven of Heavens," the capital center of all creations.

If a circle were to represent the dividing line between "time" and "eternity," then inside of the circle, where all of the heavens are located, is "time " with its limitations. The outside of the circle would represent the infinity of eternity where God's creations are outside of the dimensions of "time."

THE CREATOR'S COATTAILS

The "Heavens of Heavens" are described in Psalm 148:4-6: ". . . the Lord . . . commanded, and they (the "Heavens of Heavens" which are presumably the residences of angels), were created, He hath also established them *for ever and ever."* The planetary heaven is not established forever (2 Peter 3:10,12).

In Proverbs, Chapter 8, a beautiful autobiography is recorded by our Lord Jesus Christ in His preincarnate existence. It was "before (God's) works of old;" that is, before the creation of the heaven and earth as we know them.

Speaking in the narrative from Proverbs, Jesus said, "When (God) prepared the heavens, I was there . . . and was by him, as one brought up with him . . . I was set up from everlasting, from the beginning, or ever the earth was."

Jesus declared His preincarnation when He said, ". . . Before Abraham was, I am" (John 8:58). He said, in essence, that before the creation "I was there" sharing the excitement, thrill, and joy in the creation of all things.

In John's Gospel, its writer, in alluding to Jesus Christ said, "All things were made by him; and without him was not anything made that was made." And the Apostle Paul said, "Who (Christ Jesus) being in the form of God, thought it not robbery to be equal with God" (Philippians 2:6).

Scripture unfolds the truth that Jesus Christ was there not only during the creation of all things, but also that He shared His expertise. He was the lodestar within the Godhead, planning and directing the development of creation to perfection.

Job (28:3) said that God "searcheth out all perfection." He is never satisfied with less than perfection in His work. Here the *wisdom of Christ* (as shown in Proverbs, chapter 8) is brought into play. He is described as "the power (and) *wisdom of God"* (I Corinthians 1:24). Paul adds, "In whom (Christ) are hid all the *treasures of wisdom* (the by-products as God's entrepreneur) and knowledge" (Colossians 2:3). And in I Corinthians 2:7, "But we speak of the *wisdom of God* (Christ is God) in a mystery (compare I Timothy 3:16), even the hidden *wisdom,* which God ordained *before* the world *unto our glory"* (I Corinthians 2:7) that is, unto our enjoyment or pleasure.

During His earthly ministry Jesus Christ was recognized as an extraordinary person. The people who listened and saw Him remarked: "Whence hath this man this *wisdom,* and these mighty works?" (Matthew 13:54. And in Mark's Gospel (6:2): ". . . many hearing him were astonished, saying, From whence hath this man these things? and what *wisdom* is given unto him, even such mighty works are wrought by his hands?"

WAS GOD SITTING ON HIS HANDS?

Eternity, before the heavens and the earth were created, was not a vacuum. What makes anyone think God was sitting on His hands doing nothing before He entered into His work of creation as we know it? Eternity, as we have seen, is a *habitation* and within it is a place called "heaven" where God's throne is located (Isaiah 58:15). It was, and is, filled with a proliferation of wonders. These creations are in the dimension of "eternity" which is beyond our comprehension. Whatever existed previously must have consisted of the eternal glorious fellowship within the Godhead.

The Psalmist, in his Messianic prophecy, tells how Jesus Christ came out of His "ivory palaces" (Psalm 45:8) within the Holy City where Mount Zion and God's throne are located: ". . . Mount Zion . . . God is known in her palaces for a refuge" (Psalm 48:1-3). "Refuge" means "high place" (compare Hebrews 12:22). All of this collaborates with Jesus' own statement: "In my Father's house *are* (in existence at the time He spoke these words) many mansions (literally "abiding places"); if it were not so I would have told you. I go to prepare a place for you" (John 14:2).

To the casual observer this text could be easily misunderstood. It does not mean that Jesus was going to prepare many mansions, but that He was going to prepare a *place* (singular) for you. The former refers to that which already existed; the latter to that which is yet to come into existence.

ANGELS' ISLANDS

The "ivory palaces" apparently were created for Christ (2 Corinthians 8:9), and not for the angels, as the palaces had a prior existence. For angels God created the "Heavens of Heavens," but both the "ivory palaces" and the "Heavens of Heavens" abide within the great habitation of eternity. These places existed in the eternal past long before the planetary heavens and the earth, as we know them, were created. To God, eternity is a "habitation;" but to creatures under "time," it is timelessness.

God did not create "mansions" (abiding places) to be vacant, so presumably He created angels such as archangels, cherubim, and seraphim to occupy these palaces within the Holy City for the purpose of worship and special ministries to God. We can only assume that the other angels were created when Lucifer, one of the archangels, was enthroned on the footstool of God and given a contingent of angels. Ostensibly, other angels were assigned to the remaining archangels, Gabriel and Michael, who have continued as Holy Angels, whereas those under Lucifer became fallen angels. The Holy Scriptures reveal that there is ". . . an innumerable company of angels" (Hebrews 12:22).

Since the "Heavens of Heavens" outside the Holy City were created for angelic beings, they may be described as angels' islands in the vastness of eternity, God's habitation, where they are profusely scattered throughout its limitless expanse. The "Heavens of Heavens" were not designed or created as suitable places for the redeemed. Something special was to be created for them, called the "New Heaven," which is located in the second heaven (2 Corinthians 12:1-5; Revelation 21:2). Jesus referred to this as the "place" that He went to prepare for the redeemed between the cosmos and the Throne of God. It will extend over the infinite length and breadth of a new dimension within the habitation of "time." Because eternity is limitless, it is also timeless, since time is reckoned by the rotation of celestial bodies.[1] Scientists speak of a superspace where they believe there is an infinite ensemble of other worlds outside of our cosmos.

THE KING'S PALACE

Psalm 45:15 identifies "the King's palace" within the Holy City as the place from which Christ, the "King Eternal" reigns (I Timothy 1:17). His palace is surrounded by numerous habitations which are called "ivory palaces" (Psalm 45:8). These, as we shall show later, contain the wardrobes of the royal vestments. The "King's palace" is an archaic name for the Holy City, the New Jerusalem.[2]

The Heavens of Heavens are universes outside of the Holy City with their comparable and habitable solar-like systems that extend far into the profundity of eternity from God's throne.

PALACES WITHIN THE ETERNAL CITY

The Psalmist (Psalm 48) refers to "palaces" within the Holy City. They are like inner chambers and some perhaps serve as inns for special guests: "Now there was a day when the sons of God came to present themselves before the Lord, and Satan came also among them" (Job 1:6). According to the Apocalypse (Revelation 4:4, 5:11; 8:3-5; 7:9; 14:3), occasionally there will be massive assemblages in the Holy City near God's throne. The prophet Ezekiel indicates that this is a place of tremendous activity (Ezekiel 1:26; 10:1).

Since the Royal Redeemed are to reign with Christ as "kings" (Revelation 1:6; 21:24) and are promised special audiences with God, as well as the honor of sitting on a throne adjacent to Christ, this has a special meaning to believers in Christ: "To him that overcometh will I grant to sit with me in my throne, even as I also overcame, and am set down with my Father in his throne" (Revelation 3:21).

The fact that the Royal Redeemed are to reign as kings does not imply that they will be confined either to the throne or within the Holy City. The title simply signifies the office or position of honor bestowed upon them. Some "palaces" represent centers of power within the Holy City.

[1] Or any point of references.

[2] When the Holy City was later created, it enveloped the "King's Palace" and the "ivory palaces."

5

The Three Heavens

The New Heaven has to be the second heaven if the Holy City is to descend out of the third heaven onto the New Earth.

In order for us to grasp the significance of what our Lord meant when He said, *"I go to prepare a place for you,"* let us hear what the Apostle Paul had to say in 2 Corinthians, Chapter 12, regarding the unusual experience he had when he was stoned and left dead outside the wall of the city of Lystra: ". . . whether in the body or out of the body, I cannot tell . . . ." But he said he was ". . . caught up (to the third heaven) into paradise . . ." There he saw the deceased Christians in occupancy.

Paradise is near God's throne as evidenced by the ensuing texts: *". . . to be absent from the body (is) to be present with the Lord"* (2 Corinthians 5:8). Jesus said to his disciples: ". . . you will see me" (John 14:19, *Phillips*. See Romans 8:34; Colossians 3:1; Hebrews 1:2).

GENESIS II

One hears little, if anything, about the second heaven. Yet if Paradise is located in the third heaven (2 Corinthians 12:1-4), then it follows that there must be both a first and second heaven.

The third heaven encompasses the location of God's throne,

which is in the Holy City, as indicated in Revelation 21:2: "And I, John, saw the Holy City, New Jerusalem, *coming down from God out of heaven* (third heaven), prepared as a bride adorned for her husband."

In reference to this Holy City settling upon the New Earth we read: "Behold, the tabernacle of God is with men, and he will dwell with them, and they shall be his people, and God himself shall be with them, and be their God."

So it ensues that the *place* Jesus went to prepare for the redeemed was the second heaven. This place will remain concealed, as it were, behind the stage curtains until the cue is given for its revelation.

For almost two thousand years Christ has been preparing this *place* with wonders that will surpass all human comprehension: "But as it is written, Eye hath not seen, nor ear heard, neither have entered into the heart of man, the *things* which God hath prepared for them that love him" (I Corinthians 2:9). Christ is still in the process of preparing the second heaven and filling it with incomprehensible surprises for the redeemed.

The first heaven is the planetary one and, wonderful as we know it to be, it is incomparable with the glories that are extended so lavishly throughout the second heaven: "For I reckon that the sufferings of this present time are not worthy to be compared with the glory which *shall be revealed* in us" (Romans 8:18).

The first heaven is expendable and will be incinerated some day as just so much rubble (2 Peter 3:12): ". . . the first heaven . . . .passed away . . . " (Revelation 21:1).

GOD'S STARS

The message of the following verses: "When the morning stars sang together, and all the sons of God shouted for joy" (Job 38:7) indicates that the imminence God places upon the the redeemed far outweighs the high-ranking positions of angels: "What is man, that thou are mindful of him? or the son of man, that thou visitest him? Thou madest him a little lower than the angels; thou crownest him with glory and honor, and didst set him over the work of thy hands" (Hebrews 2:6,7).

Some day the Royal Redeemed will "judge angels" (I Corinthians 6:3) and will become their superiors. This grace of God which will be extended to them astounds angels, to wit: ". . . which things the angels desire to look into" (I Peter 1:12).

Jesus Christ, when He was incarnated in flesh, was made lower than the angels, according to Hebrews 2:9,10. This was designed to elevate the Royal Redeemed to the highest position of honor which God could bestow on any of His created beings: "For it became him, for whom are all things, that by whom are all things, *in bringing many sons into glory* (that is, to be like Christ and share mutually in His glory) . . ." (Hebrews 2:10; Revelation 21:7).

JESUS CHRIST THE ENTREPRENEUR

For two thousand years Christ has been planning, designing, and creating the second heaven for His own future corporate wife (Revelation 19:7,8). He still has one thousand years of construction work to be completed during the Millennium. The unveiling and occupancy are scheduled to occur when Planet Earth is transplanted to become the New Earth in the New Heaven. The Royal Redeemed will ascend with the earth together with the redeemed from the Millennium.

The newly created heaven has to be the second heaven as the Holy City is to *descend out of the third heaven* onto the New Earth. This will set the stage for the unveiling of the completed project. It is in this Holy City, the New Jerusalem, where the throne of God will be established among the Royal Redeemed in immortal bodies.

When Jesus ascended from the earth, the Apostle Paul said He ascended, ". . . that he might fill all things" (Ephesians 4:10). The only heaven not yet filled with the incredible things from his creative hands is the second heaven. What a magnificent place it will be!

You can rest assured that nothing is being overlooked. As Creator in the past, He has filled the first heaven with planets, stars, galaxies, and other supernatural splendors. The third heaven is resplendent with all that envelops the Father's throne—Paradise and its extended proscenium. The "Heavens of Heavens," the abiding places of angels, undoubtedly are ecstatically satisfying to those extra-celestial occupants: "It is he that builds up the *stories* (universes) in the heavens . . ." (Amos 9:6).

In order for the Heavenly Father to justify the transfer of His central throne of power to the redeemed earth within the New Heaven (second heaven), it must meet the most demanding standards of all of God's creations.

No one has seen the effulgence of the second heaven, future home of the Royal Redeemed and the redeemed other than the Royal Wife (Queen of Christ), but God Himself. It has been hidden from all peering eyes, but it will be a place of overwhelming ecstasy (I Corinthians 2:9-12; 13:12).

NO MANSIONS IN THE SKIES

Many Christians are under the impression that they will inherit a physical structure in heaven called a "mansion." This is totally inconsistent with Scripture. A building will be as unnecessary for survival in heaven as it was for Adam and Eve before their sin in the Garden of Eden. First, if there were such an abode, it would be *roofless* and without *walls,* just as the Holy City has no roof over it. Also a *heating* and *airconditioning system* would be superfluous; second, there need be no *doors* to lock because of fear; third, a *bathroom* is unnecessary since the Royal Redeemed will no longer possess mortal and corruptible bodies; fourth, there is no need for a *bedroom* since there will be no marriages in heaven according to Jesus, and there will be no night there. Since there will be no beds to make or linens to clean, consequently there will be no *laundry room.* Furthermore, the garments to be worn, such as the robe of righteousness and the garments of salvation, will remain spotless; fifth, there will be no *kitchen* or *dining rooms* since the menu will be from the trees of life and not from the corruptible foods that require *cooking,* thus there will be *no dishes and utensils to be washed.* In addition, there will be *no garbage or trash* for disposal; sixth, there will be no need of a *living room* since the people will be living comfortably in the open. Of course, there will be no *broom closets* for dusting and housecleaning; and seventh, no *garages or stairs* are necessary since the immortal bodies will possess a new type of mobility.

If, for some reason beyond our present knowledge, any mundane labor will be required, we are reminded that angels are "...all *ministering spirits,* sent forth to *minister* for them who shall be heirs of salvation" (Hebrews 1:14). The Royal Redeemed in their glorified bodies will not require any physical structure. The Holy City will provide whatever will be necessary.

A PLACE—NOT A HOUSE

What a *place* Christ has gone to prepare—a macrocosm with enormous galactic habitations for the multitudes who are destined to occupy it, such as the millions of Royal Redeemed, the redeemed from the present earth, and those to be redeemed in the New Heaven. This includes those prematurely cut from life whose names are still in the book of life. Besides these countless numbers, it includes innumerable people to be born of natural parents in the extended habitations of the New Heaven before death is destroyed and before all the redeemed come into immortal bodies. The Royal Redeemed will minister as "priests" to these people (Revelation 1:6, I Peter 2:5,9).

We must not narrow our vision and limit the potential number of redeemed to a few religious denominational adherents, or even to seven thousand years of human existence on our planet in its present cosmic location. As we shall later see, millions of redeemed will emerge from the Millennium as well as untold trillions within the New Heaven.

It will take one thousand generations of one thousand years each, or a total of one million years, before the full fruit of redemption will be realized (Deuteronomy 7:9; I Chronicles 16:15; Psalm 22:20; 105:8; 112:2; Isaiah 60:22; Revelation 20:4). Only then will these words have full meaning: "... Then cometh the end, when he shall have delivered up the kingdom *to* God. even the Father; when he shall have put down all rule and all authority and power" (I Corinthians 15:24).

WHEN DEATH WILL BE BURIED

The word "end" means the termination of mortality as the following text shows: "For he must reign, till he hath put down all enemies under his feet. The last enemy *to be destroyed is death"* (I Corinthians 15:25,26).

When death is destroyed, immortality will be shared by all of the redeemed. No unredeemed will then remain in the New Heaven. No one will continue confined within a mortal body, or be confined even to the second heaven for they "... shall inherit all things..." (Revelation 21:7). The redeemed will have the capability of taking full possession of all the ever-expanded and expanding Kingdom of God. They will enjoy that glorious magnitude throughout all eternity: "...of his kingdom there shall be *no end"* (Luke 1:33).

One envisions the Almighty as a great God with magnificent plans. It is inconceivable that such a generous and all-powerful Creator will narrow His program to the short-sightedness of a few religionists who have "...limited the Holy One of Israel" (Psalm 78:41).

6

Generations of the Heavens

The originals are eternal. That which is duplicated under creation is mortal.

The Bible speaks of the "generations" (plural) of the Heavens (Genesis 2:4). The word "heavens" includes both the celestial and the extraterrestrial heavens, but let us first consider the original heaven which God created. The word "generations" suggests many acts of creation had been in progress. Jesus as much as declared this when He said, "My Father worketh hitherto, and I work" (John 5:17). The Father never grows weary nor faint (Isaiah 40:28). He may pause in order to reflect upon his achievements (Genesis 2:2,3), but His energies are abounding and continuous. So in the aeons of the eternal past, constant creations must have been occurring. God must have been originating new and exciting things continually, including creatures that would baffle the imagination and inanimate life that would defy description.

WINGLESS SPACE FLEETS[1]

Revelation 19:11-14 refers to Christ and the Royal Redeemed riding astride white horses as they descend to the earth for the Millennium. What kinds of steeds are these, that without wings can prance, defy gravity, and gallantly parade down the celestial stardust pathway toward the earth? What kinds of ranches must exist to produce and accommodate such magnificent equines?

GIANT OYSTERS AND STRANGE GOLD

How do we explain Revelation 21:21: "And the twelve gates were twelve pearls: each several gate was of one pearl?" The sizes of these gates are enormous. Are there planets or places in the fourth dimensional heavens, or elsewhere, large enough to be inhabited by such giant oysters?

Revelation 21:21 refers to the "street of the city (that) was pure gold, as it were transparent glass." Who is familiar with this type of gold? It is the original. That which we have on earth is only a copy.

Before we ridicule the above as impossible, may we be reminded that astrophysicists believe there exists new elements on other planets or asteroids within our own galaxy. The recent discovery of the strange behavior of baryons as they escape to form their own matter is causing new interest among astrophysicists.

A SUMMARIZED HISTORY OF THE UNIVERSE

A delineation of the cosmos projects the following: First, harmony of energy (the Supergrand Unified Force); second, a series of big bangs; third, mortal particles; fourth, mass; fifth, the four fundamental forces.

Quarks are the building blocks of protons and neutrons which are held together by a strong nuclear force that forms the nucleus or nuclei. In the universe it is gravity, the weakest of the four forces, that holds the large objects together. The atom, like the Universe, consists mostly of empty space.

Out of the original harmony of energy, a series of violent explosions sprouted the tiny seeds of quarks, gluons, neutrons, gravitons, and other exotic particles which gave birth to the universe. Once mass was in the process of formation, the four fundamental forces that govern the universe became the laws of the cosmos (or order). First, the strong nuclear force; second, the weak nuclear force that causes the atom's nuclei to decay into radioactivity. Radiation separates from matter so that photon (light) particles can travel through space; third, electromagnetism and the strong and weak nuclear forces which unite into a single force at ultrahigh energies; fourth, gravity. Mortality is universal as evidenced by the fact that the cosmos has a way of erasing its history.

DENIZENS OF THE ATOMIC DEEP[2]

Swarms of neutrinos, or "ghost particles" (meaning "little neutral ones") are being made continuously in the stars. They are "rest" particles; that is, they are not in constant motion. As mass compresses, neutrons can cause other particles to disappear into a flash of high-energy gamma rays, then into pure invisible energy.

Neutrinos have great penetrating power. Billions of them continually pass through the earth's mass as if nothing exists to impede their penetrating power. The reason is that most of earth's mass is empty space. It is therefore consistent that the soul and spirit entities, because of the unique structure of their properties, can pass through the earth's crust, mantle, and outer and inner cores into the inner chambers of "Hell" for confinement. Scientists tell us that probably about fifty percent of the earth's volume is made of materials unknown to man. Why can't these materials effectively serve as retainers with "gates" and "locks" adequate to hold the soul and spirit entitles (Matthew 16:19; Luke 16:19-31; I Peter 3:19; Revelation 1:18; 9:1; 20:1)?

The fact that neutrinos are "rest" particles does not mean they are less mass, but they exist in such great abundance that ninety percent of the universe is made up of them. They have a strong gravitational effect on other particles. For every proton in the universe there are no fewer than three billion neutrinos. Each group of three billion has a total mass equal to 165 protons. Thus neutrinos within each atom keep all subatomic particles (the strong nuclear force) within the confined limits of the atom. For this reason, astrophysicists have deduced that neutrinos account for a "closed" universe, thereby limiting its expansion. This gives us a curved or "bowed" universe (2

Samuel 22:10; Psalm 70:2; Isaiah 1:27), which is consistent with Einstein's theory of a curved universe.

Neutrinos resemble electrons, but they rarely interact with matter; therefore, matter has fallen into its gravitational trap. Some scientists believe that the universe is "closed" because at the precise center is the missing mass of pure energy, called "quarks," with its powerful gravity. Quarks are believed to be located at the center of each of the billions of galaxies to hold them together, which gives the universe and the galaxies their symmetry and equilibrium. Some space scientists believe these central points may be either black or white holes, maybe both (See Romans 1:20; Job 26:13).

Astronomers have theorized that the centers of galaxies, as well as the larger clusters of galaxies, are about ten thousand times brighter than the sun. Therefore, at the centers of large galaxies may be the quasars, the brightest known objects in the universe.

FACTORIES IN SPACE

With the discovery of the "Z-zero" particles in 1983, physicists believe they have uncovered the last of the twelve smallest known building blocks of matter. Electromagnetism, and the weak nuclear force are two manifestations of a single force called "electroweak." It is now possible to consider the theory of a single unifying force that drives all nature. This eluded Einstein as he pursued his concentrated search for the "unified field theory."

In the intense heat of pure energy, protons and neutrons are formed. They in turn, produce hydrogen and helium, in that order, which are the second greatest abundant elements in the universe. About 400 million tons of helium are sent from our sun into space every second. The solar factories that God created from pure energy have set in motion the mechanism of the laws of chemistry and physics in our third dimension, thus assembling gases and particles which form spiral galaxies and terraformed solar systems. When that which is pure energy turns into matter, we have the intangible becoming tangible, and the invisible becoming the visible.

A white dwarf star cannot produce new energy because it acts as a cosmic thermos bottle holding heat and remaining white-hot for aeons.

BRIGHTNESS OF PURE ENERGY

". . . God is light, and in him is no darkness at all" (I John 1:5). "Who only hath immortality, dwelling in the light which no man can approach unto: whom no man hath seen, nor can see . . . " (I Timothy 6:16). "Now unto the King eternal, immortal, invisible, the only wise God, be honour and glory for ever and ever" (I Timothy 1:17).

Pure energy is invisible and is part of the Creator Who also is invisible. One manifestation of pure energy is "light," which is invisible to man until filtered through creation and made visible as photon particles.

Cosmologists have discovered bright radio spots in the center of black holes which may be coming from the white holes behind them. The hyperspace within those holes contains pure energy which changes the properties of matter randomly with each genesis explosion into various particles.

White holes, which produce copies of many of eternity's originals, must be thinned out in order for created living things and beings to exist and survive. The originals are eternal. That which is duplicated under creation, such as water, light, and plants, is mortal.

A DATELESS UNIVERSE

Cosmologists have approximated that if one traces the motions of the outward moving galaxies backward in time, they all would have come together approximately 15 billion years ago. Cosmologists also speak of first and second generation stars with each star having an approximate life span of 10 billion years. The problem with this theory is that if the entire intergalactic system is moving in a circular, clock-like motion, there is no way to determine its number of revolutions in order to establish its starting point.

In addition to the existence of novas, one reason why astronomers believe that the age of the cosmos is over 10 billion years is that seven galaxies hold the distance record of being 10 billion light-years from the earth.

The Bible speaks not only of "bowed" heavens but also of its "circuits" (Job 22:14); and "the orbs of the heavens" (Mark 13:25, *Moffatt*). All particles of matter in the atom have a circular motion — a model of the universe.

The center of the universe with its vast white hole of pure energy could be the very heartbeat and pulse that holds all of the intercelestial galaxies in their present rhythmic rotations. It is believed that about ninety percent of all invisible mass is located here, which accounts for the enormous gravitational power that prevents the outer fringes of galaxies from spinning away.

THE COSMIC INCUBATOR

The cosmos acts like an incubator in which there are nests of dense dust and electrically charged hydrogen gas clouds. A nearby supernova or radiation from a neighboring star can ignite these clouds and the invisible hot core will begin to eat its way through the clouds like a brush fire and thus a new star is born. Some astrophysicists compare these cluster clouds to cocoons.

When Scripture can be accurately applied to the wonders and mysteries of the universe, then cosmological theories will become more then mere hypothesis.

[1] See page 514.

[2] There are some 200 subatomic particles, plus their antiparticles, in an atom.

7

God's Cyclotron

God took pure energy . . . and as a sower scattered it like seed within the vast universal field of blackness.

God has decreed the boundaries of darkness: ". . . where is the place thereof, That thou shouldest take it to the bound (literally, boundaries) thereof . . . " (Job 38:19,20). "He setteth an end to darkness . . . " (Job 38:3). The cylindrical curtain of blackness served as God's laboratory: "He made darkness his secret place . . . " (Psalm 18:11).

WHEEL WITHIN WHEELS[1]

The prophet Ezekiel had a vision of the wonders of God's creation of the heavens: ". . . the heavens were opened, and I saw visions of God . . . behold one wheel upon the earth" (Ezekiel 1:1,15). The "wheel" within "wheels" took Ezekiel beyond the celestial heaven to see the "Heavens of Heavens" which were teeming in the habitation of eternity.[2] The "rings" were the "glory of God" suspended over each of the heavens like enormous halos: "Through faith we understand that the *worlds* were framed by the word of God . . . " (Hebrews 11:3). These heavens are far beyond the universe and completely out of the range of cosmologists.

According to Ezekiel, eternity is as a huge sea in which are scattered innumerable island universes (heavens). God has entrusted these universes to four living creatures (cherubim) that serve as His appointed guardians over all of the wondrous creations (Ezekiel 10:1).

CELESTIAL WEAVERS[3]

God took pure energy within His hand, as it were, and as a sower, scattered it like seed within the vast universal field of blackness. The four great cherubim under the direction of the Holy Spirit were guiding the emerging stars, planets, and other celestial objects into their respective places where they were woven into beautiful patterns and designs: "By his Spirit he hath garnished the heavens . . . " (Job 26:13). " . . . Whither the (Spirit) was to go, they went . . . And the living creatures ran and returned as the appearance of a flash of lightning . . . " (Ezekiel 11:12,14). " . . . The fire was bright, and out of the fire went forth lightning" (verse 13).

These wheels were "lifted up," suggesting the galaxies were placed in assigned locations. The solar systems within our Milky Way galaxy could look to the prophet like "a wheel in the middle of a wheel."

BIG BANGS[4]

The theories concerning the manner in which the celestial heavens came into existence have been raging for some time. The "big bang," as it is called, has been accepted with reservation by many cosmologists. The basic problem has been the fact that there is an uneven distribution of galaxies in the cosmos. Space scientists are coming to the concept that at the hub of every galaxy is a huge black hole and only recently has the theory of "white holes" been accepted. White holes are the wombs purportedly giving birth to new galaxies and solar systems. If it were not for them, this universe would come to an end with the death of the stars. This is consistent with Scripture which teaches " . . . the generations of the heavens . . . " (Genesis 2:4).

It is conceivable that the Spirit of God not only moved over the submerged earth to reactivate its atoms, but also germinated the seeds of pure energy scattered in the darkness of the cosmos, thereby providing a bountiful harvest of celestial glories and establishing the laws of physics and chemistry in the third dimension.

The huge cyclotron of blackness, the cosmic space, was the birthplace of countless billions of galaxies and solar systems, with their celestial children guided in rotation within the fringes or apogee of the universe.

[1] Astronomers recently have discovered several gyroscope-shaped galaxies. An incadescent ring surrounds each of the galactic bodies spinning in vertical orbits.

[2] "He that sitteth in the heavens..." (Psalm 2:4). God is surrounded by the "Heavens of Heavens" as He sits on His throne in the "Heaven (Capital) of Heavens."

[3] It is absurd to believe that the universe came out of nothing. The "Spirit-Being" cannot be described as a non-entity simply because physicists or chemists cannot analyze ultra-properties in their laboratories. They are confined to researching third-dimensional particles.

[4] A number of voids recently have been located in space that extend as far as 300 million light-years across, giving credence to the concept of many big bangs. See Isaiah 40:22; 44:24; Jeremiah 10:12; 51:15; Psalms 104:2, Zechariah 12:1. Uniform motion is inert. Only through bombardments is there activity and results, possible only under multiple big bangs. Astrophysicists say a bubble (seed) of pure energy one trillionth the size of a proton is adequate to initiate the beginning of the universe. This is near the fourth dimension—the realm of the Creator.

8
Graveyards in the Cosmos

If all matter in the entire universe were crushed into pure energy it could all be held within a person's hand.

The conservation laws of energy have been a foundation for physicists for hundreds of years. Now with the discovery of the radiation decay of the nucleus in the atom this fundamental symmetry has been demolished; therefore, the mortality of the universe is proven.

When a star explodes it is called a nova; or a supernova, if it is a very large star. Some day the entire universe will revert back into pure energy, back to the Creator Himself who first released it.

Quarks, which scientists now believe could be pure energy, are presumed to be the first building blocks of atoms. Here is contained the germ of life passed from the Creator into the atom! If there were no life the atom would be "stillborn;" the soil that is made up of atoms would be sterile and nothing would grow. This principle is true throughout the physical earth.

PINHOLES IN SPACE[1]

Black holes are called "pinholes" or "wormholes" in space because they are scattered throughout the universe. They result from the collapse of huge stars into black holes whose gravity is so strong that they form neutron stars, resulting in the breakup of stars into fragments that quickly die. The black hole, which is in rotation, collects all matter by its gravity within the sphere of its influence. The acceleration is increased as matter, whirling like eddies, gets closer to the core of the black hole and sinks into it to be crushed and disappear into pure energy.

Consider what will ultimately happen to Andromeda, the Milky Way's sister galaxy. It is roughly one hundred thousand light-years in diameter. By the time its apogee reaches the central point of its collapse into a black hole, its eddy will be spinning at the unbelievable velocity of four and one-half million miles an hour. Like water rushing down an open drain, the black hole will attract all of the solar systems containing thousands of millions of times more mass than that of the sun, yet so small in relationship to the entire core region that they will be like grains of sand in the center of a large watermelon. There is such an extremely powerful gravity near a black hole that nothing is left but a void where once was the existence of matter.

Black holes have remarkable digestive systems, transforming all matter back to its primordial state of pure energy! This pure energy is then passed on into white holes which convert it back into matter, but with the possibility of a different geometry, an entirely new sort of physics, a zero-second. This new arrangement of particles is in a different dimensional law, an astrolaw of physics and chemistry beyond our third dimension.

Many scientists believe in a participating universe in which some outside force had to make possible the elements constituting the "big bang." Einstein's theory teaches that there was nothing before the "big bang;" that time is transcendent; and that laws are mutable. The energy seed of our universe had to come from outside the sphere of our cosmos.

Gravity can reduce matter into a solid ball of neutrons weighing billions of tons, yet containable in a teaspoon. All matter is 99.9% space. Black holes have been called the "gates or exits of time" and the rehearsal of the ultimate cataclysmic end of the universe.

A HANDFUL

If all matter in the entire universe were crushed into pure energy, astrophysicists inform us it all could be held within a person's hand. Some astrophysicists believe it is reduced to *spirit* and is completely invisible. Matter of an average collapsed star before it is reduced to pure energy has enormous weight—a matchbox full will weigh ten tons. As someone has so aptly stated, "the Creator can roll the great ball of the universe around in His fingers."

Atoms, in a pound of matter changed into energy, are equivalent to ten million tons of TNT or enough coal to produce an equal amount of energy to fill nineteen railroad freight cars. A little over two pounds of matter could be converted into 25,000,000,000 (billion) kilowatt hours of electric energy. No wonder the nuclear destruction of the universe is described in this manner: " . . . the heavens will pass away with a *great* noise . . . " (2 Peter 3:10).

Many stars end their existence in an explosion of hydrogen, helium, and heavier elements which can become material to form new stars, called second generation stars. Astronomers speak of new generation stars, like our sun, as Population I stars, and the older ones as Population II stars.

WHY THE EARTH MUST BE REDEEMED

Every particle has a time cycle because it is mortal. Pure energy is invisible and therefore is not a particle at all; it is eternal. Albert Einstein said that all matter returns to Spirit. God, as Spirit Being, is eternal Light; not mortal, as is photon light. Photon light is sometimes produced when antiparticles strike certain particles; therefore, photon light is really a third generation particle and is in a class by itself. When photons collide they produce mesons which then decay in less than a millionth of a second.

The only way pure energy can be converted back into matter is by the Holy Spirit because pure energy is not a particle. If this did not perpetually occur, all matter eventually would cease to exist: "Thou sendest forth thy Spirit, they are created: and thou *renewest the face of the earth"* (Psalm 104:30. Compare Job 26:13).

The planet earth must be redeemed and changed from mortality into immortality so that it may be eternally preserved as the New Earth. The universe will come to its demise. It will not be made immortal.

ONLY THE ORIGINALS ARE ETERNAL

How can something emerging from the Eternal Creator be mortal? Simply because it is but a shadow of the real. Mankind was created *in the image and likeness of God* (Genesis 1:26,27). Only the originals are eternal. The copies can become eternal only when transformed into immortality by the power of the Spirit of God (See chapter on Immortality, "The Perfect Gift").

A nova occurs when the hydrogen is about half spent. The star first shudders, then expands. As it does so, it becomes brighter, a thousand times more luminous, for a period of seven to fourteen days. As a nova it will reach a temperature of sixty million degrees Centigrade, (127,000,000 degrees Farenheit), three times hotter than the center of the sun, and ten thousand times its surface heat. At this point, it will turn much of its solar system into gas particles.

As a nova begins to contract by implosion it becomes cooler but its gravity increases as it impacts. Pressure builds as matter is drawn in from its solar system, increasing both its gravity and temperature, then it begins to crush and consume all matter out of existence back into pure energy (Spirit).

A star that is too small to exert enough gravity to become a black hole becomes a neutron star. It fluctuates between dying and reviving. When such a star rotates at the fantastic speed of twice every second it is called a pulsar.

[1] The term "black hole" was first coined in 1968.

9

Wombs of the Universe

The word "heaven" comes from the Greek word ouranos *from which we have derived the word 'uranium." It could literally read, "the powers of* uranium *shall be broken (or fissioned)."*

"Yea, the darkness hideth not from thee; but the night shineth as the day; the darkness and the light are both alike to thee" (Psalm 139:12).

God requires no light in His laboratory, for He is light Himself. He created the planetary heavens within this sphere of blackness (Isaiah 45:7; Job 26:7; 38:19).

". . . The Lord thy maker, that hath stretched forth the heaven, and . . . plant(ed) the heavens, and laid the foundations of the earth . . ." (Isaiah 51:13,16). Imagine the Creator gently planting the cosmic eggs of pure energy throughout the globe of blackness, each with the potential of a solar and galactic system. Then picture the Holy Spirit brooding over these cosmic eggs and germinating them as He did the chaotic earth when He resurrected it from its dismal grave (Genesis 1:2),

Space scientists have now found that the solar systems are encased in a magnetic bubble, as discovered by Pioneer 10. This shell is called the heliosphere and is shaped like an aerodynamic teardrop. It is located five to ten billion miles from the suns (stars). It is believed that the fourth dimension lies beyond this bubble where speed and time have new meaning. Scientists refer to universes beyond the third dimension as tachyon universes. Here antielectrons (positrons) travel backward in time, and are alluded to as "parallel universes" to explain the paradoxes. The heliosphere breathes like our lungs, expanding the bubble once every eleven years from the streams of ionized particles, like gusts that blow across the solar system from the solar flares erupting on the sun's surface.

"By his Spirit he hath garnished the heavens . . . lo, these are parts of his ways, but how little a portion is heard of him? but the thunder of his power who can understand?" (Job 26:13,14). White holes are the nests of life, not the graves of matter.

The solar and galactic systems did not come into existence by chance. Each was designed and carefully placed with loving concern according to an intricate plan. God is the caring and providing Heavenly Father of all He has created. His celestial children are as precious as the rest of His creation.

He names the stars and knows their total number (Psalm 147:4). He frequently call them His "host" (Psalm 33:6): the same way that He addresses His angels and the descendants of Adam and Eve (Luke 2:13; Genesis 2:1). Even the fallen angels are called His "host" (Daniel 8:11,12) and presumably all are named (Revelation 9:11).

Just as the Creator knows about the death of every bird (Matthew 10:29,30), so He knows about the death of every star even to the second generation and beyond; and all about His celestial grandchildren in the cosmos.[1] He knows the divisions of the celestial heavens: "Thus the heavens and the earth were finished, and all the *host* of them" (Genesis 2:1).

While atoms are the building blocks of matter, spiral galaxies are the building blocks of the universe. Galactic collisions occur quite frequently on the cosmic calendar. Although galaxies were originally believed to be spiral in shape, collisions among galaxies have caused elliptical, toadstool, ring, globular, pinwheel, needlespike, arc, and other varying distortions or patterns. We have an orderly rotating universe, not an escalating or expanding one.

CELESTIAL GARDEN

To create the celestial garden the Holy Spirit compressed and heated the pure energy seeds until they burst forth like beautiful empyreal flowers with petals spiraling around the galactic centers in assigned plots of space. The universe now looked like a gorgeous supernal garden with many designs of galactic spirals like "wheels within wheels" interspersed with comets, shooting stars, asteroids, meteors, and other cosmic phenomena.

COSMIC CHAMBER

". . . I (God) *plant* the heavens . . . " (Isaiah 51:16); ". . . *he prepared* the heavens . . . " (Proverbs 8:27); ". . . hath *spanned* the heavens . . . " (Isaiah 48:13); ". . . *spreadeth out* the heavens" (Job 9:8); ". . . he hath *garnished* the heavens . . . " (Job 26:13); ". . . who stretcheth the *line* upon it" (Job 38:5).

Astrophysicists inform us that globular clusters, which contain many galaxies, show variation of age; and because they are very old, they help cosmologists uncover secrets about the birth of the universe. The areas of uneven nothingness conflict with the theory of a single birth explosion. Moreover, the concept of many "big bangs" helps to explain the startling differences in development of the universe.

"He stretcheth out the earth over the *empty place* . . . " (Job 27:7).[2] In this specifically prepared prodigious chamber, surrounded by the beauty of the cosmic garden, the Creator occasionally enjoys solitude, contemplation, and reflection: "He maketh darkness *his secret place* . . . " (Psalm 18:11). ". . . The darkness and the light are both alike to thee" (Psalm 139:12).

CELESTIAL NURSERIES

Space scientists have observed immense jets of gas clouds shooting forth from the centers of galaxies. We submit the possibility these may be issuing from white holes which are the wombs of second and third generation galaxies. Black holes only come into existence with the death of stars.

Astronomers have discovered scores of these massive molecular clouds at the outer fringes of our galaxy which combine to form spiral arms each stretching up to thirty thousand light years. These clouds are the stuff out of which other solar systems are made.

Primordial white holes are the wombs in which pure energy is the germ of life that eventually gives birth to the universe. Astrophysicists believe a massive white hole also may exist at the precise center of the universe.

Physicists are of the opinion that the first particles to emerge were the positive charged protons. These God took and combined with various subatomic particles thus creating over one hundred different kinds of atoms, each a world in itself. These atoms are the building blocks of matter and final creation. They are miniature models of the universe. Gravitons keep electrons orbiting like small planets. They bend the photons, or particles of light, into circular motion, holding every particle within its sphere of the atom and the universe. Scientists are hopeful that in the unified field theory, and through their experiments, they will be able to reduce the third dimensional physical world to two particles and one force.

CELESTIAL TIME BOMBS

White holes may not necessarily have black holes adjacent to them from which to replenish pure energy. If this is true, then the lack of pure energy within the white holes could reach a critical point, which would make them potential time bombs in the universe. As their supply of pure energy reaches a point of diminishing returns, they begin to consume their matter and turn themselves into cannibalistic black holes.

In the light of this, Scripture predicts a time when the white holes will coordinate with the black holes and simultaneously begin to devour all the galaxies. Perhaps the Creator may synchronize these celestial time bombs with the massive one at the center of the universe. This would produce a chain reaction of such magnitude as to turn the universe into one gigantic explosive inferno and huge black hole as predicted in 2 Peter 3:7,12 and Hebrews 1:10-12.

GOD'S ATOM SMASHER

Jesus Christ in Luke 21:26 said, "Men's hearts failing them for fear, and looking after those things which are coming on the earth: for the *powers of heaven shall be shaken."*

The word "heaven" comes from the Greek word "ouranos" from which we have derived the word "uranium." This verse could literally read ". . . the powers of *uranium* shall be broken (or fissioned)." Coincidence? Hardly!

Since the atom is a miniature universe consisting of elements similar to those of the cosmos, the laws that govern the atom also govern the heavens. Collisions occur when the atom is fissioned, resulting in a chain reaction. It is the disorientation of the particles that results in the death of the atom *per se.*

A CRISIS GENERATION

Jesus stated in His Olivet discourse: "And when these things *begin* to come to pass, then look up, and lift up your heads; for your redemption (the pre-Tribulation Rapture of believers) draweth nigh" (Luke 21:28). "Verily I say unto you, *This generation* (normal life span at the time) shall not pass away, till *all* be fulfilled" (verse 32. See also verse 26).

The earliest of these predictions concerns the splitting of the atom (uranium 238) on December 2, 1942. It was on that date in Chicago, Illinois, that man's first reactor successfully split the atom. The first explosion of an atom bomb was at Alamogordo, New Mexico, on July 16, 1945.

"Generations" (life spans) have been various lengths throughout history. Christ's return to earth can be expected within the life span of the generation that witnessed the above events: "And then shall they see the Son of man coming in a cloud with power and great glory." The fact that God has set the "time" (generation) does not justify man's setting dates.

EACH ATOM AN ARMED CAMP

An atom has approximately 200 elementary particles, each with its own counterpart, or antiparticle, going in opposite directions at the incredible speed of a billion circuits around the nucleus in a millionth of a second. Any collisions between the protons and the antiprotons would be so great that a thousandth of an ounce could take a spacecraft to Mars. Only electromagnetism keeps them from colliding, else they would destroy each other in a flash into pure energy. Every atom is literally an armed camp comparable to 200 marauders with loaded pistols. And to think there are more than 100 billion billion atoms in a drop of water!

What sagacity the Almighty had in creating the complex atom with its many subparticles! Most of the atom consists of space with each particle having its own distinctive characteristics, functions, and purposes.[3]

This summation must be carefully observed: First, the *disordered universe.* "... The foundations of heaven *moved and shook* (past tense), because he (God) *was* wroth... He *bowed* (meaning, "to bend") the heaven also..." (2 Samuel 22:8-10). These facts emerge: 1) God *shook* the universe ("shook" from the Hebrew root meaning, to agitate violently or to toss); 2) God then *moved* the universe ("moved," meaning to fall out) into a new location; 3) He *"bowed"* (that is, bent; put the cosmos into an orb or global contour). It is no problem for the Almighty Creator to move and transplant any one or more of His universes within the "Father's House."

We see evidence of this tremendous alteration in the cosmos: 1) the *scattering* of the celestial galaxies that created collisions among them, changing their once circular shape into a variety of distortions; 2) the *breaking off* of fragments from planets causing meteorites and comets to bombard celestial objects; 3) the *death* of stars into black holes, to occasionally release invisible antimatter particles that roar out of these huge black hole cannons, burning as they bore their ways deep, and, sometimes, completely through planets. Many scientists believe this is what happened in Siberia in 1908; 4) the mortality in the universe that is causing it to become like a worn-out garment. Therefore, God must ultimately terminate its present death pangs in a nuclear furnace (Isaiah 50:9; Hebrews 1:11).

Second, *the disordered atom.* "Men's hearts failing them for fear, and for looking after those things, which are coming on the earth: for the powers of heaven (Greek, "ouranos," from the etymological root, "uranius") shall be shaken" (Luke 21:26). This prophecy concerns an event that will cause apprehension and fear, because of the discovery of nuclear power within the atom.

Third, *the disordered earth and heaven.* ". . . Yet *once more* I shake not the earth only, but also heaven" (Hebrews 12:26). This prophecy must yet be fulfilled as follows: 1) the earth must first undergo violent *changes* in its topography; 2) the Pangaea is to return; 3) the universe will finally end its history *in* a series of gigantic nuclear explosions.

[1] Some astrophysicists believe that when a star disappears into a black hole it may reappear beyond the white hole in another galaxy. According to Scripture, it cannot reappear outside of our cosmos in one of the fourth-dimensional universes.

[2] Since angels are fourth-dimensional spirit beings, they have the capacity to instantly materialize into the complexity of a physical body. God, as *the* Hypostasic Spirit Being, with substance, should have no difficulty bringing from Himself the existence of a materialistic third-dimensional world.

[3] The Letter "o" is halfway between an atom and the earth in size. The nucleus is 100,000 times smaller than the atom itself. The smallest visible objects are about the size of light's shortest wavelength—one-sixteen millionth of an inch.

10
Flying Mountains

Celestial dominions are habitable places suitable only for spirit beings. *Jesus Christ referred to them as "dry places."*

"And the second angel sounded, and as it were a *great mountain* burning with fire was cast into *the* (Mediterranean) *sea:* and the third part of the sea became blood" (Revelation 8:8). "And the third angel sounded, and there fell a great star (meteor) from heaven, burning as it were a lamp . . . " (Revelation 8:10).

Space scientists have discovered that in our solar system the distant worlds are composed of hard rock, ice, stormy gases, whistling plasmas, methane oceans, sulfurous volcanoes, and other strange unexpected substances.

Scripture refers to asteroids as "flying mountains" which are wayward space rocks that wander into our solar system from interstellar space. There are approximately 1500 asteroids one-half mile or larger in diameter and they are generally irregular in shape as a result of weak gravitation. These are congregated primarily in the area between the orbits of Mars and Jupiter. There are also about 200 thousand asteroids circling the sun, many of which are believed to have an abundance of diamonds. An estimated 560 meteorites strike the earth each year, and one thousand tons of cosmic dust fall to the earth each day.

When a killer asteroid impacts the earth, burning as it enters the atmosphere, its destructive power is equivalent to one million Hiroshima-type bombs. One can readily visualize the tremendous tidal waves that will result as Jesus predicted in Luke 21:25: ". . . the sea and the waves roaring."

The scenario shows the asteroid first striking at the center of the kingdom of Antichrist, followed by showers of meteors bombarding his Babylonian Empire, as well as much of the rest of the earth. Then, according to Revelation 8:10,11, a comet will sweep across the earth. Astronomers indicate that comets are full of poisonous cyanides, so if one should sweep or plunge onto the earth its cyanogenous fallout would pollute the waters. The Scriptures refer to this as: "Wormwood . . . and many men died of the waters, because they were made bitter" (verse 11).

The headquarters of Antichrist as the "Beast" will be Rome, the megalopolis of his world commercial trade center, which is located in the heart of the Mediterranean Sea (once called "The Sea" and "The Great Sea"): "And I stood upon the sand of *the sea,* and saw a beast rise up out of *the sea* . . . And all that dwell upon the earth shall worship him, whose names are not written in the book of life of the Lamb slain from the foundation of the world" (Revelation 13:1,8).

INTERGALACTIC PLAZAS

"For we wrestle *not against flesh and blood,* but against principalities, against powers, against the rulers of the darkness of this world, against spiritual wickedness *in high places"* (Ephesians 6:12; Colossians 1:16).

The material emanating from the white holes in space could conceivably have its own molecular structure because of its unique arrangement of molecules and atoms. God, by combining atoms in various ways has created mass of various physical characteristics, as exemplified in wood and iron. The combination of these atoms in the fourth dimension (or fifth) in the intergalactic heavens is not necessarily the same as this combination on earth in the third dimension. Therefore, those principalities may have an entirely different character of matter. They would be adequate for the existence of spirit beings, but not necessarily for the survival of mankind.

God has allowed the cosmodominions to be inhabited by Satan and his legions (Colossians 1:14). Thus we note: ". . . the heavens are *not clean* in his (God's) sight" (Job 15:15). Satanic forces seek to control and influence entire kingdoms. In Daniel 10:12,13, the archangel said to Daniel, "But the prince of the kingdom of Persia (a fallen angel) withstood me one and twenty days: but, lo, one of the chief princes came to help; and I remained there with the kings of Persia. . . ."

Celestial dominions are habitable places suitable *only for spirit beings*. Jesus Christ refers to them as the "dry places" to which he expelled the exorcised evil spirits (Matthew 12:43; Luke 11:24), and as such they could not provide for the survival of human beings.

When Satan and his legions are finally cast from these astrodominions, they will be replaced by holy angels who occupy other dominions in the cosmos. From these stations or worlds in the universe will then come the resounding and victorious words: "Therefore rejoice, ye heavens, and ye that dwell in them . . ." (Revelation 12:12).

There are "things" in these astrostations according to 2 Peter 3:10-13: ". . . seeing then that all these *things* shall be dissolved. . . ." This signals the end of all existing materials in any form in the galaxies and habitable places in space which were once occupied by angelic beings.

11

Vanguard of the Godhead

Without the Holy Spirit as the Vanguard of the Godhead, God would still exist in eternity without any creatures enjoying His creation outside of Himself.

There came a moment in the eternal past when the Godhead decided to move into a new direction — to bring into existence other beings to share the abundant glories and accomplishments of the Triune Being:

"... the things which God hath prepared for them that love him ... God hath revealed them unto us by his Spirit ... the things of God knoweth no man, but the Spirit of God" (I Corinthians 2:9-11). Without the Holy Spirit as the Vanguard of the Godhead, God would still exist in eternity without any creature enjoying His creation.

OASIS IN ETERNITY

As one gazes toward heaven and sees the glories that span the evening sky on a clear night, one observes the genius of the creative power of the Holy Spirit. These stars and planets would not be there except by Him: "By his Spirit he hath garnished the heavens ... " (Job 26:13).

As one beholds the magnificent vistas on earth, he observes an abundance of trees, flowers, mountains, rolling plains, streams, and seas, which all became reality because of the Holy Spirit: "When thou sendest forth thy Spirit, (thou) renewest the face of the ground" (Psalm 104:30, *Revised Standard Version*).

The good earth owes its very existence to the Spirit of God. As it once lay submerged in the murky depths, and ice held it in its grip under the darkness in a chaotic condition, the Spirit of God exercised His power of resurrecting the terrestrial: ". . . And the Spirit of God moved upon the waters" (Genesis 1:2).

One can visualize the Holy Spirit caressing the frigid world as it begins to rise from its deep shadows and the darkness begins to fall away like bands that are broken: "When I made the . . . thick darkness a swaddlingband for it" (Job 39:9).

When God was about to create mankind, He turned to the Holy Spirit as His Vanguard: "The Spirit hath made me, and the breath of the Almighty hath given me life" (Job 33:4).

CRADLE THRONE

The Written Word is the by-product of the Holy Spirit: "Every prophetic writing . . . and prophecy did not come by the will of man . . . but by the Holy Spirit" (2 Peter 1:20,21, *Lamsa*).

The dynamic power going forth from the Written Word is "the sword of the Spirit" (Ephesians 6:17).

When the time came for the Redeemer to arrive on the earth the Holy Spirit accomplished this great miracle: ". . . Mary was found with child of the Holy Ghost . . . " (Matthew 1:18). ". . . And thou shalt call his name JESUS: for he shall save his people from their sins" (Matthew 1:21).

As Jesus grew into manhood and the time came for Him to fulfill His mission, the Holy Spirit chose a moment to thrust Him forth on His redemptive mission. Jesus, quoting from Isaiah, declared: "The Spirit of the Lord is upon me, because *he* hath anointed me to preach the gospel to the poor; *he* hath sent me . . . " (Luke 4:18). He followed this with these words: "This day is this scripture fulfilled in your ears" (verse 21).

When Jesus faced the crisis moment of His life by having His expendability tested and by going through the gory ordeal of Calvary, it was the Holy Spirit again Who was there: "How much more shall the blood of Christ, *who through the eternal Spirit, offered himself* without spot unto God . . ." (Hebrews 9:14).

Then came the decisive hour for Christ's resurrection when the Holy Spirit brought Him forth from the grave: "For Christ . . . being put to death in the flesh, was quickened by the Spirit" (I Peter 3:18). ". . . The Spirit of him (God) that raised up Jesus from the dead . . ." (Romans 8:11).

SPIRITUAL REBIRTHS

With redemption completed and now in the background, the harvest of souls must become paramount, so we read that the Spirit of God was ". . . reproving the world of sin, and of righteousness, and of judgment" (John 16:8).

Then must come the transformation of lives: ". . . Ye are washed, (ye) are sanctified, (ye) are justified in the name of the Lord Jesus *by the Spirit of our God"* (I Corinthians 6:11). ". . . And it is the *Spirit* that beareth witness, because the *Spirit* is truth" (I John 5:6).

SPIRITUAL DEVELOPMENT

Once the process of the new birth has been accomplished, the Christian faces the ever-presence of sin and the encroachment of wickedness. Again it is the Holy Spirit that comes to the believer's rescue: ". . . ye through the *Spirit* do mortify the deeds of the body . . ." (Romans 8:13). ". . . When the enemy shall come in like a flood, the *Spirit* of the Lord shall lift up a standard against him" (Isaiah 59:19).

WITNESSES FOR CHRIST

The believer in Christ has a message to proclaim and share. The Spirit of God is his ever present enablement: "Who hath made us able ministers of the new testament; not of the letter, but of the *Spirit,* . . . for the *Spirit* giveth life" (2 Corinthians 3:6). "But ye shall receive power, after that the *Holy Ghost* is come upon you: and ye shall be witnesses unto me . . ." (Acts 1:8).

When the early apostolic church was limiting its witness because of prejudice, the Spirit of God had to break it out of its shell to get the message to the world. First, Peter was sent by the Spirit to the Gentiles: "While Peter thought on the vision, the *Spirit* said unto me... go with them..." (Acts 10:19,20).

When Peter arrived at his destination, Gentiles surrounded him. As Peter addressed them, "... the *Holy Ghost* fell on them which heard the word" (Acts 10:44).

Peter then faced the deeply prejudiced church in Jerusalem where he gave an account of his actions. He replied, "... the *Spirit* bade me go... the *Holy Ghost* fell on them, as on us... what was I, that I could withstand God? When they heard these things, they (said), Then hath God also to the Gentiles granted repentance unto life" (Acts 11:12-18). However, this was not adequate. The apostolic church must understand why the Gentiles were receiving the gospel, so God raised up a persecutor of the church, transformed his life, and sent him off to Arabia for a fourteen-year course in a theological seminary under the direct tutorship of the Holy Spirit. (Galatians 1:15-17; 2:1)

Paul, this one-time persecutor, now a graduate from this unique school in Arabia said, "Nevertheless, brethren... because of the grace of God that is given to me of God... to be the minister of Jesus Christ to the Gentiles... by the power of the *Spirit* of God" (Romans 15:15-19).

THE TRUE CHURCH—WHERE?[1]

There is one area in which God has not entrusted anyone but the Spirit of God. This is in the area of calling and making members of the Body of Christ—the Church. This is the sole function of the Spirit of God: "For by *one Spirit are we all* baptized into one body... whether we be bond or free, and have *all* been made to drink into one *Spirit*" (I Corinthians 12:13).

"The *Spirit* itself beareth witness with our spirit, that we are the children of God" (Romans 8:16). "And it is the *Spirit* that beareth witness, because the *Spirit is truth*" (I John 5:6). "For by one offering he hath perfected for ever them that are sanctified. Whereof the *Holy Ghost* also is a witness to us . . ." (Hebrews 10:14,15).

In order to enrich the believer's life, the Spirit of God has made available the fruit and gifts of the Spirit (I Corinthians, chapters 12 and 14; Galatians 5:22,23).

EARTH GETS A FACELIFT

Once the redeemed have been gathered with Christ to receive the rewards and to be eternally united with Him in a proper ceremony (Revelation 19:7-9), they will return to earth. The Antichrist and his followers, unlawful squatters on the earth, will be dislodged. There will be a massive environmental and topographical change.

The shambles of Armageddon and the debris from the judgments of God will be cleared away. The earth will assume a glorious transformation which will be the prelude of the New Earth as follows: "Upon the land of my people shall come up thorns and briers . . . *until the Spirit be poured upon us from on high,* and the wilderness be a fruitful field . . . and my people shall dwell in a peaceable habitation and in sure dwellings, and in quiet resting places" (Isaiah 32:13,15,18).

So as we review the operations of the Holy Spirit we are faced with the magnificence of His work as the Vanguard of the Godhead: *"Who hath directed the Spirit of the Lord?* . . . lift up your eyes on high, and behold who hath created these things . . . by the greatness of his might . . . he is strong in power; not one faileth" (Isaiah 40:13,15,18).

[1] "Church" (ekklesia). *Ek* means "out of" and *kaleo* "to call." Literally, "called out ones."

12

The Angel of the Lord[1]

If "the angel of the Lord" were the preincarnation of Jesus Christ as a theophany, then after His incarnation there would be no more such appearances.

"Behold, I send an Angel before thee, to keep thee in the way, and to bring thee into the place which I have prepared. Beware of him, and *obey his voice,* provoke him not; *for he will not pardon your transgressions:* for *my name is in him.* But if thou shalt indeed obey his voice, and do all *that I speak* then I will be an enemy unto thine enemies, and an adversary unto thine adversaries. For *mine Angel* shall go before thee . . . " (Exodus 23:20-23).

"Therefore now go, lead the people unto the place of which I have spoken unto thee: behold, *mine Angel* shall go before thee . . . " (Exodus 32:34). It is easy to conclude as so many have, that this Angel is the theophany of Jesus Christ, but obviously this is not so according to Luke 2:9-12, because here Christ is in the manger at Bethlehem while "the angel of the Lord" is addressing the shepherds in the field.

Further evidence indicates that if it were the preincarnation of Jesus Christ as a theophany, then after His incarnation there would be no more appearances of "the angel of the Lord." However, we see proof quite to the contrary: The angel of the Lord at Christ's tomb said, "He is not here; for he is risen . . . " (Matthew 28:1-7). And in

Acts 5:19: "But the angel of the Lord by night opened the prison doors, and brought them forth." Acts 12:23 reads, "And immediately the angel of the Lord smote him (Herod), because he gave not God the glory....."

THE UNFORGIVING ANGEL

That the words "Angel" and "Holy Spirit" are synonymous is indicated by the fact that "Angel" is capitalized in Exodus 23:20,23. In order to more fully understand that this Angel of the Lord is the Holy Spirit, observe the following Scriptures:

"... He will *not pardon* your transgression..." (Exodus 23:21).

"Wherefore I say unto you... blasphemy against the Holy Ghost shall *not be forgiven unto men*..." (Matthew 12:31).

"... His countenance was... very terrible..." (Judges 13:6).

"... My name is in him..." (Exodus 23:21).

"But Peter said... Why hath Satan filled thine heart to lie to the *Holy Ghost*... thou hast not lied unto men, but unto *God"* (Acts 5:3,4).

"And the *Spirit* bade me go... what was I that I could withstand *God?"* (Acts 11:12,17).

"... *Obey his voice*..." (Exodus 23:22).

"Wherefore (as the Holy Ghost saith), Today if ye will *hear his voice,* Harden not your hearts, as in the provocation, in the day of temptation in the wilderness... Wherefore *I was grieved* with that generation... *So I sware in my wrath,* They shall not enter into my rest..." (Hebrews 3:1-11,17-19).

"... *provoke him not*..." (Exodus 23:21).

"And grieve not the holy spirit of God, whereby ye are sealed unto the day of redemption" (Ephesians 4:30).

A SECRET NAME

"And the angel of the LORD said unto him, Why askest thou after my name, seeing *it is secret?"* (Judges 13:18).

Have you ever wondered why the surname of the Godhead has never been revealed in Scripture? The Heavenly Father's given name is JEHOVAH (Psalm 83:18); the given name of the Son of God is JESUS (Matthew 1:21); but there is *neither a given name nor a surname for the Holy Spirit.*

This mystery of the surname of the Godhead is conveyed in the formula for baptism: "Go ye therefore, and teach all nations, baptizing them in the *name* (not names) of the Father, and of the Son, and of the Holy Ghost" (Matthew 28:19). *That name is secret.* Jesus is not that secret name as some would imply on the strength of Acts 2:38; 19:5; for His name is not secret (Matthew 1:21). So by using the "name" in baptizing we recognize the unity and mystery of the Godhead.

THEOPHANY OF THE HOLY SPIRIT

It usually is not difficult to envision the embodiment of the Heavenly Father and His Son, the Lord Jesus Christ; but to many the Holy Spirit remains an enigma. The Holy Spirit has various manifested forms as follows: "The Angel of the Lord," dove, fire, wind, and water.

"But David... was afraid because of the *sword* of the angel of the Lord" (I Chronicles 21:30). In Ephesians 6:17: "And take... *the sword of the Spirit,* which is the word of God."

[1] The Holy Spirit is also a Guardian Angel.

13

What is Eternity?

If we could obliterate all points of reference that mark time, such as the stars . . . we would have eternity.

In order to comprehend eternity, we first must understand that time requires at least two points of reference in the universe, and these God created when He formed the planets, satellites, and stars (suns). Time is calculated according to the rotation and the revolution of these individual bodies in relation to one another in their respective galaxies. In our solar system, the length of days, nights, and years is determined by the sun and moon: "And God said, Let there be lights in the firmament (literally expanse) of the heaven to divide the day from the night; and let them be for signs, and for days, and years" (Genesis 1:14). If we could obliterate all of the points of reference that calibrate time, such as the sun, the stars, and the planets in the universe, we would have eternity.[1]

There is also another method to help us understand eternity. Envision the outer rim of our universe, or any other universe an incalculable distance away; beyond that furtherest apogee is—eternity, or nothingness! Time has many potential dimensions; eternity has none.

CURVATURE OF THE UNIVERSE

It is an incontrovertible fact that the universe, with all of its galaxies and inherent solar systems, rotates in a circular motion. The vastness of the cosmos gives the appearance of an expanding universe moving away from the Planet Earth, yet circling within the outer limits of its orb.[2] Dr. Albert Einstein discovered this curvature of the cosmos, and his General Theory of Relativity suggests that the straightest of the lines, from the International Meter Bar to beams of starlight, are not straight at all but are bent along the subtle curvature of the space-time continuum. Even light bends to form an arc because it is made up of photon particles which are subject to the same laws as all other matter. Further evidence of a curved globe-like universe is that everything which is large enough will curve in upon itself because of gravity. The Bible supports this theory and speaks of the heaven as being "bowed" (2 Samuel 22:10; Psalm 18:9; 144:5).

QUASARS

For some time space scientists admitted they knew little or nothing about quasars or how they produce such enormous amounts of energy; however, at this point in time about 1,500 have been discovered. Quasars (quasi-stellar sources) are the most luminous objects in space and pour out energy equivalent to that of 100 trillion stars. Some quasars are thousands of times brighter than our Milky Way galaxy and they are estimated to be eight to twelve billion light years from the earth, which is regarded as being very near the edge of the universe. This confirms the theory that the cosmos is at least that old (see "let" in Genesis 1:14,15). The estimated size of the quasars is about the diameter of a solar system and they appear to be moving like gyroscopes at speeds greater than light. This can be possible only if they consist of pure energy or are fourth-dimensional lights not composed of photon particles. Cosmologists only recently have concluded that the energy of the quasars could be coming through portals or gateways at the fringes of our universe from universes outside our own cosmos. Scriptures refer to them as "doors" or "gates." The fascinating conclusions present two very interesting observations from the Biblical point of view: first, quasars lend strong support to the scientific theory of extra-cosmological universes which Scriptures call the "heavens of heavens." Also, these openings out of our universe could prove to be God's created doorways in and out of our universe through which heavy traffic passes to the throne of God via the beams of eternity's light.

"Lift up your heads, O ye gates; and be ye lift up, ye everlasting doors; and the King of glory shall come in. Who is this King of glory? The Lord strong and mighty, the Lord mighty in battle. Lift up your heads, O ye gates; even lift them up, ye everlasting doors; and the King of glory shall come in. Who is this King of glory? The Lord of *hosts,* he is the King of glory. Selah" (Psalm 24:7-10).

"Open to me the gates of righteousness: I will go into them, and I will praise the LORD: This gate of the LORD, into which the righteous shall enter" (Psalm 118:19,20).

Astronomers regard quasars as very bizarre, not only because of so much power packed into such a relatively small space, but because their brilliance changes so quickly. Astrophysicists call this cosmological red shift.

Light coming into our universe from outside universes (or from the light of eternity) converges near the fringes of our cosmos from the various openings. Since the entire vast universe is rotating, this perhaps could account for the shifting brilliance of the quasars, if its North Pole is pointed toward God's throne (see chart "New Heavens").

UNLIMITED RICHES

Christ, the "King Eternal" who reigned in the infinite past from His King's palace within the Holy City, was endowed with unbelievable riches: ". . . though *he was rich,* yet for your sakes he became poor, that ye through his poverty *might be rich"* (2 Corinthians 8:9). The Royal Redeemed shall "inherit all things" (Revelation 21:7). The Psalmist said in his prophecy: "Wealth and riches shall be in his house (Heaven of heavens); and his righteousness endureth for ever" (Psalm 112:3).

The inheritance of the Royal Redeemed will not be limited to the New Heaven and to the New Earth. For their exclusive pleasure they will have access to explore all of the glorious residences of the angels and all of God's infinite, boundless creation (Psalm 68:33; 148:4).

How tangible and how real are the existing things which God created in the eternal past? God produced for our enjoyment that which is perceptible to our senses. Heavenly things are corporeal and material: ". . . that I may cause those that love me to *inherit substance;* and I will fill their *treasures*" (Proverbs 8:21). And what about the things that the Godhead created for His exclusive pleasure in the remote past previous to the existence of any creatures? These would exceed our wildest comprehension! Some day, as the Royal Redeemed, we shall be in a position to evaluate that which God brought into existence beyond the horizons of what we know exists. In Hebrews 11:3 we read: ". . . things which are seen *were not made of things which do appear,"* a clear statement that *things seen* in our third-dimensional world came from *things unseen* in the fourth-dimensional world.

HEAVENLY COSMETICS

Substances and treasures are part of eternity. In connection with the preincarnate existence of Christ we are told that all of His ". . . garments smell of myrrh, and aloes, and cassia . . . " (Psalm 45:8). We are therefore led to believe that herbs, flowers, and many exotic plants existed in the eternal past before our cosmos was ever created. Why spiritualize or symbolize such statements from God's Word? Such deviations discredit the reliability of God's Word and accomplish nothing.

God created our planet as a very authentic place. We are real. It is therefore consistent to believe God would create tangibility and reality both in the eternal past and future. For what reason would God not do so?

If we can fathom the literal as coming from the hand of the Creator, why shouldn't God elsewhere in His domain bring into existence that which is corporeal? "For the *invisible things* of him from the creation of the world are clearly seen, being *understood by the things that are made,* even his eternal power and Godhead..." (Romans 1:20).

THE BIBLE SAYS WHAT IT MEANS

There was a time when prophecies relating to Israel's return to its homeland were rejected as being too literal and impossible because from a natural standpoint there was no evidence that the Turks, who held possession of Palestine, would ever relinquish it to Israel. So what did the theologians do? They spiritualized those prophecies, but how wrong they have proven to be!

There is no limit to the degree of one's speculation when he pursues such a dangerous course; thus it is possible to open a Pandora's box of strange and weird beliefs. Persevere in the exact sense of God's Word (except where specifically stated otherwise) and the future will vindicate the predictions, irrespective of how impossible they may appear temporarily. God means what He says, and He is fully capable of altering events which may momentarily seem incredulous.

The believer in Christ eventually will discover that his future home is precisely as God described it in His Word—exciting and aesthetic! He will possess an immortal body that has substance. Christ's immortal body after His resurrection was very tangible, but yet could transcend terrestrial laws of physics and chemistry. Natural barriers and third dimensional laws offered no restrictions to His activities and movement.

[1] Earth's revolutions around the sun are a perfect 500,000,000 miles in exactly 365 days, 5 hours, 48 minutes and 49 seconds.

[2] Dr. Einstein referred to our cosmos as, "...three-dimensional spherical shape, or rather of three-dimensional spherical geometry (and) finite." He called all such possible universes outside our own "rigid spheres."

14

Light in its Natural Environment

Since eternal light is not created, there are no restraining influences upon its speed.

"Where is the *way* (originally "home") where light dwelleth? and as for darkness, where is the place thereof,

"That thou shouldest take it to the bound thereof, and that thou shouldest know the paths to the house thereof?" (Job 38:19,20).

When God said, "Let there be light: and there was light" (Genesis 1:3), we know this was not solar light as it did not arrive until the fourth day of creation. From where then did it come?

". . . God is light, and in him is no darkness at all" (I John 1:5). God, as light, is faster than a light-second, which is the distance photon light travels in a second. Proof that God is not limited to the speed of photon particles is evidenced by His ability to give instant response to our prayers and needs: "And it shall come to pass, that before they call, I will answer; and while they are yet speaking, I will hear" (Isaiah 62:24).[1] God is a "Spirit" Being (John 4:24) Who is not restricted to the known laws of physics and chemistry." X-rays are limited to the speed of light but God has ultra-speed, hyper-penetration, and super-observation; He is not restricted.

"And the (holy) city had no need of the sun, neither of the moon, to shine in it; for the glory of God did lighten it, and the Lamb is the light thereof" (Revelation 21:23).

The light with which mankind is familiar is mortal because it consists of photon particles. Evidence that created light consists of photon particles lies in the fact that it has weight which exerts two pounds of pressure upon the earth for every square mile. Further confirmation that photon light is made of particles (quanta) is the fact that it cannot penetrate objects to eliminate shadows. Conversely, the glory of God will cast no shadows on the New Earth because it lacks created particles.

Created light cannot move superluminally because photon particles must pass through "rest" particles such as protons, gravitons, and neutrinos. In contrast, eternal light has no third-dimensional particles to impede its velocity, therefore, it is instantaneous. Since God is both "Light" and "Spirit," created particles are no deterrent to the Creator because He operates under a dual law of physics.

"And the Spirit of God moved upon the face of the waters" (Genesis 1:2); first, upon the earth to free it from its icy grip and send up a canopy of water; and second, above the earth to allow the molecule atoms to emit photon particles.

When God said, "Let there be light: and there was light," the Spirit of God, as energy, excited the molecule atoms in the canopy above the earth, thus creating photon particles to illuminate the earth from the glowing sky.

Since God is a consuming fire, He could excite the atoms to emit photons by heating; or He could let the light of His glory strike the atoms so they could absorb the energy, and then in turn, reflect their own light, which is similar to what the moon does.

WHY LIGHT BENDS

Physicists have determined that light bends by refraction and is affected by "rest" particles whose gravity attracts photons; for example, the enormous number of neutrinos in the universe (three billion to one photon) can gravitationally attract the photon particles and bend them. Because of the rotation of all particles, the photon is also compelled to follow the same circular movement. Thus no created light particles can escape through the fringes of the universe. They must circle within the outer limits of the cosmos.

SOURCE OF LIFE WITHIN PARTICLES

All particles are matter in atomic or subatomic form (except those beyond the third dimension which have their own distinctive properties under different laws of physics and chemistry). Atoms require a source of life or they are nothing more than carcasses. Thus a Creator, as the source of life, is essential to impart intelligence to each particle so that it may perform according to design.

A NONCONSUMABLE CREATOR

The unique characteristics of the eternal light and fire indicate that they have no photons. Fire associated with Deity has its own strange qualities and behavior such as being eternal, self-perpetuating, and inexhaustible (Deuteronomy 33:27; Romans 1:20; I Timothy 1:17; Hebrews 9:14; Genesis 21:33; Psalm 103:17; Isaiah 40:28). This explains why it is distinctive from combustion which requires something consumable (Deuteronomy 4:24; Hebrews 12:29).

Consider these examples: the burning bush which was unconsumed in the presence of Moses (Exodus 3:3); and the day of Pentecost when one hundred and twenty persons experienced "... cloven tongues like as of fire (which sat) upon each of them" (Acts 2:2). No physical harm or hurt was experienced by any of them; rather, they were infused with an uplift of spirit and spiritual dynamics not previously possessed. And the people became very bold in their witness and were willing to become martyrs for their Christian faith.

The irradiance of Deity is manifested as either fire, light, or glory. While having the unusual distinct trait of being harmless when imparting elevating benefits, it also has a searing effect unlike any known characteristics of fire. In contrast with light and fire, it demonstrates intelligence.

We read in I Kings 18:38, that while in the presence of Elijah and the prophets of Baal on Mount Carmel "... the fire of the Lord fell and *consumed* the burnt sacrifice, and the wood, and the stones, and the dust, and licked up the water that was in the trench."

There are mysteries about the Creator which are humanly inexplicable. God possesses capacities that defy human comprehension as displayed in His abilities, intelligence, wisdom, and perfection. Let us not limit the Creator in any area; God in action is the Almighty moving in unlimited dimensions.

GAZE AND BLAZE

Moses had another experience of interest besides that of the burnlng bush. In this instance we read: "And he (God) said thou (Moses) canst see my face, for there no man shall see me, and live." To assure his protection God said, "... when my glory passeth by... I will cover thee with my hand while I pass by" (Exodus 33:20-23).

For fear the Israelites might inadvertently behold His eternal glory and meet with disaster, God told Moses, "Go down, charge the people, lest they break through unto the Lord to gaze, and many of them perish" (Exodus 19:21).

SONLIGHT

The manifestations of God by fire; light, and glory are responsible for everything that has come or will come into existence. Besides having the characteristics of consuming or discharging benefits to God's creatures, they also have the capacity to bring into existence tangibles which beautify and extol the magnificence of the Almighty.

In seeking to explain the mystery of the incarnation of Jesus Christ, John referred to Him as the Light of Deity and that "All things were made by him; and without him was not anything made that was made" (John 1:3).

The Apostle Paul described His appearance in this fashion: "And without controversy great is the mystery of godliness: *God was manifested in the flesh,* justified in the Spirit, seen of the angels, preached unto the Gentiles, believed on in the world, received up into glory" (I Timothy 3:16). Again in I Corinthians 2:7: "But we speak of the wisdom of God in a *mystery*, even the hidden wisdom, which God ordained before the world unto our glory." Paul calls this "... the mystery of Christ" (Colossians 4:3).

THERAPEUTIC LIGHT

In John 3:34 we read: "... God giveth not the Spirit by measure unto him (Christ)." In Colossians 2:9: "For in him (Christ) dwelleth all the fulness of the Godhead bodily."

The expression of God was revealed in the manifestation of Jesus Christ. Before His appearance, God was known only in the abstract.

This "Light," John tells us, had extraterrestrial therapeutic effects, such as "eternal life," on those who received it in the Person of Jesus Christ: "In him was life; and the life was the light of men... That was the true Light, which lighteth every man that cometh into the world" (John 1:4,9).

Any attempt to articulate this mystery of God in the flesh flounders unless it is fortified with faith. It works, and that should determine its true value.

FAITH IS NO COPOUT

We must give credit to God that there are some areas about His Divinity which we cannot comprehend; that is why He is God. If we could understand everything about Him, we would equate with Him.

Some proclaim that faith is a copout. It would be a sad commentary if our limited thoughts became the sole indicators as to whether or not something exists.

"Hereby know we the Spirit of God; every spirit that confesseth that Jesus Christ is come in the flesh is of God: And every Spirit that confesseth not that Jesus Christ is come in the flesh is not of God; and this is that spirit of antichrist..." (I John 4:2,3). "Therefore if any man be in Christ, he is a new creature: old things are passed away; behold, all things are become new" (2 Corinthians 5:17).

The incarnate Christ brings therapeutic value from God, and it is for all who will by faith receive Him. He is the Physician with the necessary remedy to cure all of the contagious and infectious ills infesting mankind because of sin.

SEVEN LAMPS OF GOD

"And out of the throne proceeded lightnings and thunderings and voices: and there were seven lamps of fire burning before the throne, which are the seven spirits of God" (Revelation 4:5).

The Apocalypse is permeated with dynamic portrayals of the Almighty God. Nowhere is this more eloquently demonstrated than when God created light. From the spectrum of light, and the electromagnetic field that surrounds it, have emerged lightning, thunder, and voices in scientific achievements. Many technological innovations have come into existence to benefit society, as evidenced in the healing arts, communications, and transportation. Unfortunately, technology also has destructive capabilities causing the creations to sometimes react violently. All of these technological advances represent the ultimate source—the Creator.

The number seven predominates throughout the Apocalypse: seals, angels, trumpets, thunders, eyes, horns, heads, to name a few. Seven portrays the perfection of God and is seen proceeding from Him. There are seven orifices in our heads which serve as receptacles: two nostrils, two eyes, two ears, and a mouth; there are seven continents divided from the Pangaea; and the Church is divided into seven dispensations which characterize the different periods of the Church as symbolized by the selection of seven local churches from Asia.

Our electronic age has benefitted from the knowledge about created light. Its electromagnetic field has produced x-rays, radio waves, heat or infrared rays, visible light waves, ultraviolet, and gamma rays.

BODIES OF ULTRA-MATTER

No man in his physical body has seen or could hope to see God in all His majesty and power—and survive. Enoch and Elijah are no exceptions, nor are those in Paradise. They have assumed paradisiacal bodies that are tuned in with the manifested glory of God (2 Corinthians 5:1-9)

Eventually, when the metabolistic structure is changed, we shall be able to survive and share fully in the wonders of Paradise that exist in the presence of the glory of God. At present the full measure of the glory of God is filtered, enabling us to live and to enjoy His creation comfortably in our mortal bodies: "For now we see through a glass, darkly; but then face to face: now I know in part; and then shall I know even as also I am known" (I Corinthians 13:12).

Neither the paradisiacal nor the immortal bodies can be antimatter for then they would destroy matter whenever they come in contact with it. A particle is regarded as antimatter when it strikes certain other particles which cannot survive upon impact, thereby causing all of them to disappear in an instant. Antimatter is used in the medical profession for destruction of antibodies that invade good cells.

Because of their superior quality, paradisiacal and immortal bodies are totally immune to all particles. They are free from any bacteria that are capable of bringing death. Such bodies are tangible even though they possess unknown elements of a higher dimension, as was the case with Jesus after His resurrection. Thomas touched Jesus and suffered no harm: "Behold my hands and my feet, that it is I myself: handle me, and see; *for a spirit hath not flesh and bones,* as ye see me have" (Luke 24:39).

In the Heavenly Father's Kingdom, when all of the redeemed possess bodies of immortality, the new atomic structure may possibly be of ultra-matter foreign to our present understanding of the composition of matter. It may be composed of textures, colors, and aromas that are beyond our present material world.

NEW DIMENSIONAL SENSES

"But it is written, *Eye hath not seen, nor ear heard,* neither have entered into the heart of man, the things which God hath prepared for them that love him" (I Corinthians 2:9).

"Beloved, now are we the sons of God, and *it doth not yet appear what we shall be . . .*" (I John 3:2).

There are eight notes in an octave which form the basis for the musical range. Out of these few notes have emerged great symphonies, classicals, sacred musicales, operas, and other musical scores. But what if the Creator expanded these beyond the eight octave notes, thereby creating entirely new sounds to enlarge and enhance the quality and style of music?

We have six basic colors: three primary (red, yellow, and blue), and three secondary (orange, green, and purple). Out of these six have emerged combinations resulting in thousands of hues. What if the Creator were to increase these basic colors beyond the six? What new beauty would unfold before our vision! This is not as impossible as it may seem. For example, think of photon light as a bundle[2] of different compressed wave lengths, each a different color, from the more compressed short wave length, blue, to the thinner wave length, red. What is to prevent the Creator from creating other wave lengths of colors?

From twenty-six letters in the English alphabet there have been produced untold volumes of fascinating and exciting pieces of literature. What if the Creator were to expand these characters? We would have a greatly increased vocabulary with which to communicate and would need an enlarged lexicon to fit into the new dimension of heaven. Such knowledge by osmosis would give everyone a grasp of the entire English language without the laborious task of having to learn it. Paul, with his limited vocabulary, encountered such extreme difficulty in attempting to express what he saw and heard in Paradise that he abandoned the idea (2 Corinthians 12:1-4). It already has been demonstrated that communication could be greatly enhanced by adding fourteen letters to our English alphabet. This is called the unifon, or phonetic alphabet.

HOSTAGES OF MORTALITY

Eventually the mortals who are redeemed will be released from their earthly confinement with its limited range of senses. Their immortal bodies will become like Christ's because of the aggrandizement of their senses, which will enable them to detect the melodic sounds of cosmos that the Creator has tuned to perfection.

"But when that which is perfect is come, then that which is in part shall be done away . . . For now we see through a glass darkly; but then face to face: now I know in part; but then shall I know as also I am known" (I Corinthians 13:10, 12).

THE MORTALITY OF TIME[3]

Time is not the distance between two points of reference; it is the rate of velocity between them. Time without energy is static, and since energy consumes matter, it is mortal. To put it another way, because energy is required to reach any given point of reference, time is composed of particles; therefore, time is mortal. Time is not an absolute quantity.

As speed is accelerated, time is reduced. The point at which they meet is called "zero-time," which is speed at the velocity of light, or the maximum speed for mass when it turns into pure energy. (See chart on "How increased velocity reduces time"). Since time can no longer go forward in order to break out of the third dimension, speed must become superluminal, thereby thrusting itself backward in time.[4] An object can move both backward and forward in "time" only after it has penetrated the barrier of the third dimension into a fourth dimension. It also must be composed of properties which can survive superluminal velocity. Because time in the fourth dimension can assume any variation (called elastic time), it supersedes all laws of physics and chemistry of a third dimension. This being true, spirit beings, for example, can traverse in a third or fourth dimension without difficulty.

In order for a *person* to enter a fourth dimension, the body must be "changed"[5] to properties that can move and survive superluminal velocity. When the physical body of the deceased Christian is discarded, his spirit and soul are no longer trapped into a third dimension; and interestingly enough, this is the paradisiacal body which enters Paradise backward in time for here time moves slower (2 Peter 3:8).

Time, like photon light, can be consumed in a black hole because when all references that mark time are eradicated, there exists only space-time.

"God hath made the world and all things therein . . . and hath determined the times before appointed . . . " (Acts 17:26). Time in the fourth dimension becomes immortal, thereby those in the heavenly Paradise have a different calibration of time. Some theologians believe that its "day" comprises one thousand years of our solar time, based on 2 Peter 3:8.

Scientists speak of elastic time that can be compressed and stretched; also, of exotic places where time stands still or even ceases to exist. They refer to subatomic particles traveling backward in time because of the laws of quantum physics.

Time is calibrated by the ratio of speed which requires that the energy supplied by matter can be warped and slowed by gravity. It is therefore mortal.

THE MORTALITY OF LIFE

The Hebrews called the soul "the breathing spirit," or "life," *per se.* The attributes of the soul are distinct from those of the spirit, as indicated by Scripture. The properties of the soul, which comprise the memories, senses, and personality of a person, cannot be dissolved when one is deceased, according to Matthew 10:28. The soul can be terminated only by what the Scriptures call the "second death." It will be resurrected eventually, along with the physical body (so it has to have some kind of properties), in order to stand in judgment before the Lord (John 5:28,29; I Peter 3:19; 4:5,6). The spirit is that God-given entity that communicates between the person and his Creator (Romans 8:16) and gives man his superiority over other creatures.

The Scriptures also distinguish between "life" and "eternal life." The unredeemed, who will be lost eternally, will have their lives (souls) terminated somewhere in the distant lake of fire as determined by the decision rendered at the White Throne Judgment under the indictment of the "second death." Eternal punishment is also indicated in specific cases (Revelation 14:9-11). Angel (spirit) beings cannot be annihilated or returned from their fallen state back to the Creator. The "spirit," like God, is eternal (Ecclesiastes 12:7; John 4:24; Deuteronomy 33:27).

DIAMONDS ARE NOT FOREVER

The calculated mortality range of our third-dimensional universe goes from the shortest life-span subatomic particle *meson* of a mere 10 millionth of a billionth of a billionth of a second to the longest life-span which is the *isotape* lead 204 of a half-life of some 14 billion billion years. The most startling discovery is that protons, once held to be the very paragon of stability, actually decay.

The new ARM (atomic resolution microscope) is now capable of giving a clear picture of how atoms are mutually bonded. Scientists have for some time understood the way atoms are arranged in lattices deep inside perfect crystals of solid matter, but nature is rarely perfect (compare Romans 8:21,22). Most materials are laced with impurities that alter atom arrangements within the lattice. Are these imperfections non-existent in fourth-dimensional structures? The ARM is yet unable, however, to image electrons. Microscopists believe this is a long way off.

REMOVING THE STINGER, DEATH

"O death, where is thy sting? . . . thanks be to God which giveth us the victory through our Lord Jesus Christ" (I Corinthians 15:55,57).[6] Inasmuch as strong nuclear forces hold the atom together, weak nuclear forces govern nuclear *decay*. Here lies the culprit, death! God needs only to remove this stinger for life to go forth untarnished and unaffected in all matter.

[1] Proof of God's foreknowledge.

[2] Each wavelength is about half a micron in width, about one hundred and twentieth the diameter of a human hair.

[3] To break out of the dimension of "time" is to enter eternity. Everything existing in the universe is under the penalty of "time," which makes it mortal. Living beings existing in eternity are not under this penalty. God, therefore, has neither a beginning nor an ending. The fact that God created the dimension of "time" in the extension of His creation does not necessarily locate nor confine Him there. It is only one phase of His creation. Therefore, "time"

is only a small segment of God's accomplishments. So to step out of the third dimension into the fourth is to step backward in time because time in the latter moves at a slower pace than in our cosmos (2 Peter 3:8). This is one difference between Eden's Paradise and the Heavenly Paradise.

[4] In an over-simplification, one might state that a person travelling faster than sunlight in a westward direction is going backward in time.

[5] See page 278, footnote 1.

[6] The deadly Type A flu virus is so small that 2-½ million in a line would take up an inch. Seasonal outbreaks of flu affect anywhere from five to twenty percent of a local population. The major pandemics in our century include the great pandemic of 1918, the Asian flu of 1957, the Hong Kong flu of 1968, and the Russian flu of 1977. Influenza usually lasts from six to twelve weeks. The 1918-1919 global pandemic affected some two billion persons, and resulted in twenty million deaths.

15

Doctrine of Unified Creation

The earth, in order to have been the footstool of God in proximity to His throne, must have been created exceptionally beautiful in its primordial state.

Creation is consistent with the second law of thermodynamics which states that the universe began in an orderly manner but is regressing into disorderliness. Everything is deteriorating, collapsing, and breaking down by itself, which is contrary to the theory of evolution that suggests that the universe began in a discordant, chaotic state and is becoming more orderly.

The universe could not have created itself naturally. The probability of getting the first cell is astronomical. If the DNA could be lined up in the right order, even then the probability of matter arranging itself as a bacterium is one chance in 100 billion zeros. The age of the universe, as generally accepted, makes this impossible because there would be insufficient time.

Evolution suffers major weaknesses, but it is the only alternative to scientists who will not accept a Creator. Thousands of scientists are convinced creationists, however, and the number grows rapidly.

In order to develop harmony in the scenario of the divine creation, it is essential that three basic scriptures[1] are understood. Once these are comprehended, science and theology will be brought into closer harmony.

THE ORIGINAL CREATION

"*In the beginning* God created the heaven . . . (Genesis 1:1).

God first brought into existence a place called "heaven" where He established His throne and center of power for all that He would be bringing into existence.

THE PRECELESTIAL EARTH

". . . And the earth" (Genesis 1:1).

God created the earth as His footstool, signifying its proximity to His throne, but subordinate to it. In Genesis 2:4 we read: ". . . the LORD God made the earth and the heavens." Chronologically, therefore, the earth came into existence *before* the present cosmos.

Moffatt translates "heaven" as "the highest heaven" and "the very heaven . . . itself" (2 Chronicles 2:6; 6:18; Deuteronomy 10:14). It is the "Heaven of heavens," the heaven that oversees all of the heavens; it is the capital of all created heavens, including the "Heaven*s* of Heaven*s*" (Psalm 148:4; 68:33; John 14:2). The future "New Heaven" and "New Heavens" also will come under eternity's central government (Revelation 21:1; Isaiah 65:17; 66:22).[2]

The earth, in order to have been the footstool of God in proximity to His throne, must have been created exceptionally beautiful in its primordial state. It apparently was nestled as a rare jewel against the background of God's throne like a rainbow halo (Revelation 4:4); a priceless gem set in a bezel of radiance.

The original earth as it came from the hands of the Creator was covered with mountains, hills, trees, orchards, and cities, with clouds floating lazily overhead (Isaiah 14:7-17; Jeremiah 4:23-25).

A DISPENSATIONAL DAY[3]

All of these habitations were created in one *dispensational* day because as yet no points of reference existed by which to mark and establish time, such as in our solar system: "These are the generations of the heavens and of the earth when they *were created* in the (dispensational) *day* (singular) that the LORD God made the earth and the heavens" (Genesis 2:4).

CREATION OF THE FIRST LIVING CREATURES

"Thus the heavens and the earth were *finished* and *all the host* of them" (Genesis 2:1).

Here we have documented for us the purpose of these created habitations. First, of course, there were the angels who were called God's "ancients" (Isaiah 24:23) and sons of the morning (or dawn) because they were the first to be created in the dawn of creation (Isaiah 14:12). They were also called "sons of God" because they were the creations of the Creator (Job 1:6; 2:1).

THE GLOWING SKY

"And the Spirit of God moved upon the face of the waters" (Genesis 1:2).

The Spirit of God caressed the earth which was embedded in its icy crypt under a shroud of blackness, which later become our solar system. This caused the water to heat and its vapors to rise and form a canopy of water in the upper stratosphere surrounding the earth like a huge grave-like bubble: "I clothe the heavens with blackness, and I make sackcloth their coverings" (Isaiah 50:3).

"The day is thine, the night is thine: thou hast prepared the light *and* the sun" (Psalm 74:16). Here we see that light did exist *before* the sun came into actuality.

"Which commandeth the sun, and it riseth not, and sealeth up the stars" (Job 9:7); also, the word "let" in Genesis 1:14-16 tells us that the beginnings of the interstellar galaxies occurred before the fourth day of renewal, during the first three days of dispensational time (which accounts for the extreme ages of interstellar objects). In order to bring into existence the first light to irradiate the renewal of the earth, God lifted the shroud of blackness sufficiently from the earth to establish a firmament (expanse): "And God said, Let there be a firmament in the *midst* of the waters, and let it divide the waters which were *under* the firmament, from the waters which were *above* the firmament: and it was so" (Genesis 1:6,7).

The glowing sky of photon particles which illuminated the earth below was the result of the Holy Spirit's agitation of the molecules in the canopy of moist clouds. Examples of how this occurred are found in Exodus 14:19,20: "... the pillar... a cloud and *darkness* to (the Egyptians), but... *light* to (Israel)...;" "Also can any understand the spreading of the clouds (Job 36:29). "... Behold, a *bright cloud* overshadowed them..." (Matthew 17:5). This canopy above the earth was laden with much of the water that helped bring about the great Flood in just forty days as evidenced in Genesis 7:11: "...all the fountains of the great deep were broken up, and the *windows of heaven were opened.*" (see Job 37:15).

Scientists have discovered that molecules in the sky can produce photon (light) particles during earthquakes of 7 or more on the Richter scale. Because of the great pressure, the rock slippages create friction in which the heat generates electric fields of photon particles to glow in the sky. This glow has been observed as far as seventy miles from the epicenter of the earthquake, indicating that the sun is not essential as a source of light upon the earth.

On the fourth day of restoration, God rent the dark veil surrounding the globe and allowed the solar system to perform its service upon the earth: "Drop down, ye heavens, from above, and let the earth open (become visible to the solar system): and let them bring forth salvation, and let righteousness spring up together; I the Lord have created it" (Isaiah 45:8).

The universe is pristine and is part of the primordial creation of the first three (dispensational) days of renewal, which accounts for at least fifteen to twenty billion years of duration, the life span of stars that have already disappeared to be replaced by second and third generation stars. Photon particles continue to travel through the space of cosmos to earth long after stars are deceased, and when they strike objects they cast shadows. Their movements are comparable to the motion of balls which are suddenly released by a pitcher. The balls do not fall to the ground at once but continue their momentum until they strike an object. The light we see now from Andromeda galaxy, our nearest full-size neighbor, probably began its journey when the earth was in its chaotic state, long before its renewal, because this light from the Andromeda is approximately two billion light-years away.

Some creationists believe that 1) distance to stars may not be very great; 2) light could take a short route through space; 3) light may have traveled faster at the time of creation; 4) the universe was created to full maturity in seconds which under normal processes would have taken years. Thus, they hold that light from the most distant star to the earth would arrive within fifteen years. But these postulations have little significance when we recognize that the primordial creation of the planetary heaven occurred under dispensational time.

WALKING GHOSTS

The words "let" and "make" in Genesis 1:26 inform us there were manlike creatures on the original earth aeons before Adam appeared on the scene. They possessed humanlike features with degrees of intelligence ranging probably midway between man and animal. At that time ". . . there was no man" (Jeremiah 4:25,29). These humanoids were called "bowmen" in the texts.

Many of these bowmen apparently joined Lucifer in his rebellion and became casualties in the form of disembodied spirits. These beings now roam the earth seeking to re-enter either *humans* or *animals*. They are called "evil spirits," "demons," "unclean spirits," and "deaf and dumb spirits" and they still create turmoil and malformations. They are *not* fallen angels because the latter have their present habitations in the cosmos or are imprisoned in the bottomless pit. Fallen angels are not loose upon the earth (Jude 6; Daniel 12:14,20; Ephesians 6:12), and furthermore, they cannot reproduce by giving birth to roaming demons on earth (Mark 12:25). If they are not the primordial humanlike creatures that fell in the rebellion, who are these demons and from where did they come?

It is entirely possible that the archaeological findings of bones from a prehistoric past could be those of disembodied spirit beings which God, in His judgement, stripped of physical substance as punishment for their participation in the rebellion of Lucifer. God created them for beneficial services upon the original earth, but because they were beyond the point of redemption, God did not want them to continue to exist in a sinful state in bodies which were in His image (but *not* in His "likeness").

". . . The angels who abandoned their *own domain* (are) within the nether gloom . . ." (Jude 6, *Moffatt*). In the writer's opinion, the scenes depicting the space beings in "Chariots of the Gods" resulted from three possible sources: 1) fallen angels who took advantage of humanoid beings on the primordial earth; 2) fallen angels who corrupted and exploited the fallen human race during the Antediluvian period; and 3) "also after that" (Genesis 6:4; compare Jude 7). These fallen angels materialized (compare Genesis 18:1-9; 19:1) in order to take possession of the earth and corrupt the human race in the same manner that they did the humanoids. Both the polluted humanoids and the human race were denied physical bodies; the former became demons *on* the earth (Matthew 8:29) and the latter were confined to the lower regions *in* the earth (I Peter 3:19,20). When these fallen angels are released upon the earth during the last day Tribulation period, they again will duplicate their wicked designs through the Antichrist and False Prophet (Revelation 9:1-11; 13:1-18; 14:9-11; Luke 17:26-30).

GENERATIONS OF THE EARTH

"These are the generations . . . of the earth . . ." (Genesis 2:4).

The earth always has received priority in either of the heavens of which it was and is a part. How do we know this? We know it because the creatures restored to the renewed earth had a prior existence as implied by the word "let." There cannot be a reappearance of that which never existed previously. Before Lucifer existed on the original earth, living creatures already had lived upon it, but they all became extinct. In the renewal of the earth only those creatures which God desired were restored. The remainder probably will be restored in the New Heaven on planetary habitations comparable with their required environments.

Before it became part of our present solar system (Ezekiel 28:16) the original earth undoubtedly was carpeted with extravagant and captivating jewels that sparkled, glistened, and glittered like *stones of fire* under the brilliance of the glory of God. The light of God's glory radiated out from this gem-sprinkled oasis with its dazzling and shimmering mirrored multicolors spiraling upward as giant prisms. It must have resembled reflecting fingers reaching toward God's throne in worship. No wonder God called it "my footstool." In order to have assumed such an honor, it must have been exquisitely luxurious in every detail. Genesis 1:14,15 does not imply the original earth was part of a solar system but rather that such a system existed in God's creation.

ANIMAL PETS

"Let," in Genesis 1:20 and 24, speaks of the replenishment of certain primordial creatures to the restored earth (Acts 3:21; see chart "The Generations of the Earth"). The word "let" is used in the same sense one would say "Turn on the light." It never means "creation" but "permission," "grant," "allowance," or "sanction." The fact that these deceased creatures from the primordial earth were returned to the restored earth is important because it is plain that the "soul" from the Hebrew "nephesh" which means "the breathing spirit" or simply "life" continues to exist and that its properties remain intact. This is confirmed in Psalm 76:6: ". . . horse (is) cast into a *dead* sleep," which negates any kind of active existence in a kennel or corral paradise.

The Scripture ". . . the soul that sinneth, it shall die" (Ezekiel 18:4) is not a reference to animals but to human beings who have the capacity to sin (verse 20). This death separates man's spirit and soul from his body, with his spirit entity always returning unto God. The soul of the *redeemed* is preserved and retained with his spirit because his soul now possesses "eternal life" (John 3:16). Before the resurrection both body and soul will be preserved for an appearance at the final tribunal (Matthew 10:28; John 5:25-30). If the name at that time does not appear in the book of life, the defendant must face the prospect of a *second death which has power over the soul.* The terminal point will be determined at the White Throne Judgment.

MANKIND IN BOTH THE IMAGE AND LIKENESS OF GOD

The word "let" is used interchangeably with "make" and "created" when reference is made to mankind: "And God said, *Let* us *make*[4] man in our image . . . So God *created* man in his image . . . " (Genesis 1:26,27. See also verses 20,21). Mankind is distinguished from all other creatures, but Scriptures do inform us there were humanlike beings which previously existed. They were devoid of the higher faculty in mankind called "likeness" ("after our likeness"), which goes beyond the physical structure. It is only in reference to mankind that we read: "And the LORD God . . . breathed into his nostrils the breath of life; and man became a living soul" (Genesis 2:7). Man possesses a quality of God that animals (or any primordial creatures) do not possess.

God told Adam and Eve to *"replenish the earth"* (Genesis 1:28), certainly evidence that the earth was occupied previously by other living creatures. And in Psalm 104:30, the Spirit of God was sent to *"renew* the face of the earth" which includes both animate and inanimate life.

There are many mysteries of the past that archaeologists have sought to explain from their discoveries of ruins, fossils, ancient monuments, structures, relics, artifacts, and hieroglyphics. These all testify to the previous existence of prehistoric creatures on the earth.

FINAL CURTAIN ON ANCIENT HISTORY[5]

Archaeological finds of many prehistoric animals and manlike creatures that predate Adam and Eve in no way diminish the Biblical record, but they must be distinguished from modern man: "I beheld, and, lo, there was no man . . . and not a man dwell therein" (Jeremiah 4:25,26). "And every plant of the field before it was in the earth, and every herb of the field before it grew: for the LORD God had not caused it to rain upon the (primordial) earth, for *there was not a man to till the ground"* (Genesis 2:5).

The problem is that dating attempts by archaeologists, geologists, and paleontologists fail to dissociate the creatures on the primordial earth from modern man.[6] There were glaciers and other gigantic upheavals on the primordial earth that indicate total annihilation of all life on the earth. Then there is the period of "rest" in earth's history when all living organisms upon the original earth came to an abrupt halt and died atomically. The length of this period of "rest" is uncertain, but because of the wide gap separating the primordial past from-the present reactivated earth, all connection of modern man with the past is impossible. Man is a unique and distinctive being who was formed by the Creator and imparted with a faculty that aligned him closely to his Creator: "But there is a *spirit* in man: and the inspiration of the Almighty giveth them understanding" (Job 32:8). "LORD . . . thou also hast wrought all our works in us" (Isaiah 26:12).

"The whole earth is at *rest,* and is quiet . . . " (Isaiah 14:7). The word "rest" means to "settle down." Zechariah 1:11 reads: " . . . behold, all the earth sitteth still and is at *rest."* There is no record that such an incident ever occurred apart from the time that the primordial earth literally atomically died! Archaeological finds of an ancient past indicate only that there were generations prior to the demise of the original earth.

Creatures on the primordial earth cannot be linked to modern man as evolutionists have attempted to do. First, those creatures literally and atomically died, so man could not have evolved from them. Second, Adam and Eve were "created," and not evolved from any prior creatures. Mankind is distinguished from any beings of the past as indicated in Job 32:8. Animals have instinct; man is created uniquely different. He possesses that special faculty called a "spirit," which links his abilities to his Creator Who is a Spirit Being (John 4:24). Scripture in this regard calls mankind "gods" (Psalm 82:6; John 10:34). He is in "the image *and* likeness" of his Creator, which intellectually at least, makes him superior to all of the creatures on earth. All creatures and things that have emerged from the Creator's hands on the renewed earth, except for mankind, are but copies of the prototypes which He created to inhabit the original place of abode.

Charles Darwin, the noted naturalist, whose research involved mainly the biological structure of man, stated that intellectual and moral faculties must have had their origin in the Unseen Universe of Spirit. When asked concerning the existence of an omnipotent God, Darwin replied: ". . . this has been answered in the affirmative by some of the biggest intellects that ever existed."

THE BIG REFRIGERATOR

There is no conflict with science regarding the longevity of the earth when one recognizes that the earth was once placed on "hold" and held in "cold storage" as it were. When the earth was entombed and held in suspension it was in a state of inertia with every particle in every atom coming to an abrupt halt. Not only did protons, gravitons, and neutrinos become rest particles, but even the fast moving photons that race at the speed of light were brought to a complete rest, and *darkness* enveloped the *earth.* The earth literally died, and was buried. Historically, this was the final colossal ice age.

ANGELIC BEINGS

In Jeremiah 4:29 we read: ". . . the whole city shall flee for the noise of the *horsemen* and bowmen; they shall go into thickets, and climb up upon the rocks: every city shall be forsaken, and no man dwell therein."

The word "horsemen" frequently makes reference to God's loyal and holy angels being called into combat against evil forces" (2 Kings 6:17; Joel 2:1-11).

LISTENING TO BONES

As previously stated, the archaeological findings of ancient bones may have been those of the disembodied spirit beings that now roam the earth as "demons" and "evil spirits." Because they possess knowledge of past events, they can transfer this information to persons opening themselves to them, such as in seances. However, as "lying spirits," their counsel may well lead the subjects to disaster.

When demons were exorcised by Jesus, He first bound them and thereby limited their movements; then He commanded them to leave the victim and join the fallen angels in the "high" and "dry" places (Ephesians 6:12; Matthew 12:43; Luke 11:24). However, they have the capacity to return to earth under instructions from Satan (Matthew 12:45; Luke 11:24).

THE MYTHICAL UNICORN[7]

". . . The unicorn . . . his strength is great . . . " (Job 39:10,11). Many creatures that once existed now are extinct. Many have become legend or forgotten because there are no records of their existence. The mythical unicorn perhaps may represent all of the creatures that once graced the primordial and renewed earth. Its return to habitable places in the New Heaven is highly probable, not only to eradicate the tragedies that follow because of sin, but to fulfill what Jesus said He would do: ". . . might fill all things" (Ephesians 4:10).

AN AUDIENCE IN HELL

Apparently fallen angels were responsible for serious acceleration of wickedness in the days of Noah, thereby forcing God to bring swift devastation upon the earth by a Flood (Genesis 5:1-7; 2 Peter 2:4,5; Jude 6). As a result, these particular angels were confined in the bottomless pit, and the Antediluvians who were caught in this immorality were visited by Jesus in the lower parts of the earth (I Peter 3:18,19). They will not be judged in the same manner as the guilty angels because Jesus said they would some day stand before Him to ". . . be judged according *to man in the flesh . . .* " (I Peter 4:6) and not as fallen angels (spirit beings). Christ preached to these Antediluvians the "Gospel," which means "Good News."

Unless their works were justification for God to remove their names from the book of life, they are assured of an opportunity after the White Throne Judgement to make their decision for Christ in the New Heaven under a more favorable climate. Otherwise, why would He preach the gospel to them if there were no hope? Christ could do nothing for them while they were "spirit" beings devoid of their bodies of "flesh." Response to the Gospel can be made only after they are resurrected into bodies of "flesh."

GOD'S FOOTSTOOL: A TOMBSTONE

When mankind fell into sin God opened separate compartments within the earth for the occupancy of the righteous and the unrighteous after their decease (Luke 16:19-31; Ephesians 4:9; Psalm 63:9; 71:20; Isaiah 26:19; Luke 23:43; I Peter 4:6).

God did not place deceased fallen men in the bottomless pit with the fallen angels. As we shall see, the bottomless pit and outer darkness are separate places located outside and below the cosmos farther from God's throne than the earth, with the earth being between God's throne and the bottomless pit.

The "bottomless pit" cannot be located inside the earth and be bottomless—it would be *all* bottom. And "*outer* darkness," by its very name, indicates it is located a greater distance from God's throne than the earth. Jeremiah 10:11 states: ". . . the gods that have not made the heavens and the earth, even they shall perish *from* the earth, and *from under these heavens."*

Until the White Throne Judgement, the earth will continue to gather into its deep hold the unredeemed souls whose names are not in the book of life: "The Lord said unto my Lord, Sit thou on my right hand, Until I make thy foes thy footstool" (Acts 2:34,35).

Those whose names are in the book of life are no longer being held in Paradise within the earth. Christ has transported all of them into the heavenly Paradise (Ephesians 4:8-10; 2 Corinthians 12:1-4). All of those now in the book of life ascend at death directly into the heavenly Paradise (2 Corinthians 5:6,8).

Paradise in the earth was emptied of all its occupants, and hell was enlarged to accommodate the growing numbers of lost souls erased from the book of life. Presumably this means that this Paradise now has been opened to receive these lost ones descending into hell: "Therefore hell hath enlarged itself, and opened her mouth without measure: and (those lost) . . . shall descend into it. And the mean man shall be brought down, and the mighty man shall be humbled: But the Lord of hosts shall be *exalted in judgment,* and God that is holy shall be sanctified in righteousness" (Isaiah 5:14-16).

[1] Genesis 1:1,2; 2:1.
[2] See chart on page 603.
[3] If the concept of dispensational days is ruled out, there is no scriptural evidence that the rebellion of Lucifer (first ruler on the earth) occurred during any of the six twenty-four hour solar days. Hebrew word *yom,* 1480 times in the Old Testament is flexible to day periods.
[4] Hebrew for "make" is "asah" meaning to make out of already existing material. It means the opposite of the Hebrew "bara" (to create) as used in Genesis 1:1.
[5] Although theologians are uniformationists, they do embrace the concept that in the ancient ages of the primordial earth, it underwent a series of major disasters and judgments.
[6] See chart on page 614.
[7] Whatever legends remain of the past, undoubtedly many have their roots in the primordial ages of the earth and some in the Antediluvian period.

16

Sons of the Morning

The earth was carpeted with gleaming and sparkling gems that appeared to radiate like fire under the light of God's glory.

"How art thou fallen from heaven, O Lucifer, son of the morning (or, son of dawn)! how art thou cut down to the ground, which didst weaken the nations!" (Isaiah 14:12).

"... The devil... abode not in the truth" (John 8:44).

God's creative masterpiece was an archangel named Lucifer, not a devil; but since his fall he has been the bearer of darkness, not a "bringer of light." He was given the title "Son of dawn" (or morning) because he was formed first at the dawn of all created living creatures.

During his long time of loyalty he gave God no occasion to anticipate anything other than his continued faithfulness. "... Thou wast perfect in thy ways from the day that thou was created, till iniquity was found in thee" (Ezekiel 28:15). Even under His perfect foreknowledge and wisdom, God would not have been deterred from creating Lucifer. Lucifer's fall was the result of exercising his own will and for this he must accept full responsibility. Lucifer's fall was not the fault of God's creation, but God permitted his existence to fit into His long-range plan of redemption.

God made the earth a kingdom and placed Lucifer in charge.

We conclude that his subjects were numerous angels and intelligent beings in physical form because many who fell with Lucifer are today disembodied spirits seeking habitations in fleshly creatures. They also will be severely judged some day: "And, behold, they cried out, saying, What have we to do with thee, Jesus, thou Son of God? art thou come hither to torment us before the time?" (Mark 1:24).

PAVEMENT WITH STONES OF FIRE

Before our solar system came into being, earth's thoroughfares were carpeted with gleaming and sparkling gems which radiated as fire under the light of God's Glory: ". . . thou hast walked up and down in the midst of the stones of fire" (Ezekiel 28:14,15). Undoubtedly edifices and promenades also were adorned richly with these precious stones.

After Lucifer fell into disgrace,[1] these elegant gems were allowed to crumble and be buried in the ground: ". . . how art thou cast down to the ground" (Isaiah 14:12). The word "ground" shows that the earth at the time was solid and not a misty cloud of gaseous vapors as is the case with many celestial bodies (Isaiah 51:12,16; 54:11; Zechariah 12:1). Lucifer also was called "King of Tyrus," meaning "rock," further evidence of earth's solid foundation (Ezekiel 28:13).[1]

There must have been areas for flowers, plants, and other lush vegetation, as well as sanctuaries for many lovely fowl and animals gracing the landscape. Paradise, where Lucifer often must have relaxed, was nearer God's throne than the earth: "Thou hast been in Eden the garden of God . . ." (Ezekiel 28:13).

Lucifer was a handsome and attractive being who was luxuriously clothed. In every sense, he appeared as regal and noble as a "prince," a title often conferred upon him in God's Word even after he transgressed (Daniel 10:12,20; Ephesians 2:2).

THE PRINCE'S WARDROBE

Prince Lucifer's wardrobe was exquisitely and ingeniously tailored. Each garment was trimmed and decked with the finest materials for His Majesty. The most perfect jewels adorned the vestment of His

Royalty to complement and befit His Honor (Ezekiel 28:15).

God warned Lucifer that if he pursued the course of rebellion that was being planned, his fall would be so disgraceful that his very features and appearance would become gruesome. Any attempt to readorn himself would be met by dismal failure: "And when thou art spoiled, what wilt thou do? Though thou clothest thyself with crimson, though thou deckest thee with ornaments of gold, though thou rentest thy face with painting, in vain shalt thou make thyself fair . . . " (Jeremiah 4:30). His extreme ugliness would never be reversed.

His followers in this rebellion were to despise him because of the eternal humiliation and disgrace brought upon them. Why shouldn't they? He was their leader and he was proven to be a liar from the beginning of this plotted rebellion (John 8:44; Revelation 12:9; 20:10).

ORCHESTRA HALL

God assembled and presented to Lucifer, son of dawn, a marvelous orchestra with instruments of the finest quality and elegant workmanship, together with top talented musicians for his pleasure and enjoyment: ". . . The workmanship of thy tabrets (musical instruments) and of the pipes (bezels) was prepared (for) thee in the day that thou wast created" (Ezekiel 28:13).

Did Gabriel play in this orchestra, or was he perhaps the maestro directing this orchestration in superb and pleasing symphonies?

EARTH'S ORIGINAL CITIES

God is a Builder of cities, as well as of the heavens with their many intergalactic systems, as is evident in Ezekiel 28:18: "Thou (Lucifer) hast defiled thy *sanctuaries* by the multitude of thine iniquities . . . ." In Isaiah 14:17 we read: "That made the world a wilderness, and *destroyed the cities thereof . . . .*" God built the Holy City, New Jerusalem. Abraham, we read: ". . . *looked for a city* which hath foundations, whose builder and maker is God" (Hebrews 11:10).

Perhaps in Abraham's day there were signs of ruined ancient

cities on the primordial earth which enabled him to recognize the existence of prior prehistoric metropolises. Within one of these cities God had erected a gorgeous palace, as well as a magnificent throne, for His Majesty Lucifer. Did not Lucifer say, "...I will exalt my throne (and) I will be like the most high" (Isaiah 14:13,14)? Lucifer knew he could not dethrone God, but it was his ambition to exalt his own throne above that of the Almighty.

A PROUD AND JEALOUS GOD

There is no indication that Lucifer intended to dethrone God. He knew that would be impossible because he was no match against the Almighty. His purpose was to approach God, backed by massive support and destructive military power, and thereby coerce God to accede to his demands of abdication, or devastation would follow. Lucifer wanted to be absolute ruler over his kingdom without any interference from the Creator. He must have felt sufficiently secure in his kingdom in order to have made such a drastic challenge to the Almighty. His pride had truly corrupted his wisdom. In some ways it paralleled the pride of King Nebuchadnezzar of Babylon in the days of Daniel, as evidenced by Daniel's warnings to the King:

"It is thou, O king, thou art grown and become strong...O king, and this is the decree of the most High, which is come upon my lord the king that they shall drive thee from men, and thy dwelling... till thou know that the most High ruleth in the kingdom of men, and giveth it to whomsoever he will... after that thou shalt have known that the heavens do rule" (Daniel 4:22-26).

God, the Creator, is an extremely proud Majesty and will not give His glory unto another, but will protect His image as the Almighty. He will not tolerate rivals who seek to equate themselves with Him. Lucifer sought to exalt himself *above* God.

"Behold, I am against thee, O thou most proud, saith the Lord God of hosts: for thy day is come, the time that I will visit thee. And the most proud shall stumble and fall, and none shall raise him up: and I will kindle a fire in his cities, and it shall devour all round about him" (Jeremiah 50:31,32).

"I am the Lord: that is my name: and my glory will I not give to another . . . Let them give glory unto the Lord, and declare his praise . . . " (Isaiah 42:8,12).

"His scales are his pride, shut up together as with a close seal. One is so near to another, that no air can come between them" (Job 41:15,16).

POISONS OF PRIDE AND JEALOUSY

God was proud of His position of power and of His achievements with which no one can compete. Why has God reserved this faculty unto Himself? It is simply because no created beings could handle it and be just in their actions. There would be disastrous effects upon other creatures and upon God's creations as we have seen demonstrated in the actions of Satan and his followers since their fall, and as portrayed in these scriptural warnings:

" . . . Jealousy is *cruel* as the grave: the coals thereof are coals of fire, which have a most vehement flame" (Song of Solomon 8:6).

"For jealousy is the *rage* of a man: therefore he will not spare in the day of vengeance. *He will not regard any ransom; neither will he rest content, though thou givest many gifts"* (Proverbs 6:34,35).

What a picture of Lucifer! When he failed to obtain his desire he flew into a rage and became destructive. His very nature was poisoned beyond recovery and God could offer him no ransom because he would have spurned and rejected it. And if God had enriched him further he would never have been satisfied. It would have become extortion by an ingrate with continuing demands and unthankful ness (See Romans, chapter 1).

Jealousy is not a desirable quality. Why then is it right for God to have it? It is because He can be jealous in *righteousness*. God is the only One who can handle it. God is not competing with someone who equates with Him or who is above Him, therefore He is justified in His jealousy.

To man, jealousy is a cancer, but God can contain it without destroying His demeanor. Satan's jealousy made him the father of lies (John 8:44; I John 2:4; 4:20) and the creator of all impurity. God is the Father of truth and there is an outpouring of love and profitable benefits issuing from Him.

[1] The expression in Matthew 13:35, "... the foundation the world..." literally means "before the overthrow of the world" ("Kosmos," meaning the social world). Also see illustration VIII.

17

Sons of God

There are angels, unfortunately, who do not represent Christ faithfully—they must be rejected.

"Now there was a day when the sons of God came to present themselves before the LORD . . . " (Job 1:6).

Angels are called sons of God because they were God's first created living beings. (Compare Genesis 6:1-5 with Jude 6).

GOD'S ONLY BEGOTTEN SON

Jesus Christ is God's only begotten Son (John 3:16) but *when* did He become God's only *begotten* Son? "God . . . hath *raised up Jesus again;* as it is written in the second Psalm, Thou art my Son, *this day* have I begotten thee . . . *he raised him from the dead* . . . " (Acts 13:33,34).

"So also Christ glorified not himself to be an *high priest;* (compare Hebrews 9:11); but he that said unto him, Thou art my Son, *today have I begotten thee*" (Hebrews 5:5). "Yet have *I set* my king upon my hill of Zion. I will declare the decree: the LORD hath said unto me, Thou are my Son, *this day* have I begotten thee" (Psalm 2:6,7).

When Christ was in the bosom of the Father (John 1:18) within the Godhead per se, He was not the begotten Son of God.[1] He became that Son intrinsically when the tomb became the womb which issued forth the Christ into a new relationship with the Godhead: "And from Jesus Christ, who is the faithful witness, and the *first begotten of the dead*..." (Revelation 1:5).

He was first to have His body of mortality changed into immortality: "Who only hath immortality, dwelling in the light which no man can approach unto (unless he is changed like Christ); whom no man hath seen, nor can see (in his mortal state): to whom be honour and power everlasting" (I Timothy 6:16).

Christ Jesus, as God's only begotten Son, has now assumed a state of unequaled glory and power: "... his countenance was as the sun shineth in his strength" (Revelation 1:13-17). To Him the Father has committed all judgment (John 5:22-29) and all power: "And Jesus came and spake unto them, saying, All power is given unto me in heaven and in earth" (Matthew 28:18).

ORIGINAL SONS OF GOD

Angels are sons of God in the sense that they are created spirit beings. God is a "Spirit Being" (John 4:24) and so spirit beings have a very close affinity to God. Their composition enables them to defy all known laws of physics and chemistry. They are creatures of great strength (Psalm 103:20; 2 Peter 2:11) possessing all of the senses of man, but with a greater degree of acuteness.

REDEEMED SONS OF GOD

Since man is also the possessor of "spirit" (I Thessalonians 5:23; Job 32:8; Ecclesiastes 3:21; 12:7), his sensitivity toward God is keener than that of animals. He has unlimited creative abilities in contrast with other creatures who are limited to instinct.

By His resurrection, Jesus Christ, the only begotten Son of God, elected to include others as sons of God in order that they might participate in an inheritance with Him: "For it became him, for whom are all things, and by whom are all things, in bringing many sons unto glory, to make the captain of their salvation perfect through sufferings" (Hebrews 2:10).

Christ is, therefore, but the "firstfruit" of a great harvest of redeemed beings who are to share in His eternal inheritance (Revelation 14:4; James 1:18; I Corinthians 15:20,23). But they must have the "firstfruits of the Spirit" (Romans 8:23) which are the first three listed in the fruit of the Spirit—love, joy and peace. These three concern a personal relationship with Christ and are the first evidences of the spiritual rebirth.

When the redeemed finally possess bodies of immortality with Christ and reign with Him, they will be revealed in the Lord's visible Kingdom under "the manifestation of the sons of God" (Romans 8:19), which will affect the entire environment of the earth. The Millennium will provide them with the opportunity of apprenticeship for service in the greater expanded kingdom in the New Heaven and on the New Earth (Luke 1:33).

MYSTIC OF ANGELS

There are angels, unfortunately, who do not represent Christ faithfully or at all. They must be rejected: "I marvel that ye are so soon *removed from him* that called you into the grace of *Christ* unto another gospel: Which is not another; but there be some that trouble you, and would *pervert the gospel of Christ.* But though we, *or an angel from heaven* (which place contains both good and evil ones), *preach any other gospel* unto you than that we have preached unto you, *let him be accursed"* (Galatians 1:6-9).

"Beloved, when I gave all diligence to write unto you of the *common salvation* (not some strange new doctrines that lead away from Christ), it was needful for me to write unto you, and exhort you that ye should earnestly contend for *the faith which was once delivered* unto the saints" (Jude 3). Adhere to the original; it is superior and it is true.

The Devil and his fallen angels are not only deceivers, but pathological liars; therefore one needs to be on the defensive constantly: "Stand therefore, having your loins girt about with *truth* . . . and the sword of the Spirit, which is the word of God" (Ephesians 6:14-17).

WHY BE A SURE LOSER?

Because man's "spirit being" belongs to God (Ecclesiastes 12:7), his soul is his most precious possession, so why jeopardize its eternity on either a sure loser or a doctrine of questionable nature? The standard (canon) for determining the inspired writings of the Holy Scriptures is clear from what the Apostle Paul had to say: "And are built upon the foundation of the *apostles and prophets,* Jesus Christ himself being the chief corner stone" (Ephesians 2:20).

The Old Testament was written by prophets of God and their works were included in the Hebrew collection of the Scriptures which fell into four divisions—the Law or Pentateuch, the Historical Books, the Poetical Books, and the Prophets. The Apocrypha was rejected from the Hebrew collection as uncanonical. Neither Jesus Christ nor His Apostles ever quoted from it, but they did quote from the four divisions of the Old Testament.

The canon of the New Testament was established by the writers themselves as they were Apostles (appointed by Jesus Christ), and it was confirmed by the Founding Fathers, who were converts of the Apostles, and who immediately followed the Apostles' ministry. Theologians believe Paul was the apostle replacing Judas Iscariot and that the early Apostolic Church had erred in appointing Matthias before the outpouring of the Holy Spirit at Pentecost. It was an appointment by lottery, not by the Holy Spirit, and no subsequent mention is ever made of Matthias. Paul became the apostle to the Gentiles when he broadened the ministry of the Church to witness to the whole world, as was the intent of the Holy Spirit in empowering the Church.

THE CANON IS COMPLETE

The Written Word is complete with the final book of Revelation. Nothing must be added nor removed from it. The canon of truth (criterion for determining what is part of the Holy Scriptures) as contained therein is concluded. Any supplement that equates itself to it must be soundly rejected as the Holy Scriptures are unique and stand singularly. They need no defending; they will defend themselves. All of men's writings, including this book, are but commentaries to assist toward a better understanding of the Written Word.

"For I testify unto every man that heareth the words of the prophecy of this book, If any man shall *add* unto these things, God shall add unto him the plagues that are written in this book; And if any man shall *take away* from the words of the book of this prophecy, God shall take away his part out of the book of life, and out of the holy city, and from the things which are written in this book" (Revelation 22:18,19).

The book of Revelation closes forever all future apocalyptic utterances that equate as God's inspired Word, either as part of, or as an essential supplement to the Written Word. All prophecies coming from any so-called prophets now fall into the category of the prophetic gifts of the Spirit: "Let the prophets speak two or three, *and let the other judge*" (I Corinthians 14:29). You do not judge the Written Word; it judges you. It must be *believed.* Prophetic utterances, whether oral or written, can and must be judged to see if they harmonize with God's Holy Scriptures in their completed form.

The last of the prophets who make proclamations or have inspirations that equate as God's infallible Word have come and gone: ". . . the mystery of God should be *finished (fulfilled), as he hath declared* to his servants the prophets" (Revelation 10:7). We are now living in the time of the fulfillment of God's Written Word.

MILLENNIUM AND PROPHETS

There will be kings, priests, lords, and judges in the Lord's manifested kingdom, but no prophets (Revelation 1:6; 19:16; I Corinthians 6:2,3). ". . . Prophecies, they shall fail (literally, pass away, be fulfilled)." In one translation we read: "Are there prophets? their work will be over" (*New English Bible*).

The prophet Zechariah, referring to the Millennial reign of Jesus Christ, stated emphatically that no prophets will be permitted to exist at that time: ". . . I will cause the prophets . . . *to pass out of the land.* And it shall come to pass, when any shall *yet prophecy (he) speaketh lies* in the name of the Lord: and... shall... *thrust him through when he prophesieth.* And it shall come to pass on that day, that the prophets shall be ashamed every one of his vision when he hath prophesied..." (Zechariah 13:2-4).

[1] The Godhead is not triple 1+1+1=3 but triune 1x1x1=1, for one must move to a higher mathematical formula.

18

Lucifer Blew It

God would not create automatons divested of such a precious commodity as volition.

Lucifer's reasoning powers became warped and bent on evil. As he surveyed God's creation in his bid for exclusive supremacy there possibly were several options which he may have considered.

First, God was such a loving Creator that He never would punish Lucifer or his followers for their brazen and open rebellion; or at the worst, He would mildly rebuke them.

Second, God would vacate or abandon His works of creation in lieu of seeing any of them tainted or destroyed by Lucifer's threats.

Third, the worst that possibly could happen would be that God would annihilate him and his comrades-in-arms. The idea that God would bring eternal punishment upon them must have seemed unthinkable as they believed it would be totally foreign to God's character. How tragically they miscalculated God: "And the devil that deceived them was cast into the lake of fire and brimstone (and) shall be tormented day and night for ever and ever" (Revelation 20:10). ". . . Everlasting fire, prepared for the devil and his angels" (Matthew 25:41). ". . . A fire not blown shall consume . . . " (Job 20:26).

ETERNAL SPIRIT BEINGS

God is a "Spirit" Being (John 4:24) and angels are "spirit" beings: "Who maketh his angels spirits . . . " (Hebrews 1:7). Since God as a Spirit Being is eternal, all spirits that come from His hand are equally eternal. There is no record that a spirit can be annihilated or deceased; therefore angels are subjects of eternal punishment. Perhaps Lucifer did not realize this fact, or was obsessed with the idea that failure in his campaign was deemed impossible.

PROPAGANDA MINISTER

Lucifer undoubtedly possessed charisma. He was to build a strong military power which would be armed with the most powerful and sophisticated weapons of destruction. These would be mighty enough to convince the Almighty that Lucifer's army was capable of totally destroying the kingdom on earth, the footstool of God. As terrorists, Lucifer and his cohorts were ready to prove their point if their demands were not accepted.

What were they demanding? They were insisting upon the right to throw off the reins of God's righteousness as sovereign beings: ' . . . iniquity was found in thee. By the multitude of thy merchandise (his rich and beautiful kingdom) they have filled the midst of thee with violence, and thou hast sinned: therefore I will cast thee as profane out of the mountain of God: and I will destroy thee, O covering cherub, from the midst of the stones of fire" (Ezekiel 28:15,16).

Discarding the restraints of righteousness was the prime objective as demonstrated in Mark 1:23,24: "And there was in their synagogue a man with an unclean spirit; and he cried out, saying, *Let us alone;* what have we to do with thee, thou Jesus of Nazareth . . . ?"

Lucifer's prescription for success depended upon his ability to sell his program as was indicated by the great number of enlistees in his audacious and infamous venture. As evidenced by Lucifer, there is a danger of misusing one's charisma for purposes of deception.

LORDS OF THE DEEP

Prophetic scriptures bear out the fact that Apollyon (or Abaddon) was appointed as the top staff officer to organize and lead the assault on God's throne (Revelation 9:1-11). He is described as "king" over his dedicated and vicious military force. That is why, after their unsuccessful coup, he and his forces were incarcerated as lords of the deep in the bottomless pit which is located far out from the universe and below "outer darkness" in the habitation of eternity (Jude 6; Jeremiah 10:11).

In Revelation 9:1-11 the degenerate and nefarious character of these forces is obvious when they are loosed upon the earth for five months and turn the entire world into a veritable insane asylum, except for those whose names are in the book of life: "And in those days shall men seek death, and shall not find it; and shall desire to die, and death shall flee from them." Any person who takes the mark of the beast will be immune to death for a period of five months (verse 5). Will the vengeance of these persons be pitted against the tribulation saints who are susceptible to death? Will this account for the martyrdom of tens of thousands?

Lucifer (now Satan) always employs others in his campaigns against God (2 Thessalonians 2:3; 2 John 7), and they frequently appear as angelic messengers of light bearing felonious and destructive communications detrimental to the souls of mankind.

What is the safeguard against Satan? It is the Word of God permeating hearts and minds with His truth which will inspire faith and loyalty to Him and disarm the Wicked One (Ephesians 6:1-18; Romans 1:17; 10:8,17). His assaults upon a person are devious ("wiles"; Ephesians 6:11). Satan seeks to drive a wedge between humans and the Almighty by casting shadows upon the integrity of God as stated in His Written Word (Romans 3:2).

WHY NOT MORE VISIONS?

The appearance of angels is not a reliable criterion for truth. The Apostle Paul, who had an abundance of visions (2 Corinthians 12:7), made this clear in Galatians 1:8,9: "But though we, or an angel *from heaven* (and there are fallen angels in "high places"), preach any other gospel unto you than that which we have preached unto you (or) ye have received, let him be accursed." Occasionally fallen angelic spirit beings are hidden in personable individuals who are ready to "... pervert the gospel of Christ" (Galatians 1:7).

Why doesn't God more frequently communicate via visions or similar manifestations to convey His position? A basic reason lies in the danger that truth can be perverted. Such sources are difficult to prove or disprove apart from the Written Word of God, and can make a person rely too heavily upon this means for spiritual direction and doctrinal instruction. This is a poor second to the constant reliability of God's Written Word which is steady and firm; it does not undulate like a cork bobbing on the waves of the sea. To turn to any secondary means for divine truth and guidance is froth with possible deceptions which can open a whole pandora of false information.

Satan considers the Written Word of God as superfluous. By making the Written Word of God secondary, Satan and his lieutenants can infringe upon and furtively lead easily influenced souls into false cults and occultism. The only stabilizing foundation for truth has been and always will be God's completed Written Word. Additions or deletions are no longer admissible as divinely inspired.

"When they tell you to consult mediums and ghosts (spirits) that cheep and gibber in low murmurs, ask them if a nation should not rather consult its God... Why consult the dead on behalf of the living? Consult the Message and the Counsel of God" (Isaiah 8:19,20, *Moffatt*).

The *Authorized Version* says: "To the *law* and to the *testimony* (God's Word): if they speak not according to this word, it is because there is no light in them" (Isaiah 8:20). "Study to shew thyself approved unto God, a workman that needeth not to be ashamed, rightly dividing the word of truth" (2 Timothy 2:15).

WHY ANGELS CANNOT BE FORGIVEN

God created beings, angels, and mankind with the full foreknowledge that they would sin and could be lost eternally. Of course, there is no doubt about the risks involved in creating beings with sovereign wills, but God would not create automatons divested of their volition. With this sovereign will imparted to His creatures, God relinquished total control over their actions, even at the risk of their going askew.

God wants a voluntary response of love, not one of coercion. The opportunity of exercising sovereign will forms the basis of divine justice in the final judgment of that being's eternal future.

Redemption is a personal matter and cannot be imposed arbitrarily against one's will: "... The righteousness of the righteous shall not deliver him in the day of his transgression ..." (Ezekiel 33:12). There must be an atonement.

Spirit beings incur a greater liability when they pursue the route of transgression: first, they have a unique relationship with the Creator who is also a Spirit Being, but more than that, He is the Supreme Spirit Being; second, angels are more knowledgeable; third, they have no tempter, but they are self-deceived; fourth, man is informed in Scripture that when he dies, even in sin, his spirit being returns to God (Ecclesiastes 3:21; 12:7). It is his "soul" that is in jeopardy: "... the soul that sinneth it shall die" (Ezekiel 18:4,20). The angels possess no soul *per se,* but they have a spirit which is their whole being and that never can be destroyed any more than God, the Spirit Being, can be destroyed. It is an eternal substance. There is no record that the angel spirit ever returns to God as does man's spirit. God will never receive the fallen angel spirit unto Himself and perhaps it would not be possible; fifth, pride put a crown on Lucifer's head as it does on all creatures: "Woe to the crown of pride ..." (Isaiah 28:1,3). Spirit beings, for whatever reason, find themselves unable to abdicate and have this crown removed; sixth, pride hardens the being. This is evidenced by the fact that fallen angels have shown no signs of change, remorse, or repentance (compare Daniel 5:20); seventh, pride can become an integral part of the being to live in self-deception: "The pride of the heart hath deceived thee . . . " (Obadiah 3); finally, for the above reasons, any offers of redemption would be rejected and scorned: "He will not regard any ransom, neither will he rest content, though thou givest many gifts" (Proverbs 6:35). Fallen angels, therefore, neither possess the capacity nor the will to repent, and without repentance there can be no forgiveness (I John 1:9; Luke 13:3,5; Hebrews 9:22). It is this sad state that the Scriptures refer to as the unpardonable sin (Matthew 12:24-32; I John 5:16; 2 Peter 2:4; Jude 6,7; Galatians 1:8).

How could an atonement even be made for bloodless spirit beings? There would be no benefit in Christ's shedding His blood for them; neither would the Holy Spirit be a likely Candidate as their Redeemer (Matthew 12:32). Transgression places them totally out of any redemption and return into the grace of God. No wonder angels look with such apprehension at this grace of God which is offered to mankind (I Peter 1:12).

Why was there no compensation to Lucifer for the period of faithful service performed in the will and plan of God? According to Ezekiel 33:12 which is equally applicable to fallen angels and to mankind, it is for the reason that "The righteousness of the righteous shall not deliver him in the day of his transgression . . . ."

Five times Lucifer defiantly said, "I will," thereby threatening the authority of God. God replied in kind with His "I wills" (Isaiah 14:13-15; Ezekiel 28:16-18). The will of a transgressor is no match against the will of the Almighty.

Fallen angels who pursue the same course as Lucifer are but carbon copies. As "spirit" beings their modus operandi equates them with their leader, Lucifer (Revelation 9:10,11; 18:2; Mark 9:17-26; Zechariah 13:2; Hosea 4:12; 5:4; Acts 19:12-16; Isaiah 19:3).

19
The Genesis War

When the primordial earth literally died every particle composing the earth's mass came to an abrupt halt.

Lucifer, in his rebellion against God, would leave nothing to chance. All eventualities must be taken into account. What if God should decide to resist and move with swift and terrible vengeance? As Commander-in-Chief of his forces, Lucifer must forge the most powerful expeditionary might possible. In the event that he underestimated the Almighty by relying too heavily upon His good nature and long-suffering, he must be prepared.

HELL ON EARTH

Apparently it became necessary for Lucifer to deploy the most feared weapon in his power, perhaps nuclear explosives. If so, it is conceivable how the primordial earth became so devastated: "That made the world a wilderness, and destroyed the cities thereof. . . " (Isaiah 14:17).

The arsenal of nuclear bombs today is equivalent to ten tons of TNT for every person on earth. The unleashing of such destructive power would be hell on earth. Most people would die, and for the survivors there would be no anesthetics to relieve the pain, nausea, or the epidemics of black plague, typhoid, tuberculosis, and dysentery that would follow. Would Satan hesitate to see these deadly weapons used today? He is just as insensitive as ever, and unless God intervenes, there could be a repeat performance: "And except those days should be shortened, there should no flesh be saved . . . " (Matthew 24:22).

NO HONOR AMONG THIEVES

God entrusted Lucifer, an archangel, with the responsibility of reigning over the earth. He was given a contingent of angels to serve with him, but Lucifer and his followers proved unfit for such great responsibilities. Pride led to the downfall of Lucifer; God did not create the Devil!

Lucifer led the revolt against the Creator, which resulted in devastating consequences for both himself and his followers, as well as for the earth itself. The wrath of Lucifer's most powerful military lieutenants turned against him when they suffered such a major defeat. Their ferociousness was so diabolical that God had to incarcerate them: " . . . thy lovers will despise thee." God warned that " . . . they will seek thy life" (Jeremiah 4:30). Therefore, God had them confined within the bottomless pit.

The extent of their vicious character is graphically demonstrated when God gives them a momentary release prior to the battle of Armageddon. Their actions are senseless and unconscionable and their atrocities are vented to the maximum that God allows: "And they had tails . . . and their power was to hurt men five months . . . " (Revelation 9:10). "And in those days shall men seek death, and shall not find it; and shall desire to die, and death shall flee from them" (Revelation 9:6).

If these vicious beings were unrestrained, they would kill the entire human race: "And to them it was given that they should *not* kill . . . " (Revelation 9:15).

Shortly after these monsters from the bottomless pit are loosed upon the earth, Lucifer (or Satan) also will be cast out of heaven onto the earth where he will directly face these denizens of the pit. Armageddon is not just a battle between human beings; it is also between evil angelic forces locked in fierce combat.

The descent of Jesus Christ with His holy angels, together with the Royal Redeemed in bodies of immortality, will bring a conclusion to the tragic and horrible casualties and to the destruction suffered from the battle of Armageddon.

Scripture indicates that the battle of Armageddon will have its repercussions and counterpart throughout the earth: "And the slain of the LORD shall be at that day from one end of the earth even unto the other end of the earth: they shall not be lamented, neither gathered, nor buried; they shall be dung upon the ground" (Jeremiah 25:33).

The prophet Isaiah reflects upon the problems that faced Lucifer after his rebellion. When God cast him into the bottomless "pit;" as his "grave," we read that he was "cast out... like an abominable branch... Thou shalt not be joined with them (his allies) in burial..." (Isaiah 14:19,20). It was reminiscent of gang war. Truly, there is no honor among thieves, nor among wicked angels.

WHEN THE EARTH LITERALLY DIED

After Lucifer's rebellion, the earth, as God's footstool, was thrust deeply into an entombing orb of blackness along with his anarchists. Wrapped like a swaddlingband, the earth was interned in history's largest mausoleum in an orb of blackness created by God (Job 38:9). For all practical purposes, the atomic clock stopped ticking and the earth lay dead in its dismal grave. Geologists who are dating their oldest rock finds are discovering a gap of 800 million years as to what they believe is the actual age of the earth.

All matter is composed of atoms and subatomic particles, so when the primordial earth literally died every particle composing the earth's mass came to an abrupt halt: "The whole earth is at *rest,* and is quiet..." (Isaiah 14:7. Compare Zechariah 1:11). Life within every particle was withdrawn. Matter became lifeless, sterile, useless, and worthless.

THE MISSING LINK IS FOUND

For seventy years Jerusalem was in ruins and unoccupied with an eerie cemetery-like atmosphere over the city. The prophet Zechariah was concerned about its future and its people, so to illustrate a point, God gave him a reflection of the past judgments which had visited the earth.

The prophet was shown the earth as it lay in its vast ruins. He saw how God had sent Michael, the archangel and leader of the holy angels, into combat. Undoubtedly, these holy angels were the same ones who contended against the rebellious legions under the influence of Lucifer and his star general, Apollyon (also Abaddon, Revelation 9:11). The archangel Michael (see Daniel 10:13-18; 12:1) directed his contingent of angels to gather intelligence information regarding the state of the earth and to report back whether everything was under control. The disclosure indicated, "... We have walked to and fro, through the earth, and, behold, *all the earth sitteth still, and is at rest*" (Zechariah 1:11).

Jerusalem lay like the chaotic earth, and now, like the earth, there would be a resurrection from the cemetery-like conditions (compare Ezekiel, chapter 37). Jerusalem would rise again: "Therefore thus saith the Lord; I am returned to Jerusalem with mercies: my house shall be built in it, saith the LORD of hosts... My cities through prosperity shall yet be spread abroad... and shall yet choose Jerusalem" (Zechariah 1:16, 17).

God made the same predictions about the chaotic earth that had lain in its grave: "He setteth an *end* to darkness, and searcheth out all perfection: the stones of darkness, and the shadow of death" (Job 28:3).

The remaining chapters of Zechariah contain great prophetic promises of the last days concerning Jerusalem, her people, and the earth at the time that Jesus Christ returns to the world to set up His manifested Kingdom.

HEAVENS' FIRST BLACKOUT

"For this shall the *earth mourn,* and the *heavens above be black:* because I have spoken it, I have purposed it, and will not repent, neither will I turn back from it" (Jeremiah 4:28). "... The heavens shall be *draped with black*" (Isaiah 50:3, *Living Bible*).

The first and foremost tragedy of all times occurred, in defiance of God's warnings, when Lucifer with his malfeasants continued to rebel against God: "...iniquity was found in thee (and) filled the midst of thee with violence..." (Ezekiel 28:15,16).

The "Heavens of Heavens," the residences of angels, became like mortuaries filled with billions of mourners. The Creator reached His hand, as it were, and turned down the rheostat to the Glory of God which enlightened the habitations of angels. The light began to dim and soon "all the heavens were dark" (verse 23, *Living Bible*).

These innumerable holy angels went into mourning because their most personable, intelligent, and charismatic archangel, Lucifer, together with millions of his compatriots, or sons of God, had sinned so grievously. The sin was tragic! God's judgment was swift and final and God said, "*I will not repent,* neither will I turn from it" (Jeremiah 4:28). They had reached the point of no return back into God's favor and they forever forfeited the glory that once was their possession.

The beautiful earth, once the luxurious footstool of God, was rent and torn as a result of the aftermath of the battle. There was unbelievable chaos everywhere and lovely cities lay heaped in massive piles of debris and total shambles.

FIRST BLACK HOLE IN SPACE

God said, "I form the light, and *create darkness*..." (Isaiah 45:7); "Thou *makest darkness*..." (Psalm 104:20); "I clothe the heavens with *blackness* (when they went into mourning), and I make sackcloth their coverings" (Isaiah 50:3); "When I *make (the) thick darkness* a swaddlingband for it (the earth)" (Job 38:9).

As we have already established, light was the normal, natural function and environment of eternity and of the heavens before God created our solar system or any similar primordial systems: "...and darkness was upon the face of the deep" (Genesis 1:2).

Darkness had to be created and then only because of transgressions. It has remained a symbol of sin and judgment ever since. The earth lay in a state of inertia deep in space far removed from its original position as the footstool of God.

The place in which the earth was later entombed became the planetary heavens or first heaven (Revelation 21:1): "And God said, Let there be a firmament (meaning expanse) (and) God made the firmament" (Genesis 1:6,7).

WHAT IS DEATH

Death is a horrendous experience when it isolates one from God, as was the case when it separated fallen angels from Him and from the "Garden of God." This was earth's first paradise and it was designed for its rulers—past, present, and future. Paradise was lost because of folly, arrogance, pride, and stupidity. The kingdom of Lucifer was the last to suffer violence and judgment from God in the primordial ages: "Thou shalt break them with the rod of iron; thou shalt dash them in pieces like a potter's vessel" (Psalm 2:9).

TIME FOR REDEMPTION

After this debacle caused by Lucifer's rebellion, God was determined to establish a foolproof plan of redemption that would prevent any repetition. The plan was as follows: 1) the retention of sovereign wills; 2) the bestowal of a Redeemer to initiate, forge, and complete such a master plan; 3) the forgiveness of sins for those who truly repented and turned to God; 4) the righteousness of God implanted to ensure eternal loyalty (Romans 4:11); 5) an assurance given that all of the names in the book of life would remain there forever (Revelation 3:5). The only justification for removing names placed there before the creation of the world would be for committing acts without conscience and demonstrating incorrigibility, as did Lucifer; 6) an assurance given that all would get a clear message of redemption under equitable circumstances; and 7) assistance from the Spirit of God in the redemptive process so as to facilitate the redeemed toward their ultimate eternal inheritance.

Redemption has been provided and is available if one acts decisively and in time (Colossians 1:14; Hebrews 9:27).[1]

[1] See chart on page 626.

20

God's Olive Branch

Temptation cannot be equated with the deed itself.

"... I *will* cast thee as profane out of the mountain of God: and I *will* destroy thee, O covering cherub, from the midst of the stones of fire... *I will* cast thee to the ground, *I will* lay thee before kings, that they may behold thee... *I will* bring thee to ashes upon the earth in the sight of all them that behold thee" (Ezekiel 28:16-18).

SCENARIO OF EVENTS

First: Lucifer reached Mount Sion, God's throne, in his rebellion and God cast him out from His presence.

Second: God destroyed Lucifer's honorable position as well as his physical attractiveness.

Third: God cast Lucifer to this cursed and mangled earth which was imprisoned under darkness.

Fourth: During the Millennium, Lucifer will be cast into the bottomless pit to join his fallen comrades (Revelation 20:1-3) in viciously tearing at one another in their hatred, but being unable to kill each other as spirit beings.

Fifth: At the end of the Millennium, the human body that

Lucifer possesses (called Gog) will lie in ashes. Of that event we read: ". . . fire came down from God out of heaven, and devoured them." The same prophecy is given in Malachi 4:1-3: "And ye (the physical redeemed saints who emerge from the Millennium) shall tread down the wicked; for they shall be ashes under the soles of your feet in the day that I will do this, saith the LORD of hosts." This could not refer to Armageddon because the destination of Satan and Antichrist are different; Satan is not incarnated in Antichrist (Revelation 19:20; 20:1-3).

The "I wills" of God were warnings which Lucifer received *before* he launched his dastardly attack against God's throne. The Almighty was very explicit about the consequences if Lucifer consummated his plans.

God's warnings to Lucifer also served another purpose. If there were any indecisiveness on the part of either Lucifer or his volunteers, this would discourage them from embarking on a catastrophic course of action. Moreover, the determination to continue this campaign demonstrates how deeply this sin of rebellion had ingrained itself into the ego fabric of the followers of Lucifer. They seemed insanely driven to this disastrous aggression; therefore, they all must share with Lucifer the guilt of this infamous act.

Likewise, concerning the final tribulation period of the Apocalypse, God again gives adequate warnings in reference to taking the mark of the beast:

"And the third angel followed them, saying with a *loud voice,* if any man worship (Antichrist) and his image, and receive his mark in his forehead, or in his hand, the same shall drink of the wine of the *wrath of God,* which is poured out *without mixture* into the cup of his indignation; and he *shall be tormented* with fire and brimstone in the presence of the holy angels, and in the presence of the Lamb" (Revelation 14:9,10). Nothing could be plainer! The nature of the warning shows that the decision will be a *voluntary* act rather than a compulsory one.

TEMPTATION AND SIN

"Let no man say when he is tempted, I am tempted of God: for God cannot be tempted with evil, neither tempteth he any man" (James 1:13). This places temptation squarely on the shoulders of Lucifer, to wit: "... that Satan tempt you not..." (I Corinthians 7:5).

Temptation in itself is not a sin, but yielding to it is a sin; however, continual meditation on the object of tempatation will inevitably result in one's inability to resist (James 1:2,12; I Corinthians 10:12,13). Temptation cannot be equated with the deed. It is "works" (deeds) that will determine the degree of judgment, not the temptations (I Corinthians 3:13; Revelation 20:12).

God extended His hand of love to Lucifer with adequate warnings before the attack was launched showing forgiveness is always in order with God if there is repentance and it occurs in time (Luke 13:3; Acts 17:30).

After death, or the literal separation of man's triune entity, must come the judgment (Hebrews 9:27): "Who will have all men to be saved, and to come unto *the knowledge of the truth*" (I Timothy 2:4). "The Lord is...not willing that any should perish, but that all should come to *repentance*" (2 Peter 3:9). "... The Lord God, merciful and gracious, longsuffering, and abundant in goodness and truth" (Exodus 34:6).

RATTLING OF SABRES

When Lucifer launched his attack, his doom was sealed, as well as the doom of those who joined in the contemptible assault on God's integrity. God's prior warnings apparently had borne some fruit because there appears to have been a strong resistance movement by angels opposing this insurrection. The following is a subsequent account by the holy angels who gathered in the Heaven of heavens after the battle had been successfully won: "... they break forth into singing...saying since thou (Lucifer) art laid down, no fellow is come up *against us*" (Isaiah 14:7,8).

This is echoed in Revelation 12:12 when Satan and his fellow angels are cast out of the heavenly universe prior to the battle of Armageddon: "Therefore rejoice ye heavens, and ye that dwell in them . . . . " To put events in perspective, the former event happened after the cessation of mourning over their fallen comrades when the habitations of the "Heavens of Heavens" were draped in blackness.

SURVIVAL OF PARADISE

Although the earth as God's footstool became a shambles, Paradise escaped unscathed as indicated by these words: "Yea, the fir trees (filled with birds) rejoice at thee, and the cedars of Lebanon (with its romping deer and playful animals) . . . " (Isaiah 14:8).

Paradise, a garden or park where rulers of the earth might relax and communicate with nature, was a sanctuary unexcelled in beauty. The only sanctuaries that were defiled were those Lucifer had created as shown in Ezekiel 28:18: "Thou hast defiled *thy* sanctuaries by the multitude of thine iniquities, by the iniquity of thy traffic . . . . "

A POWERFUL KINGDOM FALLS

God had said that He will "destroy them which destroy the earth" (Revelation 11:18). Like a confirmed and incorrigible criminal blinded by his own pride, his corrupted wisdom, and a false sense of security, Lucifer planned to destroy and plunder if God would not accede to his demands. He conceived of his strategy as being brilliant and successful.

Filled with an insatiable appetite for increased power, Lucifer's greed ultimately and tragically thrust him into direct confrontation with the Almighty. He literally brought his world of great achievements crashing down upon himself and his loyal subjects as he destroyed his glorious realm by instituting a scorched-earth policy. This is comparable to a young child angrily destroying the structure he has carefully and painfully erected because all did not go according to plan.

CONFRONTATION OF ANGELS

Presumably the first assault upon God's throne was made by the angels, who as spirit beings, could ascend directly to God's throne.

After the defeat of the fallen angels, the holy angels made an assault against the "bowmen" (humanoid creatures) in a ferocious battle upon the earth, culminating in the fallen angels and bowmen retaliating with a scorched-earth policy when the battle turned against them. Thereupon, many fled to other parts of the earth as they were pursued by the holy angels.

The extent of the ferocity between God's holy angels and Lucifer's army is described in Jeremiah 4:29: "The whole city (presumably the capital of Lucifer's kingdom) shall flee for the noise (of the confrontation) of the horsemen and bowmen (compare Revelation 6:15-17); they shall go into thickets and climb up upon the rocks; every city shall be forsaken . . . ." "That made the world as a wilderness, and destroyed the cities thereof . . . " (Isaiah 14:17). "Thou shalt be joined with them in burial (Lucifer's allies), because thou hast destroyed *thy* land . . . " (verse 20).

Like abandoned victims exposed to the wrath of God, the creatures in physical humanlike bodies on the primordial earth apparently scrambled like frightened rabbits to the rocks and mountains. This situation will be repeated during God's final judgments in the last tribulation days, and it will be called "the day of the Lord:" " . . . Fall on us, and hide us from the face of Him that sitteth on the throne, and from the wrath of the Lamb: For the great day of his wrath is come; and who shall be able to stand?" (Revelation 6:12-17).

Lucifer's army will be unable to evade the army of God's holy angels in "the day of the Lord" as indicated in this prophecy: "A fire devoureth before them (holy angels); and before them a flame burneth: the land is as the garden of Eden before them, and behind them is a desolate wilderness; yea, and nothing shall escape them. The appearance of them is as the appearance of horses (compare Revelation 19:11-16); and as horsemen, so shall they run . . . They shall run like mighty men; they shall climb the wall like men of war . . . and when they fall upon the sword, they shall not be wounded . . . The earth shall quake before them . . . " (Joel 2:8).

21
The Wrath of God

Lucifer... denuded of his former superior beauty, now bears the visible hideous forms from God's wrath.

After pursuing the fleeing remnants of Lucifer's forces and overpowering them, God took firm action. He said, "... I will cast thee as profane out of the mountain of God..." (Ezekiel 28:16). "I will cast thee to the ground" (Jeremiah 4:17). Lucifer and his hordes were driven from the profusely luxurious Garden of God (Paradise) that was enveloped by the aura of God's throne.

Upon suffering such humiliating indignities, Lucifer's wrath waxed near insanity. Infuriated, he and his armies, in effect became terrorists. Unleashing his full arsenal of destructiveness upon his own kingdom, his raging and uncontrollable temper revealed his deranged, demeaned, and diabolically twisted character.

EVOLUTION IN REVERSE GEAR

Incredible as it may seem, this once beautiful being was fated to a gruesome physique. His appearance was like unto that of dragons, serpents, and other horrific creatures. Even his vicious followers acquired heinous features, resembling scorpions with tails (Revelation 9:1-11; 12:9). At times they seemed to take on the appearance of "angels of light" (2 Corinthians 11:14,15).

These are disfigurations that are totally compatible with the inner evil designs and incorporeal self. Could these frightful features likewise befall all of those who must finally be dispatched into eternal punishment? We know what sin does to handsome physiques and undefiled souls who are given over to satanic influences and immoral rubbish. The "spirit" returns to God (Ecclesiastes 12:7) so the lost souls are equated with animals not having that God-given attribute.

Antichrist, who is a normal human being before being incarnated with evil power, is described as a "beast;" so is the False Prophet (Revelation, chapter 13).

As we shall see later, Adam and Eve regressed to the level of the animal kingdom when they sinned, thus becoming carnivorous. Previous to that their bodies were quickened or kept alive by the pulsating inner power from the tree of life because they came as "living soul(s)" fresh from the hand of the Creator (Genesis 2:7), but now these souls were to face the trauma of death (Ezekiel 18:4, Genesis 2:17).

FRANKENSTEIN'S LABORATORY

Lucifer, stripped of his beautiful and powerful kingdom, as well as denuded of his once superior beauty, now bears visible hideous disfigurations, roars like a ferocious lion (I Peter 5:8), and devours like a dragon (Revelation 12:11,17; 13:2,4; 16:13; 20:2).

Infuriated at what happened, and as retaliation against God, these debauched fallen angels continue their destruction and devastation by inflicting humanity with diseases. Even the redeemed are unexempt while dwelling in their fallen mortal bodies, except when through prayer and faith God extends His healing hand.

Weak and mutilated bodies could never have existed before Adam and Eve sinned. In demonic obsessions, depressions, and possessions there is pain, deformation, and sometimes fearful contortions, so it is conceivable that those lost in eternity could have a sinister and ugly appearance. Sin molds to its own loathsome nature.

The disasters that Lucifer and his hench-fiends have inflicted upon mankind and the earth characterize how wars since the beginning of time have been fostered and motivated. Behind the scenes Lucifer creates situations that inevitably lead into confrontations between individuals, races, nations, and society in general.

GLOBE-TROTTING GHOSTS

"Which shaketh the earth out of her place (as God's footstool), and the pillars thereof tremble" (Job 9:6).

"Hell from beneath is moved for thee to meet thee at thy coming: it stirreth up the dead for thee, even all the chief ones of the earth; it hath raised up from their thrones all the kings of the nations. All they shall speak and say unto thee, Art thou also become weak as we? art thou become like unto us? Thy pomp is brought down to the grave, and the noise of thy viols: the worm is spread under thee, and the worms cover thee" (Isaiah 14:9-11). This text reveals those in hell are in a state of consciousness.

"And the LORD said unto Satan, Whence comest thou? Then Satan answered the LORD, and said, From going to and fro in the earth, and from walking up and down in it" (Job 1:7). This reveals Satan does have access to the throne of God but only alone.

". . . Be vigilant; because your adversary the devil, as a roaring lion, walketh about, seeking whom he may devour" (I Peter 5:8). He is not confined to the habitations in the planetary heaven because on occasion he does make solitary visits to the earth. His fallen angels are confined to either the bottomless pit or the planetary heavens.[1] The disembodied beings, who once lived in their physical bodies on the primordial earth, and who later joined Lucifer in his rebellion, are those that roam the earth as demons or unclean and evil spirits. Because their punishment was denial of a physical body, they crave to occupy one. They create certain physical and mental diseases, as well as non-organic disorders, perhaps as a form of retaliation against God and against those still possessing bodies in the "image" of God.

Since these disembodied spirits are the lowest rank of beings in the rebellion with Lucifer, a child of God has superior power over them in the name of Jesus Christ and through the power of the Holy Spirit: "... In my name shall they cast out devils..." (Mark 16:17). "Behold, I give unto you power... over all the power of the enemy... the spirits are subject to you..." (Luke 10:19,20). "And having spoiled principalities and powers, he made a shew of them openly, triumphing over them in it" (Colossians 2:15).

The above verses reveal how well organized the kingdom of Lucifer was on the primordial earth, as well as subsequently, and the recovery from its chaotic condition.

Because of these evil forces above, within, and on the earth, the whole creation of God is waiting for that day when all these evil forces will be judged, forever removed from the earth, and confined in eternal punishment outside the cosmos: "For we know that the whole creation groaneth and travaileth in pain together until now" (Romans 8:21,22). All of creation is waiting to be "... delivered from the bondage of corruption into the glorious liberty of the children of God." That will be the redemption of the earth!

[1] Those in the primordial earth who helped Lucifer in the construction of cities and who built various enterprises are called "hewers" in Isaiah 14:8, *SRV*.

22

Resurrection of the Earth

As the Spirit of God moved over the earth, the land started to rise from its frozen quarters.

"He setteth an end to darkness, and searcheth out all perfection: the stones of darkness, and the shadow of death" (Job 28:3).

". . . And the Spirit of God moved upon the face of the waters" (Genesis 1:2). The word "moved" literally means "to brood:" that is, to warm and bring into existence in the same manner that a hen incubates her eggs to hatch chicks.

The planet earth probably was plunged deep into the orb or tomb of blackness. This would place it inside of the cosmos at the furthest point from God's throne. Below this, and outside of the cosmos still further from God's throne is located the outer darkness, the bottomless pit, and the lake of fire.

STONES OF DARKNESS

The extent of the chaotic condition upon the earth is dramatized by the words "stones of darkness." The land mass was covered with frozen wastes which lay completely buried in cold storage except for an occasional mountain peak exposed above the frozen debris. Those few land peaks above the congealment were so sparse and convulsed that they gave no clue as to the former topography of the earth. There were no definable outlines of earth's previous appearance. There were no landmarks to indicate earth's former configuration because they were too deeply entrenched and solidly submerged. Atop the few peaks jutting upward were the many precious and beautiful rare jewels that then were merely "stones of darkness" because there was no light to enhance them.

As the Spirit of God moved over the earth, the land started to rise from its frozen quarters. Its rigid chains slowly melted away and the waters began to flow, thus releasing the earth from its icy hold. This so beautifully illustrates the manner in which the Spirit of God effectively moves upon cold and unresponsive hearts to begin the melting process toward redemption.

After the Holy Spirit made His beachhead on the once submerged earth and the icy refrigeration was loosened, God said, "Let there be light: and there was light" (Genesis 1:3). This was the first time since its chaotic condition that the earth was once more bathed under the warm glow of the brightness of God's glory. Light comes from two sources as the Psalmist declared, ". . . thou hast prepared the light *and* the sun" (Psalm 74:16).

EARTH IN SUSPENSION

There is no scientific conflict concerning the earth being older than the universe when we accept the premise that the earth was once placed on "hold," as it were. When it was entombed and held in suspension it was in a state of inertia. Not only did the neutrons, protons, and gravitons become rest particles, but all particles came to a full stop. Even photons, racing at the speed of light, were brought to a complete rest and darkness enveloped the earth. The earth literally died. All mass was a carcass of useless atomic particles.

The resurrection of the earth precedes its eventual total redemption at which time it finally will regain its original beauty and be relocated as the footstool of God's throne. It then will become the predicted New Earth with a new King and Royal Redeemed occupants to brighten the heart of God. When it becomes the New Earth the plan of redemption will be in its final stage in the New Heaven.

INCORRIGIBLES

When God incarcerated Lucifer and his fallen compatriots, the essential ingredient for forgiveness and restoration was missing—repentance. That is the common denominator that God requires of all transgressors in order to regain favor with the Creator. As previously stated, Lucifer would not have accepted redemption even if it had been offered him. God said, ". . . he will not regard any ransom" (Proverbs 6:34,35). He became incorrigible!

GOD'S TERRESTRIAL CHILD

God had not forgotten the earth which lay embedded in its depressive funereal as evidenced by His plans for its recovery: "For thus hath the Lord said, The whole land shall be desolate; *yet will I not make a full end*" (Jeremiah 4:27). He would restore it as His magnificent footstool.

The good earth owes its very existence to the Holy Spirit: "Thou sendest forth thy spirit, they are created: and *thou renewest the face of the earth*" (Psalm 104:30).

As the earth lay chaotically submerged in the icy clutches of blackness, the Spirit of God exercised His power of terrestrial resurrection: "And the Spirit of God moved upon the waters" (Genesis 1:2).

One can almost visualize the Holy Spirit caressing the cold terra firma until the earth slowly rose from its "valley and shadow of death" and the shroud of blackness fell away like bands that are broken (Job 38:9).

The rebellious ones locked within the bowels of the earth must have wondered what was happening! The earth began to move as its metabolistic structure of atoms and molecules generated renewed life.

SPOTLIGHT FROM HEAVEN

"And God said, Let there be light; and there was light" (Genesis 1:2).

The word "let" here does not imply reference to our present solar system but it does indicate that there were also other solar systems. The light now coming upon the earth, as mentioned in verse

2, was not coming from any solar system, but from the Spirit of God, thus causing the molecules in the water canopy to glow and light up the earth. The sun and moon did *not* shine upon the earth until the fourth day of the earth's renewal. It was not necessary in these initial stages: ". . . thou hast prepared the light *and* the sun" (Psalm 74:16).

Some theologians believe this light on the first day of the earth's renewal came from the Glory of God directly, citing Revelation 21:22: "And the city had no need of the sun, neither of the moon, to shine in it: for the glory of God did lighten it. . . ."

The word "let" in Genesis 1:14-19, offers the probability that the universe, including the primordial earth (or the footstool of God), was removed to be held in an impenetrable shroud of blackness until the fourth day of restoration.[1] This could account for the vast age of our universe, as well as that of our planet Earth. The restoration of the Earth, together with its "rest" period, would make our planet appear younger than the cosmos.

[1] See illustration XII.

23

The Book of Life

God extended forgiveness to Lucifer and his followers before *the fact; but to humanity . . .* after *the fact.*

"Remember the former things of old: for I am God, and there is none else; I am God, and there is none like me,

"Declaring the end from the beginning, and from ancient times the things that are not yet done, saying, My counsel shall stand, and I will do all my pleasure . . .

"I have spoken it, I will also bring it to pass; I have purposed it, I will also do it" (Isaiah 46:9-11).

This faculty of God known as foreknowledge is beyond our comprehension, but His foresight is as perfect as His hindsight. Let us examine this amazing ability of God.

HANDFUL OF SHADOWY FIGURES

"Where wast thou when I laid the foundation of the earth?" (Job 38:4).

" . . . I have covered thee in the shadow of mine hand, that I may plant the heavens, and lay the foundations of the earth, and say unto Zion, Thou art my people" (Isaiah 51:16).

Even before God created the planetary heavens and placed the earth in its orbit in the solar system, He held all of the potentials for redemption as shadowy figures in His hand according to the vision of His perfect foreknowledge. What a beautiful handful of priceless jewels!

GOD'S TELESCOPIC VISION

God has penetrating telescopic vision.[1] His perfect foreknowledge established the basis for inscribing into the book of life the names of unborn billions. Contrary to that which many believe, a name is not placed in this book when one accepts Christ as his personal Saviour. There are no supporting texts for it. If this were the case, then innocent deceased children would be lost: "And whosoever was not found written in the book of life was cast into the lake of fire" (Revelation 20:18).

In Exodus 32:33 Moses received this message from God: ". . . Whosoever hath sinned against me, him will I blot out of my book." How can a name be blotted out if it never was inscribed? God has given everyone an equal start by placing his or her name in the book of life.

God has possession of the instrument that inscribed every person's name in the book; also He has possession of the eraser which will remove the name. He never will entrust either of them to the wisdom of men! Whether or not one believes in the doctrine of perennial grace, the fact remains that only God can delete or retain the names in the book of life.

MAN'S SELF-DESTRUCTIVE FACULTY

God's creative ability and plans would not exclude anyone *without* a sovereign will; however, with this inherent power, a person possesses the potential of self-destruction. Therefore, one cannot blame God when a person by his free choice determines his own destiny: "And the Spirit and bride (redeemed) say, Come. And let him that heareth say, Come. And let him that is athirst come. And *whosoever will,* let him take of the water of life *freely"* (Revelation 22:17). *"My soul is continually in my hand"* (Psalm 119:109).

GOD HOLDS AN ELECTION

"*Elect* according to the foreknowledge of God the Father, through sanctification of the Spirit, unto obedience and sprinkling blood of

Jesus Christ . . ." (I Peter 1:2).

This indicates that the members of the Godhead instituted a redemptive plan in which each committed Himself to the fruition of the master plan.

The Godhead is clearly stated in Romans 1:28; Acts 17:29; and Colossians 2:9. Examples of the plurality of the Godhead is seen by such words as "us" and "our" in Genesis 1:26; 3:22; 11:7; Isaiah 6:8.

There are only three members in the Godhead—the Father, Son, and Holy Spirit—and they equate as *one* (John 17:11), which makes them a compound unity. The redeemed, in spite of their future great inheritance and transformation into immortality (I Corinthians 15:33; Revelation 21:7), will never become members of the Godhead. They will share in the Glory of God, but always will remain "adopted" until their marriage to Christ Jesus as His corporate wife. Jesus Christ is God's *only* begotten and He will always be "King" of kings and "Lord" of lords. The redeemed will forever remain His loyal subjects as "kings" and "lords." The redeemed in the Church, even as the wife of Christ, can only be God's "daughter-in-law."

ELECTION OF GRACE

God extended forgiveness to Lucifer and his followers *before* the fact; but to humanity *after* the fact. This is the "election of grace" and the angels yearn to explore this new turn of events: ". . . which things the angels desire to look into" (I Peter 1:12).

CHOSEN ONES

God began His plan of *redemption* by providing it to *all:* "According as he hath *chosen* us in him *before* the foundation of the world, that (prophetically) we *should be* holy and without blame before him in love" (Ephesians 1:4. Compare Romans 3:22).

Paul the Apostle said, "For ever *before I was born God had chosen me* to be his . . ." (Galatians 1:15, *Living Bible*). In this respect we are all exceptions and privileged beings, as was Paul. To be "chosen" by God is not arbitrary; it is still contingent on the "will" of a person to accept. God by foreknowledge knows what that decision will be, but it is not a compulsory one. Because names are initially inscribed in the book of life, a follow-up plan is required to preserve as many as possible in that book according to the redemptive plan of God.

THE CALLING OF GOD

"For the gifts and calling of God are without repentance" (Romans 11:29). Phillips translates it, *"God does not withdraw his gifts or calling."* They are irrevocable.

This calling involves two basic responsibilities that God has totally assumed, as best expressed in these scriptures: "Who will have *all men* to be saved, and to come unto the *knowledge of the truth"* (I Timothy 2:4). "The Lord is not slack concerning his promise, as some men count slackness; but is longsuffering to us-ward, not willing that any should perish, but that *all should come to repentance"* (2 Peter 3:9). The word "should" clearly shows that this is not universal salvation, but that it is an invitation to repentance.

God knew by foreknowledge that all men were potential sinners, so to accelerate the redemptive process, Christ was *chosen before* the foundation of the world to pay that price for *all:* "Even the righteousness of God which is by faith *of* Jesus Christ *unto all* and *upon all them that believe:* for there is no difference: For all have sinned (in God's foreknowledge), and come short of the glory of God" (Romans 3:22,23). Any alternative not only would have delayed the redemptive process, but also would have blunted any initiative on the part of individuals to seek redemption. God's love and grace had to be enacted by the death of His Son, the Lord Jesus Christ, in order to demonstrate and clarify the issue.

The two redemptive options afforded mankind as provided by Jesus Christ are clearly stated: "For the *wages of sin* is *death;* but the *gift* of God is *eternal life* through Jesus Christ our Lord" (Romans 6:23). The options are death or eternal life.

The Holy Spirit assumes a blanket responsibility in directing the three basic essentials toward redemption: "And when he (the Holy Spirit) is come, he will reprove (or convince) the world of *sin,* and of *righteousness,* and of *judgment"* (John 16:8). The context shows these categories as originating from three fundamental problems: *unbelief;* man's *sinful nature;* and the *serious consequences of transgression.*

PREDESTINATION

"According as he hath chosen us in him before the foundation of the world, that we *should be* holy and without blame before him in love: Having predestinated us unto the *adoption* of children by Jesus Christ to *himself,* according to the good pleasure of *his will"* (Ephesians 1:4,5). The emphasized words reveal the entire text is prophetic only to the redeemed.

Predestination takes us one step beyond God's plan of redemption. He is no longer addressing the unredeemed under foreknowledge. God is now concerned about the future *inheritance* of those already converted to Christ under His foreknowledge.

Foreknowledge opens the door of redemption.

Predestination opens the door of inheritance to the redeemed.

The immortals whose names are in the book of life are assured that their names will never be removed and that their eternal inheritance is secure: "He that overcometh . . . I will *not blot out* his name out of the book of life . . . " (Revelation 3:5).

HOPE OF THE REDEEMED

The point at which we will be conformed to the image of God's Son is clearly shown in I John 3:2,3: "Beloved, now are we the sons of God, and it doth not yet appear *what we shall be:* but we know that, *when he shall appear, we shall be like him;* for we shall see him as he is. And every man that hath this *hope* in him purifieth himself, even as he (Christ) is pure."

Predestination is, therefore, the "hope" of the redeemed. It involves the believer's future *inheritance: "To an inheritance* incorruptible, and undefiled, and that fadeth not away, *reserved in heaven for you"* (I Peter 1:4); *". . . they which are called might receive* the promise of eternal inheritance" (Hebrews 9:15).

As redeemed we are *now adopted* sons of God, but we will be "conformed to the image of his Son" by the Spirit of God, to the degree unto which we can qualify maturely as His eternal corporate wife.

A MATURE BRIDE

When a person is "born again," he is spiritually immature; and as far as his personal transformation is concerned, he still is not like Christ to the degree of His full stature: "Till we all come in the unity of the faith, and of the knowledge of the Son of God, unto a perfect man, unto the measure of the stature of the fulness of Christ" (Ephesians 4:13). That will require some doing: "For when *that* (not "He;" i.e. Christ) which is perfect is come, then *that* which is in part shall be done away" (I Corinthians 13:9,10).

Christ will not wed an immature child bride. Only when we achieve a certain degree of maturity (I John 3:2) will we become part of the corporate wife of Christ. This new status will become a reality when the redeemed receive their bodies of immortality by the Holy Spirit.

"For whom he did foreknow, he *also* did predestinate *to be conformed* (not transformed, as in the new birth) to the image of his Son, that *he might be* the firstborn among many brethren" (Romans 8:29).

HAVE FAITH IN GOD

"And Jesus answering saith unto them, Have faith in God" (Mark 11:22). Faith involves trusting in the totality of God—His perfect wisdom, foreknowledge, and love—not just His omnipotence. If we are trusting in an isolated prayer, we are not letting all things work together for our good (Romans 8:29), because our faith is in God's power alone, and not in His wisdom to put things together; thus we send ourselves on a guilt trip and lose our peace of mind (Colossians 3:15). An example of having faith in both the Lord's power *and his wisdom* (in timing the answer) is found in Matthew 8:2,3: "And, behold, there came a leper and worshipped him, saying, Lord, *if thou wilt,* thou canst make me clean. And Jesus (said), *I will,* be thou clean . . ." (Compare Acts 12:1-16). It is one of the major Biblical truths that defies logic rather than accedes to it.

[1] Without God's foreknowledge, Isaiah 65:25 would be impossible: ". . . *before* they call, I will answer . . . ."

24
God Ties His Own Hands

By its own actions Israel made possible the expansion of the Abrahamic covenant to include the Gentiles.

God chose Abraham with whom to make an unconditional and unilateral covenant: "As for me," God said, "behold my covenant is with thee . . . and to thy seed after thee . . . *and I will be their God* . . . And I will establish my covenant between me and thee and thy seed after thee in their generations for an *everlasting covenant,* to be a God unto thee, and to thy seed after thee" (Genesis 17:4-8).

The purpose of this agreement was to restrain God from destroying the human race because of sin, thereby jeopardizing the plan of redemption: "Except the Lord (of hosts) . . . hath left us a seed, we had been made like unto Sodom and Gomorrah" (Romans 9:29). This covenant served to stave off the wrath of God upon sinful humanity and resulted in God's patience and long-suffering coming into focus: ' . . . behold, the husbandman (God) waiteth for the precious fruit of the earth (redeemed), and *hath long patience for it,* until (it) receive the early and latter rain" (James 5:7). These spiritual rains cover the periods of the Church Age (the mystery of the kingdom) and the Manifested Kingdom of Christ which will begin with His return to the earth.

Prior to this unconditional and unilateral covenant with Abraham, several devastating events occurred. The flood in Noah's time nearly wiped out all vestige of life upon the earth. Sodom and Gomorrah and their neighboring cities were totally destroyed by divine conflagration. Only a few humans were spared in both instances.

The covenant with Abraham pertained to the birth of Isaac who typified the coming Redeemer, so to punctuate His forebearance at that time, God demonstrated how the patience of the redeemed would be tried: "And so, after (Abraham) had *patiently endured,* he obtained the promise" (Hebrews 6:15). However, chapter 11 is filled with examples of the redeemed having their faith tested in an unresponsive world which periodically has been very hostile to righteousness.

UMBRELLA OVER THE GENTILES

"I say then, Have (Israel) stumbled that they should fall? God forbid: but rather *through their fall salvation is come unto the Gentiles*... For if the casting away of them be the reconciling of the world, what shall the receiving of them be, but life from the dead?... Thou wilt say then, The branches were broken off, that I might be grafted in" (Romans 11:11,15,19).

When God said, "So all Israel shall be saved" (verse 26) undoubtedly He made stipulations regarding those who adamantly refused his redemptive plan. Is God to be satisfied with merely a residue of redeemed ones, a partial fulfillment of the covenant? To complete the fulfillment of His undertaking, God had to retain the covenant throughout His redemptive plan.

God now would have to graft Gentiles outside of the commonwealth of Israel into the covenant in order to accomplish His objective. Because Israel, as a nation, did not unilaterally receive God's redemptive plan, He would provide the promised Messiah to atone for the sins of all people. By its own actions, Israel made possible the expansion of the Abrahamic covenant to include the Gentiles.

This covenant will remain in effect until Christ returns to earth:

"For I would not, brethren, that ye should be ignorant of this mystery, lest ye should be wise in your own conceits; that blindness in part is happened to Israel, until the fulness of the Gentiles be come in. And so all Israel shall be saved: as it is written, There shall come out of Sion the Deliverer, and shall turn away ungodliness from Jacob" (Romans 11:25,26).

The ultimate objective of the plan of redemption in the Lord's Manifested Kingdom will become His total responsibility as He assumes the reins of His government: "And he (Christ) shall reign over the house of Jacob for ever; and of his kingdom there shall be no end" (Luke 1:33). It will be an ever-expanding kingdom even as it emerges into the Father's kingdom.

25

God's Arsenal

God established an ideal climate with a built-in weather control system which consisted of a canopy.

"And God said, Let there be a firmament in the midst of the waters, and let it divide the waters . . . which were under the firmament from the waters which were above the firmament: and it was so" (Genesis 1:6).

As God's Glory penetrated the darkness that enveloped the earth, it warmed and evaporated the chilly waters, sending billions of silvery threads spiraling upward in shimmering rays from the reflection of the rising vapors. Clouds formed and were held in suspension by divine decree: "He bindeth up the waters in his thick clouds; and the cloud is not rent under them" (Job 26:8). A firmament was then placed between the waters upon the earth and those above it.

EARTH'S CANOPY

God established an ideal climate with a built-in weather control system which consisted of a canopy of water above the earth that enveloped and hermetically sealed in the atmosphere. In addition, it shielded out the dangerous lethal rays which cause decay and decomposition in flora and fauna.

As a result of the Flood the canopy above the earth was depleted, thus the harmful ultraviolet rays were able to penetrate the atmosphere to cause decay and decomposition. Noah was unaware of this environmental change after the Flood so when he drank from the processed fruit of his vineyard, he fell victim to alcohol, a new chemical element.

In the Antediluvian period the earth had a comfortable greenhouse climate with moderate temperatures and diffused sunlight which contributed greatly to the health and longevity of those living at the time. Isaiah 30:26 predicts the return of this water canopy during the Millennium: "Moreover, the light of the moon shall be as the light of the sun, and the light of the sun shall be sevenfold, as the light of seven days, in the day that the LORD bindeth up the breach of his people, and healeth the stroke of their wound." Unless such light is diffused by the canopy which acts as a heavenly reflector and screens out the ultraviolet rays, the heat will be intolerable. However, we know that the atmosphere will be similar to that enjoyed under the shade of a great tree with its fine mist cooling the air.

MILLIONS OF ARTESIAN WELLS

". . . All of the fountains of the great deep were broken up, and the windows of heaven were opened" (Genesis 7:11). It was as though countless artesian wells of massive proportions gushed forth throughout the earth. Evidently, the combination of these gushers, together with the release of the water canopy high above the earth, quickly caused the inundation of the earth's terrain.

When God was ready to drain the water from the flooded earth, these artesian wells undoubtedly turned into vast underground drainage systems and surplus water was diverted to the polar regions. This would have been accomplished simply by God shaking the earth, perhaps in the form of an earthquake: "Whose voice then shook the earth . . ." (Hebrews 12:26). ". . . Who shut up the sea with doors, when it brake forth, as if it had issued out of the womb . . . I made the cloud the garment thereof (canopy), and thick darkness a swaddlingband for it (ice), And brake up for it my decreed place (Flood), and set bars and doors (imprisoned Flood waters at the poles), And said, Hitherto shalt thou come, but no further: and here shall thy proud waves be stayed" (Job 38:8-11).

In the lower areas God immediately congealed the waters, like jello, by tilting the earth twenty-three and one-half degrees. This would hold the frozen waters in place during its six months of darkness and bitter cold, but there would not be sufficient time to freeze the waters. This congealment of the waters, therefore, began from the *inside*—similar to what occurred in Exodus 15:8: "And with the blast of thy (God's) nostrils, the waters were gathered together, the floods stood upright as an heap, *and the depths were congealed in the heart of the sea."*

GOD'S FUTURE ARSENAL

"Hast thou entered into the treasures of the snow? or hast thou seen the treasures of the hail, Which I have reserved against the time of trouble, against the day of battle and war?" (Job 38:22,23).

Why has this mass of frozen water been retained at the polar regions instead of having been restored as the canopy? Could it be that this is God's future arsenal to topple the regime of Antichrist and his vast Babylonian Empire in the last days of Gentile rule? ". . . And the end thereof shall be with a flood, and unto the end of the war desolations are determined" (Daniel 9:26).

Jesus predicted that there would be vast tidal waves (Luke 21:25) leaving Antichrist's world trade center in shambles: "And they cast dust on their heads, and cried, weeping and wailing, saying, Alas, alas, that great city, wherein were made rich all that had ships in *the (Mediterranean) sea* by reason of her costliness! for in one hour is she made desolate" (Revelation 18:15-19).

The great heat plagues that God will send during the last tribulation judgments upon the earth will coincide with the predicted great earthquake, thus causing a breakup of the polar ice mass. In turn, the cascading ice floes will cause the oceans to rise (Revelation 8:7-11; 16:8,9; 6:12).

The canopy presumably will be restored in time for the Millennial reign of Christ on earth and vast resources of minerals and productive lands will be uncovered in the polar regions (Job 38:22).

One of the most feared future events is the collapse of the West Antarctic ice sheet containing more than 90 percent of the fresh water on the earth. This sheet would return ten percent of the Antarctic water to the oceans and cause sea levels to rise twenty feet. The resulting flood would wipe out most of the world's port communities, inundate the greater portion of Florida, and probably alter climatic patterns. Presently the Thwaites Glacier, which is moving at the rate of almost two miles a year, is snagged on islands. An increase in temperature could cause an uncontrollable surge to the sea by the enormous ice sheets congealed there since the Flood in Noah's time. This could result in the predicted worldwide tidal waves (Luke 21:25; Daniel 9:26; Jeremiah 51:42,43). An earthquake also could cause the islands which are upholding the ice sheet to sink and thus create the prophesied catastrophe. The discovery of identical fauna fossils in the Antarctic and in the United States confirms that the Flood in Noah's time was worldwide.

26

The Craftsman at Work

We see care and intricate precision in every detail of that which God has scrupulously, meticulously, and ingeniously created.

"These are the generation*s* (plural) of the heavens and of the earth *when they were created,* in the *day* that the LORD God made the earth and the heavens" (Genesis 2:4). However, in contrast, the heavens and the earth in the *restoration* were *not* brought into existence in a single day.[1] Our text shows two different generations, one which existed on the original earth and one which inhabited the restored earth.

RESURRECTION OF THE ANIMAL KINGDOM

The first four days during this renewal of the earth were dispensational days, with the solar system assuming its role on the fifth day, in order to accommodate the plant life and the living creatures that were to be restored to the earth.

The fifth day, according to Genesis 1:20-23, was resurrection day in the animal kingdom as indicated by the word "let," which denotes *permission* or *allowance* in the sense that we would say, "Turn on the light." This infers that the primordial earth must have existed under an environmental condition similar to ours at one time because it had provided a life support system for these creatures (Genesis 1:15). The terms "generations of the . . . earth" and "replenish the earth" imply that both mortality and judgments existed on the primordial earth, but apparently not in the heavenly Paradise, which evidently was adjacent to it since it is called the "footstool" of God.

Not all of the creatures which existed on the original earth were restored during the renewal of the earth. There was selectivity, just as there will be at the time of the resurrection of believers. The original creatures that were not restored during the renewal of the earth perhaps will be resurrected and reactivated at the time of the Millennium, the New Earth, or in the New Heaven, and transported to a natural habitat in the Lord's expanded Manifested Kingdom. Did not Jesus say He would "fill all things"? (Ephesians 4:10).

ATOMS PUT TO REST

All matter is composed of atoms, and within the atoms are particles, as well as subatomic particles, which combine in such a way as to impart distinctive characteristics to plants and living creatures. This intrinsic *personality* is within the *life* of every particle and without it each would be dead and useless.

Atomic physicists have discovered that at the core of atoms, the nuclei of protons and neutrons which collectively are called "nucleons," oscillate as if they were breathing. These "nuclear vibration modes" are known as monopole, dipole, quadrupole, sexupole, and octupole, and their palpitations coordinate with their internal forces. The sizes of the nuclei determine the ratio of the pulsations.

The central mass of the nucleus remains at rest. Perhaps here in the eye of the atom hurricane resides the spirit-life that causes the vibration modes. The electrons form a wall about the nucleus and the outer shell of the atom by their fantastic rotation speed which is propelled by strong nuclear forces. Collision or spinning away from the orbit is prevented by the electromagnetic force.

When the earth literally died, each particle in every atom of all mass became a "rest" particle when it came to an abrupt halt in its motion. According to Isaiah 14:7, all living plants and creatures on the original earth encountered the same consequences, as indicated by the word "rest," which means "to settle down."

"Thou sendest forth thy spirit (Holy Spirit), they are created: *and* thou renewest the face of the earth" (Psalm 104:30). Thus the renewal was instantaneous as it will be with the multiplied millions of believers who will be resurrected some day in the "twinkling of an eye" (I Corinthians 15:52).[3]

IN THE LIKENESS OF GOD

As denoted by the word "let," God replenished the renewed earth with primordial creations of His selection. The manlike creatures existing on the primordial earth were not returned to the renewed earth except as disembodied spirits. Instead, He created Adam and Eve who were distinctly different, being in the *likeness* as well as in the image of God: "And God said, Let us make man in our image, after our likeness . . . male and female created he them" (Genesis 1:26,27; 2:7).[2] To them alone the Creator imparted His Spirit with its Godlike characteristics and superior intellectual capacity: "But there is a spirit in man: and the inspiration of the Almighty giveth them understanding" (Job 32:8). As Jesus quoted the Psalmist: " . . . Ye are gods" (Psalm 82:6; John 10:34).

CRAFTSMAN AT WORK

We see care and intricate precision in every detail of that which God has scrupulously, meticulously, and ingeniously created.

Why were we given the record of this creation on the original earth? First, that is the way it happened; second, it is to help us understand the Creator's thoroughness and patience in achieving perfection which involved planning, designing, and imparting life to all living things and creatures. Every plant, herb, and creature was given personal attention to enable it to function and survive in its environment. We have a tendency to think of God as being interested only in general things and not being concerned with specific matters and problems in our lives as well. How wrong we are!

The Creator operated under the time schedule of eternity. He never created in a rush. He planned, designed, and perhaps occasionally paused to reflect upon His production. He worked on the premise of bringing into existence a perfect result: " . . . he searcheth out all perfection . . . " (Job 28:3). "Thus saith the LORD the maker thereof, the LORD that formed it, to establish it; the LORD is his name; Call unto me, and I will answer thee, and shew thee great and mighty things, which thou knowest not" (Jeremiah 33:2,3). "And it shall come to pass, that before they call, I will answer; and while they are yet speaking, I will hear" (Isaiah 65:24), which is a clear proof of God's foreknowledge.

GOD THE PHYSIOLOGICAL BOTANIST

"And *every* plant of the field *before* it was in the earth (the original one), and *every* herb of the field *before* it grew; for the Lord God had not caused it to rain upon the earth, and *there was not a man to till the ground"* (Genesis 2:5).

Visualize the Creator contemplating the appearance, function, and contribution that every species of plant and herb would make to the balance of nature. There were myriads of varieties and each was given His personal attention concerning climate and soil. The Creator anticipated how these plants were to serve the needs of the living creatures He would bring into existence.

Within each of the different kinds of living plants and creatures God implanted seeds to assure their continuation and dispersal over the earth. Locked within each seed were all of the distinctive invisible characteristics for future generations.

Life consists of two entities: the invisible and the visible. Because something is visible we have a tendency to magnify its importance, but actually it is the invisible that contains the whole personality of that which is to become visible. See illustration VII.

DELIVERY FROM THE CURSE

Sin has brought great problems to the earth itself, but some day this will be rectified: "For the creature was made subject to vanity, not willingly, but by reason of him who hath subjected the same in *hope,* Because the creature itself also shall be delivered from the bondage of corruption into the glorious liberty of the children of God. For we know that the whole creation groaneth and travaileth in pain together until now" (Romans 8:20-22).

Sin has enabled destructive insects to corrupt the earth; however, during the Millennial reign of Christ the curse will be lifted from the earth. The biological appetites of over one million species of insects either will be changed to consume unwanted plants or they will be exterminated under the judgments of the heat plagues.

"At that day shall a man look to his Maker, and his eyes shall have respect to the Holy One of Israel . . . " (Isaiah 17:7,11). "Seek ye out the book of the Lord, and read: not one of these shall fail, none shall want her mate (no guns and hunting permitted): for my mouth it hath commanded, and his spirit (Spirit) it hath gathered them" (Isaiah 34:16). "The wilderness and the solitary place shall be glad for them (Royal Redeemed returning to earth with Christ); and the desert shall rejoice, and blossom as the rose" (Isaiah 35:1). "Then the eyes of the blind shall be opened, and the ears of the deaf shall be unstopped. Then shall the lame man leap as an hart, and the tongue of the dumb sing" (Isaiah 35:5,6). "And the ransomed of the Lord shall return, and come to Zion with songs and everlasting joy upon their heads: they shall obtain joy and gladness, and sorrow and sighing shall flee away" (Isaiah 35:10).

[1] See Illustration XII, page 614.

[2] See pages 393-394.

[3] Word "and" denotes creation *before* renewal.

27

Name Calling

God took sufficient time to work, to rest, and to enjoy His achievements.

"And out of the ground the LORD God formed *every* beast of the field, and *every* fowl of the air; and brought them unto Adam to see what he would call them: and whatsoever Adam called *every* living creature, that was the name thereof" (Genesis 2:19). When the Creator brought into existence animate creatures, He continued with the same meticulous procedure and dedication as He did with the inanimate world. In the days of renewal prior to the creation of Eve, Adam named every fowl and animal as God "*brought them unto Adam* (one at a time) to see what He would call them." This would have required many days which means that Eve's creation came some time after Adam's.

What effect did all of this have upon Adam? As he looked at each creature with a discerning eye, the overwhelming feeling of attachment which he undoubtedly had for each creature must have brought a deep sense of love and admiration for the Creator.

For God to have reached out over the earth and to have brought all creatures to Adam required His knowledge of their location: "I know all the fowls of the mountains: and the wild beasts of the field are mine" (Psalm 50:11). The Psalmist added, "Such knowledge is too wonderful for me; it is high, I cannot attain unto it" (Psalm 139:6).

CLONINGS

Not only did the Creator operate under dispensational time to give each plant, herb, and creature His concentrated attention, but in order to assure perpetuation and survival of the fowls and animals, He inseminated hormones with an affinity to each of their own kind. To prevent cross-breeding, God had to establish barriers between the different species, and this purpose was accomplished by injecting the sperm with a mystical element.

To enhance their survival, God imparted various creatures with sonar devices and other mysterious characteristics, some of which are produced not only within very minute simplistic creatures, but also within those with structural complexities. This scientists cannot comprehend.

Science is just beginning to appreciate some of these wonders: "For the invisible things of him from the creation of the world are clearly seen, being understood by the things that are made, even his eternal power and Godhead; so that they are without excuse" (Romans 1:20).

We know the Creator could have created everything in one moment of time, but because He has a personal attachment to His work and an interest in the organization and usefulness of each species in the balance of nature, He chose not to do it in that way. The Mastermind of the Creator gave sense and meaning to both the invisible and visible world, leaving nothing to chance.

God must have paused occasionally to view with admiration the beauty, design, and distinctive features of each entry He produced. This should teach us something about the Creator besides his ability to bring things into existence. He took all the time necessary to work, to rest, and to enjoy His achievements, and as God's Word informs us, He is ever ready to give His personal attention to any particular problem or need that concerns us.

Let us not underestimate the Creator's patience that flows out from His Being as He shows personal care and love for each project on His agenda.

"His eyes behold . . . the children of men . . . His countenance doth behold the righteous" (Psalm 11:4-7).

28
Principalities of Angels

Jesus Christ is "Prince of princes" over all angels, "King of kings" over the Royal Redeemed, and "Lord of lords" over all creation.

"For we wrestle not against flesh and blood, but against *principalities,* against powers, against the rulers of the darkness of this world, against spiritual wickedness *in high places"* (Ephesians 6:12).

"To the intent that now unto the *principalities* and powers *in heavenly places* might be known by the church the manifold wisdom of God" (Ephesians 3:10).

GATES OF HELL

". . . I (Jesus) will build my church; and the gates of hell shall not prevail against it" (Matthew 16:18). The principalities given over to Satan in the celestial heavens are not the gates of heaven but the "gates of hell." The Church has been given the "keys" so that ". . . whatsoever thou shalt bind on earth shall be *bound in heaven:* and whatsoever thou shalt loose on earth shall be *loosed in heaven"* (verse 19).

That which plagues and harasses the Church and the believers comes from Satanic celestial stations referred to as "high places" and "heavenly places." The Church is called upon to understand from where its opposition really comes: "To the intent that now unto the principalities and powers in heavenly places might be known by the church and the manifold wisdom of God" (Ephesians 3:10).

CELESTIAL DESERTS

How can these principalities express the wisdom of God? They show the distinctive areas of operation of the spiritual forces, and their limitations, as well as the extent of the spiritual forces weighed in favor of righteousness.

The believer can bind the disembodied spirits on earth that are seeking to reinhabit a physical body either by oppression, depression, or possession, the first being a milder form of impairment. In binding these spirit forces, we not only can contain their ability to add further difficulties; but we also can prevent their stronger allies in the heavenly (cosmos) domains from coming to the aid of the demons loosed on earth.

When a person who is in need of deliverance is "loosed," the holy angels resident in their heavenly principalities rush to the aid of the believer because opposing spiritual forces cannot be defeated with "flesh and blood" tactics. Jesus informs us that the exorcised evil spirits are sent to "dry places" which presumably are above the earth (Matthew 12:43; Luke 11:24). These habitations are suitable only for spirit beings and not for mankind. The plurality of the word "keys" suggests that there is the right "key" to bind and loose in any given situation that arises.

PRINCE OF PRINCE

"God . . . by him were all things created, that are in heaven . . . visible and invisible, whether they be thrones, or dominions, or principalities, or power: all things were created by him, and for him" (Colossians 1:13-16). Since God created the dominions He can limit the mobility of those who occupy them and give them a disadvantage with the believer (I John 4:4). The head of all principalities is Jesus Christ: "And ye are complete in him (Christ), which is the head of all principality and power" (Colossians 2:10). In this position, Christ is the Spoiler of Satan's dominions.

When Satan steps on the toes of the Church on earth, Jesus Christ is ready to step on Satan's head and defeat him for he has offended the Head as well as the Body: "And he said unto them, I beheld Satan as lightning fall from heaven, Behold, I give unto you power to tread on serpents and scorpions (see Revelation 12:9; 9:1-11; Genesis 3:15), and over all the power of the enemy: and nothing shall by any means hurt you" (Luke 10:18,19). "And having spoiled principalities and powers, he made a shew of them openly, triumphing over them in it" (Colossians 2:15).

Christ Jesus identified these evil powers during His earthly ministry and delivered those oppressed and possessed by them. He effectively used the "keys" of the kingdom which he later gave to the Church to use in releasing the holy princes (angels) from their principalities to come to His aid: "And he saith unto him, Verily, verily, I say unto you, Hereafter ye shall see heaven open, and the angels of God ascending and descending upon the Son of God" (John 1:51). These are ministering spirits ". . . sent forth to minister for them who shall be heirs of salvation" (Hebrews 1:14).

One of these holy angels prevented the premature death of Christ in Gethsemane (Luke 22:43,44). Here Satan sought a counterfeit and useless atonement: ". . . *his sweat as it were drops of blood* falling to the ground." Nothing but the actual blood of Christ, incorruptible and precious, can atone for sin.

29

Celestial Hatcheries

Astronomers tell us that a symphony of sounds which is called "cosmic harmony" reverberates throughout the uhiverse.

"The heavens declare the glory of God; and the firmament sheweth his handiwork" (Psalm 19:1).

"He made darkness his secret place . . . " (Psalm 18:11).

"Yea, the darkness hideth not from thee: but the night shineth as the day: the darkness and the light are both alike to thee" (Psalm 139:12).

The earth had long been buried near the apogee of the orb of blackness in relation to God's throne. The Creator held within His hand pure invisible energy, which He scattered like seed, thus transplanting the universe carefully in a predesigned plan for a great celestial harvest. The Holy Spirit completed His germination, and like billions of chicks bursting from their eggs, there were big bangs everywhere (not just one) which were the embryos of galaxies and their inherent multitudinous solar systems.

Space scientists are observing within galaxies very bright sources whose nuclei are shooting out from the core matter at incredibly high speed as if a million nuclear warheads are being detonated each second. Powerful radio rays are also emitting from these central cores.

Again we see the Creator under a dispensational calendar interweaving every constellation, galaxy, and solar system, as well as naming every star: "He telleth the number of the stars; he calleth them all by their names" (Psalm 147:4). Presumably this also includes those stars in the "Heavens of Heavens" and the "Heaven of heavens."

The solar system was not ready to assume its function on the renewed earth until the fifth day of restoration. It was not necessary until then because the living creatures were not resurrected until the fifth and sixth days. It is believed by some that the whales were an addition to the restoration because of the use of the Hebrew word "bara," implying an original creation.

Adam and Eve were an exception as they were *created.* Eve came on the scene after Adam had named the creatures, as evidenced in Genesis 2:21-25: ". . . the heavens and the earth were finished, and all the host of them." The animal kingdom and mankind appeared on the scene only *after* the solar system (of which the planet earth is a part) was permitted to exert its influence upon the earth, thus providing the proper environment for their comfort and existence.

STARS IN YOUR EYES

Man never has been able to calculate all of the billions of stars and the galaxies even by the most modern sophisticated radio telescopes and detection instruments. Only recently have the distant quasars been located.

In the ever-changing patterns in the universe there are deceased stars and new generation stars coming into existence. What perfect organization the Creator has in order to be fully cognizant of them and their constant changes! They are part of His family, and no new star, regardless of its size, fails to be named at birth—just as mankind names hurricanes and typhoons as they emerge. The Creator is fully aware of all happenings and changes taking place in the universe.

IN THE THROES OF DEATH

An example of God's concern for each individual entity of His creation is demonstrated in Matthew 10:29: ". . . Are not two sparrows sold for a farthing? and one of them shall not *fall* (in the throes of death) to the ground without your Father." Just a farthing to man, but God is emotionally involved in each creature regardless of its human value. In the light of this revelation, how can anyone minimize the worth of man's soul? Will God cast aside man's soul for any trivial reason? If this were so, we would have to assume that God loves the animal kingdom more than He does mankind because death to an animal carries no future punishment, but man, who is of inestimable value to God, is demeaned and cast into eternal punishment; in some cases through no fault of his own.

How much greater than sparrows are we in God's eyes! (2 Peter 3:9). But why the loss of any creature at all? Death did not arrive on the scene by divine initiation. It came as an intruder to spoil the work and plan of God. Sin is the underlying culprit. Redemption is the final solution.

God never could allow mankind to continue indefinitely in sin in the light of His mercy. This would make the earth a veritable hell and totally uncontrollable. So death was permitted to remain and to serve a purpose to God's advantage. Thank God, tyrants and evildoers do not live forever on earth! Unfortunately, the righteous also face the same fate, but their resurrection and future are eternally assured under a perfect environment.

With the demise of each wicked person, that population is reduced. Sin, therefore, is restrained to some extent. There is always hope that the succeeding generation will have learned its lesson and have profited thereby, but history too often seems to repeat itself. Eventually God must lower the curtain, call a halt to the travesty of evil, and bring all into judgement.

CASTING SHADOWS

All things, animate or inanimate, are but shadows of the originals and naturally have a limited life expectancy. They also suffer from the backlash of prevailing sin: "For the creature was made subject to vanity, not willingly, but by reason of him who hath subjected the same in hope, Because the creature itself also shall be delivered from the bondage of corruption into the glorious liberty of the children of God. For we know that the whole creation groaneth and travaileth in pain together until now" (Romans 8:20-22).

Even the celestial heavens have a limited life expectancy for the same reasons. Satan has polluted certain areas within his principalities: "... yea, the heavens are not clean in his sight" (Job 15:15).

At the close of the fourth dispensational day God rent the swaddlingband which shielded the earth from the radioactivity of exploding galaxies and solar systems. In full view of all, the angels of God stood on a massive observation platform suspended on high and beheld the vista of trillions of glittering stars displayed before them.

The entire celestial heavens burst into all of their glorious beauty. It was as if they had issued out of a womb, according to the Psalmist's description. There were billions of solar systems within the innumerable galaxies, with stars, planets, satellites, streaking comets, wandering meteorites, asteroids, and cosmic dust filling the interior of the globe of blackness.

SOUNDS OF SILENCE

"When the morning stars sang together, and *all* the sons of God (angels) shouted for joy..." (Job 38:7, Compare Psalm 148:3-11; 69:34). "... I tell you that, if these should hold their peace, the stones would immediately cry out" (Luke 19:40).

Astronomers tell us that there are extraterrestrial multichromatic tones, a symphony of sound reverberations throughout the universe, which they call "cosmic harmony." For instance, the sun rings like a bell, chiming out acoustical waves at regular intervals. Voyager Pioneer spacecraft have detected a variety of radio waves which are inaudible to the human ear because the sounds are unable to travel through the vacuum of space. Each star, solar system, and galaxy has its own distinctive sound ranging within periods of two to eight minutes. These audio frequencies are well within the auditory range of the Creator and angelic beings. Scientists believe that "quarks" are the root common denominator for all musical sounds. Since they are believed to be pure energy, or the first elements, music must have its origin in the Creator.

The earth also is a musical instrument, but its sounds, like all celestial sounds, are twenty octaves lower than the pitch one can hear. Protons are tenors; ions, the heaviest particles, are the basses. Each atom produces a song, but it is twenty octaves above our range of hearing and is produced by the changing of chemical structure. The electronic ears which "listen" are very sensitive seismographs detecting what scientists call "exotic resonance." Saturn's magnetosphere sounds, played through a music synthesizer, produce a slow, dreamy melody.

Ordinary gas rising from the earth to a height of 90 miles is heated enough to produce plasma in which all atoms are charged. This produces wonderful and mysterious musical sounds which astrophysicists are now recording on tape from Voyager Pioneer. It is believed that the energy stored for these sounds is located in our planet's magnettails, with the frequencies varying from low to higher decibels. These sounds in the universe are being called "Music of the Sphere." They are described as the complex symmetry of matter. When Jesus said that "the stones would immediately cry out," all the Creator had to do was to increase the sounds to an audible range. Human hearing is between 20 and 20,000 vibrations per second. By comparison, the piano is between 27 and 4,000 vibrations per second.

When the earth was put to "rest" after Lucifer's rebellion, all music came to complete silence as the earth lay in its lonely cemetery. The entire orchestra of creation must have ground down its melodic sounds temporarily into a mournful dirge; then silence. But soon the music in the heavens resumed its rejoicing tempo: ". . . they break forth into singing. Yea, the fir trees rejoice . . . and the cedars . . . saying, Since thou art laid down, no fellow is come up against us" (Isaiah 14:7,8).

According to Job 38:7, the celestial decibels were tuned by the Creator into harmonious vibrations which so excited the angels when the universe was completed that their emotions gave way to shouts of joy because of what they saw and heard.

HEAVENLY MEDLEYS

Is it possible that all of the universes (Heavens of Heavens) filling the habitation of eternity are like one giant music hall? The Creator and His angels have the capacity to tune these sounds in or out at their desire, as will the redeemed in their future bodies of immortality.

What a delight it will be to receive supersonic hearing one day and be able to listen to those heavenly orchestrations and angels' chorales: "But as it is written, Eye hath not seen, *nor ear heard,* neither have entered into the heart of men, the things which God hath prepared for them that love him" (I Corinthians 2:9).

God loves the aesthetic. Because these sounds have a restrictive range, they do not interfere with man's own ability of producing music within the range of his hearing. Sounds are the indication of life. All creation is alive with the breath of the Almighty God pulsating within. "... His eye seeth every precious thing" (Job 38:10).

30

Princes in Heavenly Principalities

Christ will be leading untold millions of redeemed out of Paradise in His descent from the Father's throne (I Thessalonians 4:14; 2 Corinthians 4:14).

In God's Word we read the following concerning Antichrist: "And through his policy also he shall cause craft to prosper in his hand; and he shall magnify himself in his heart, and by peace shall destroy many: he shall also stand up against the Prince of princes (Jesus Christ); but he shall be broken without hand" (Daniel 8:25).

Christ Jesus is called "Messiah the Prince" and also "Prince of life" (Daniel 9:25; Acts 3:15). He is "Prince of princes" over all angels; "King of kings" over all the redeemed; and "Lord of lords" over all created places and beings.

GOD'S ENVOYS

While all angels are "ministering spirits" either for God or Satan, the archangels, who are higher in rank, are those in command of great contingents of angels. Satan apparently was an archangel because so many angels fell into transgression with him (Revelation 12:3-9). Theologians follow the assumption that all angels were divided among three archangels—Lucifer, Gabriel, and Michael—because there are no others given in Scripture. However, this would not account for the cherubim and seraphim which God delegated to Himself. Gabriel appears to have the assignment of God's direct envoy to deliver messages of a personal nature (Daniel 8:16; 9:21; Luke 1:19,26).

THE GREAT PRINCE

When there were satanic forces that needed to be overcome, curtailed, or defeated in some specific area, the assignment was delegated to Michael and his angels who had been given tremendous power. No satanic forces could withstand or hold a line against him and his angels for long (Daniel 10:12,13; 12:1; Jude 9; Revelation 12:7).

The "voice of the archangel" (in I Thessalonians 4:16) at the time of the rapture of the redeemed is presumed to be that of Michael (Daniel 12:1), who is summoned whenever a passageway through cosmos or satanic dominions must be opened with great dispatch. He undoubtedly will be involved when Christ descends from the Father's throne (I Thessalonians 4:14; 2 Corinthians 4:14) with the untold millions of redeemed out of Paradise. These are they who will receive their bodies of immortality and then join the redeemed who will be raptured simultaneously from the earth.

The "trumpet," in Biblical terminology, always has been the signal for warfare or for a great gathering. The sound of the trumpet will be followed instantly with a surprise thrust through the heavens. The entire operation will take place "in the twinkling of an eye" (I Corinthians 15:52) before Satan and his troops can do anything to prevent it (Hebrews 10:37).

The angels in the Apocalypse with the seven trumpets cannot be any of the archangels, consequently their last trumpet does not equate with the "last" trumpet that Michael will sound. The seven trumpets are signaling judgments. The first and last trumpets are for the rescue of the redeemed.

The sound of Michael's trumpet will be the declaration of a rescue operation, for all hell is about ready to break loose upon the earth—devastations, wars, earthquakes, and tidal waves. However, before the ceiling can fall to the floors, the Royal Redeemed will have been "caught up" and gone.

LIVING LUMINARIES

Christ, the "Prince of princes," is referred to as "the bright and morning star" (Revelation 22:16). Similarly, angels are identified not only as "princes" but also as "stars" because of their luminous appearance. In Revelation 20:1, an angel is coming down from heaven with "the key of the bottomless pit" (just *one* of the keys of the kingdom of heaven) to incarcerate Satan for the duration of the Millennium. The same angel previously opened the bottomless pit with the identical key (Revelation 9:1). He is described in this manner: "I saw a star fall from the heaven unto the earth. . . ."

Jesus gave personality to these transcelestial stars of God because He is called a "Star." Christ, as the bright and morning star, is called the "Sun of righteousness" (Malachi 4:2).

STAR OF BETHLEHEM

Archaeoastronomy has failed to reveal any solid evidence that the Christmas star (Matthew 2:9-12) was a celestial object. There are many reasons for this. First, a star is a sun and each has its own solar system, thus it is unlikely to enter our solar system. This star led the wise men approximately 500 miles to the Christ Child and it was sufficiently close to the earth to direct them to a specific location; moreover, it showed intelligence and was not seen by anyone but the Magi.

The redeemed are told in Daniel 12:3 that they "shall shine as the brightness of the firmament; and . . . as the stars for ever and ever." In Revelation 12:1 the redeemed are depicted as being clothed with the sun and as having a crown of twelve stars. In Christ's hand are the "stars" symbolizing the seven church pastors of Asia (Revelation 1:16). All of this lends support to the concept that certain stars have personality.

AN ANGELIC CHOIR

"Whereupon are the foundation thereof fastened? or who laid the corner stone thereof (of the earth); When the morning stars sang together (celestial orchestration of the galaxies of stars), *and* all the Sons of God shouted for joy?" (Job 38:6,7).

Angels frequently manifest themselves in translucence symbolized by stars, but they also sometimes materialize and reveal themselves in the likeness of men, to wit: "...behold, two men stood by them in shining garments" (Luke 24:4). But Daniel 10:16,18 states: "And, behold, one like the similitude of the sons of men...like the appearance of a man...." To Abraham they appeared in the same manner: "And he lifted up his eyes and looked, and, lo, three men stood by him..." (Genesis 18:2). This is the reason for these words of warning in Hebrews 13:2: "Be not forgetful to entertain strangers: for thereby some (such as Abraham) have entertained angels unawares."

MORNING STARS

The brightness of angels is indicated in these passages: "And he (Jesus) said unto them, I beheld Satan as lightning fall from heaven" (Luke 10:18).

"And no marvel; for Satan himself is transformed into an angel of light" (2 Corinthians 11:14). "He (Cornelius) saw in a vision... an angel of God...stood before me in bright clothing" (Acts 10:3,30). "...The cherubim there appeared over them as it were a sapphire stone, as the appearance of the likeness of a throne" (Ezekiel 10:1).

CHRIST OUTSHINES THEM ALL

Christ Jesus is described in the following manner: "That was the true Light, which lighteth every man that cometh into the world" (John 1:9). Regarding the transfiguration we read: "And was transfigured before them: and his face did shine as the sun, and his raiment was white as the light" (Matthew 17:2).

When He returns to earth to reign we note: "For as the lightning cometh out of the east, and shineth even unto the west; so shall also the coming of the Son of man be" (Matthew 24:27). His brightness will destroy the Antichrist: "And then shall that Wicked be revealed, whom the Lord . . . shall destroy with the brightness of his coming" (2 Thessalonians 2:8). As recorded in the Apocalypse, His radiance will frighten the unredeemed when He is revealed to them: "And said to the mountains and rocks, Fall on us, and hide us from the face of him that sitteth on the throne, and from the wrath of the Lamb: For the great day of his wrath is come; and who shall be able to stand?" (Revelation 6:16,17).

Christ's irradiant appearance as He comes to earth to reign will destroy one-half of those remaining on earth after the final Tribulation judgments: "And the slain of the LORD shall be at that day from one end of the earth even unto the other end of the earth: they shall not be lamented, neither gathered, nor buried; they shall be dung upon the ground" (Jeremiah 25:33). "Remember Lot's wife (who was destroyed at the time of Sodom's destruction). Whosoever shall seek to save his life shall lose it: and whosoever shall seek to lose his life shall preserve it (martyred redeemed during the final Tribulation days). I tell you, in *that* night (when Christ returns to earth) there shall be two men in one bed; the one shall be taken, and the other shall be left . . . Wheresoever the body (carcass) is, thither will the eagles be gathered together" (Luke 17:32-37). This gluttonizing by the vultures who devour the slain bodies is called "the supper of the great God" (Revelation 19:17,18).

COSMIC WAR

"And it waxed great, even to the host of heaven; and it cast down some of the host and of the *stars* to the ground, and stamped upon them" (Daniel 8:10). By this violent action Abaddon (not yet incarnated in the man, Antichrist) challenges Satan to come down from his lofty perch and meet him in vicious and deadly combat. But Satan has no desire to do so, and furthermore cannot, until God evicts him and his fallen comrades. It appears by this action that Abaddon (Apollyon) will kidnap and take hostage one-third of Satan's forces and torture them. As we shall show later, Armageddon is not just a battle of flesh and blood, but it is a conflict among wicked angels who are in a state of anarchy and are unable to organize themselves against the holy angels that come to apprehend them and properly dispatch them until the day of judgment.

Has Satan had access to God's throne since his fall? Certainly. According to Job 1:6-12, he made such an appearance, but without any of his comrades.

31

Original Paradise[1]

Paradise . . . was created expressly for divinely appointed rulers of the earth.

"Thou (Lucifer) hast been in Eden in the garden of God" (Ezekiel 28:13). Paradise means "a garden or park." It was created expressly for divinely appointed rulers of the earth and it is apparently a planet located in the third heaven (2 Corinthians 12:1-4). It appears to be nearer the throne of God than the earth (separated by galaxies of stars Isaiah 14:13), which was God's footstool. Paradise is in the proximity to God's throne like a dais and is enriched with the choicest of colorful ultra-transcendent flora in a vast dimension of wonders. Its expanse must be enormous to accommodate the vastly growing numbers of inhabitants. The only restraint to approaching God's throne is His awesome Majesty and Glory.

Lucifer and his staff of rulers must have retired to this place on occasion to reflect and to enjoy its superbly relaxing environment. Here were exquisite animals and birds moving gracefully among the trees and bushes. There were lovely glens, streams, and waterfalls. How frequently Lucifer must have gathered with his lieutenants in strategy sessions with plans to enhance their wonderful kingdom on the primordial earth. Important decisions must have been made amid these glorious settings.

Only by invitation would anyone consider nearing God's throne. Perhaps there was a promenade extending from Paradise toward God's throne. Was there a huge dais before the throne where angels assembled for conferences? According to Job 1:6; 2:1 such gatherings were held: "Now there was a day when the sons of God (angels) came to present themselves before the Lord . . . Again there was a day when the sons of God came to present themselves before the Lord . . . " (Compare Revelation 5:11; Daniel 7:9,10).

IN FULL VIEW OF GOD'S THRONE

"Blessed are the pure in heart: for they shall *see God*" (Matthew 5:8). David said, ". . . thou shalt make me full of joy with thy *countenance*" (Acts 2:28). Jesus said, ". . . you will see me" (John 14:19, *Phillips*). The Apostle Paul added, "We are confident, I say, and willing rather to be absent from the body, and to be *present* with the Lord" (2 Corinthians 5:8).

A SANCTUARY

We can comprehend something about this Paradise near the throne of God by noting its analogue created for our first parents: "And the LORD God planted a garden eastward in Eden . . . And out of the ground made the LORD God to grow every tree that is pleasant to the sight, and good for food; the tree of life also in the midst of the garden . . . And a river . . . " (Genesis 2:8-10). "Praise him, ye heavens of heavens, and ye waters that be *above the heavens*" (Psalm 148:4). "And he shewed me a pure river of water of life, clear as crystal, proceeding out of the throne of God and of the Lamb" (Revelation 22:1).

During the renewal of the earth the word "let" is used repeatedly, implying a previous existence. The fact that the word "Eden" is applied in both Genesis 2:8 in the renewal of the earth, and Ezekiel 28:13 in the primordial creation shows these Paradises to be very tangible. The same Creator who made our world also created heaven. Is it therefore out of context to recognize a similarity?

FIRST HUMAN ARRIVALS

Paul the apostle said, "I . . . was caught *up* into paradise . . ." (2 Corinthians 12:1-4). This was aeons after Lucifer was cast out from the "mountain of God" (Ezekiel 28:14,16) situated on the paradisiacal globe in the third heaven. It remains a most attractive, real, and wonderful place reserved for the gathering of potential rulers of the earth who will reign with Christ during His manifested Kingdom, beginning with the Millennium. He also added: ". . . this man was caught *up* to paradise and heard sacred secrets which no human lips can repeat" (Moffatt).

All of the believers who were deceased during the Old Testament economy went to Paradise *within* the earth. The first two redeemed human beings to occupy the heavenly Paradise after the fall of Lucifer were Enoch and Elijah who were translated. Their bodies were simply enveloped by the Glory of God in the translation, in contrast to Christ, who after His resurrection possessed a body of immortality (I Timothy 6:16).[2]

It is conceivable that Paradise is in a solar system not too unlike the one in which we now live because the word "let," as applied to the animal kingdom in the renewal of the earth, shows that the restored creation requires a similar environment.

[1] Paradise is the first and oldest of all God's creations revealed in Scripture. Here God established his throne in the Heaven of Heavens.

[2] Since only Jesus Christ possesses an immortal body, Enoch and Elijah must possess a body similar to Adam in his paradise.

32

Paradise on the Earth

The idea of tangibility beyond death escapes the average laymen.

"And the LORD said, Behold, the man is become as one of us, to know good and evil: and now, lest he put forth his hand, and take also of *the tree of life,* and eat, and live for ever . . ." (Genesis 3:22).

". . . To him that overcometh will I give to eat of *the tree of life,* which is in the midst of the paradise of God" (Revelation 2:7).

ALL FLESH IS NOT THE SAME

"All flesh is not the same flesh . . .": "So also is the resurrection of the dead . . ." (I Corinthians 15:39,42).

Scripture shows there are three kinds of bodies affecting the redeemed—paradisiacal, mortal, and immortal. The source of life for each differs. The paradisiacal body requires the tree of life; the mortal body needs the blood; and the immortal body receives its life from the Holy Spirit, after which it becomes perpetuating.

PARADISIACAL BODY

The same conditions that prevailed in Eden's paradise on earth also exist in heaven's Paradise, with the tree of life in both being the lifeline to the residents. However, when Adam and Eve sinned their bodies were altered biologically, thus requiring a different source of life. The blood within them became the fountain of life, and as a consequence, their bodies were more fragile and terminal (Psalm 90:10; Leviticus 17:11; Genesis 3:22; Revelation 2:7).

Adam said of Eve, "This is now bone of my bones, and flesh of my flesh . . . " (Genesis 2:23). No mention was made of "blood" as it was totally unimportant at the time. Blood in the circulatory system was only secondary to life. Perhaps it might be classified as an auxiliary resource which God, by foreknowledge, knew would be required after mankind sinned.

CORPOREAL BODY

Job (19:26) said, " . . . in my *flesh* shall I see God." Those in heaven's Paradise are in the Lord's presence (2 Corinthians 5:6,8), and from their vantage point they can see the Lord Jesus Christ on His throne (Psalm 16:11). Adam and Eve, as well as many others, saw God while in their flesh, but only Adam and Eve in their paradisiacal bodies observed Him in His *full* majesty (Genesis 3:8; Judges 13:22; Exodus 33:20-23). Because of their biological composition, the bodies of Adam and Eve were able to survive under the intensity of the Glory of God. When Enoch and Elijah were translated to Paradise in their paradisiacal bodies they also could see God in all of His Majesty and survive (2 Kings 2:11); but no mortal man can (I Timothy 6:16; Exodus 33:20).

David said, " . . . *my flesh* shall rest *in hope*" (Psalm 16:9). When the Old Testament saints retired to *paradise within the earth,* their bodies remained at "rest" in the ground since the tree of life was not in lower Paradise (Revelation 2:7). Those whom Christ addressed were spirit beings; that is, they were devoid of their physical bodies: " . . . but live according to God in the spirit" (I Peter 4:6). Their faculties (souls) were retained, as evidenced by the fact that they were able to hear what Christ had to say to them.[1] Angels devoid of physical bodies also possess all of the senses but they are more acute.

DEATH IN LIFE

God said to Adam, "... for *in the day* that thou eatest thereof thou shalt surely *die*" (Genesis 2:17), but Adam and Eve did not die that very day in the usual sense. However, their bodies were replaced by less resilient ones whose source of life must come from the blood within them.[2] Even the source for good health, the *leaves* of the tree of life, was denied them (Revelation 22:2; Ezekiel 47:12). As a result, corruption set in and excrement resulted. God made no provision for physical waste disposal from their original bodies since none was required nor will there be in the heavenly Paradise.

Science is beginning to attribute the aberrant behavior of mankind to his blood. Latest findings believe that hypoglycemia, a disorder from low blood sugar, could be responsible even for serious crimes. With the absence of all blood from a paradisiacal body, such irrational behavior would be eliminated, which could contribute to harmony in the heavenly Paradise.

God also cursed the ground from which food for survival must come (Genesis 3:17-19), thus compounding the new physical problems.[3] The food not only was inferior in quality to that which they were accustomed, but it was more difficult to produce and its preservative qualities were limited.

LIFE IN DEATH

The fact that those in Paradise (both Eden and heaven) require the tree of life and its leaves for medicinal purposes indicates that even though their bodies are eternal (I Corinthians 5:1), they are not yet comparable to the resurrected immortal bodies. The latter will be totally autonomous and severed from all prior and outside life support as well as from confinement in Paradise. The resurrected body will possess characteristics far excelling both the mortal and paradisiacal bodies.

While the paradisiacal bodies are "eternal," God never intended that those in Paradise be confined *there* forever. The resurrected body will break away from all restraints to "inherit all things" (Revelation

21:7) and eventually every one in Paradise will leave when it is emptied at the time of the White Throne Judgment.

In Romans 8:11 we read: "... the Spirit... shall *also* quicken (bring to life) your *mortal* bodies by his Spirit that dwelleth in you." This text clearly indicates that those in Paradise have retained their individual personalities just as have those who are living in mortal bodies.

The final change of the body will be into immortality at the time of the resurrection. This is called "the redemption of the body" (I Corinthians 15:23). Actually, it is a complete redemption for the whole person. Since the transfer of the redeemed to the heavenly Paradise, God has designated that all whose names are in the book of life possess a paradisiacal body to cover their soul and spirit. At both judgments of mankind, the redeemed as well as the unredeemed will appear in physical bodies (I Peter 4:6; John 5:22-30).

RESURRECTION OF THE BODY

"For we know that if our earthly house of this tabernacle were *dissolved,* we have... an house not made with hands, eternal in the heavens... to be clothed upon... we shall not be found naked" (2 Corinthians 5:1-4).

The word "dissolved" means to be loosened, disintegrated, or destroyed. Because man's body is primarily space, a minute combination of atoms is all that will be necessary as the embryo for a resurrected body. These particles are now at "rest" or "asleep" (Psalm 16:9; Revelation 6:11; 14:13; I Corinthians 15:6,18; I Thessalonians 4:13), but the soul is not asleep. It does not consist of atomic particles but of other properties. Its composition does not follow the physics of elementary particles; it is guided by its own laws of physics.

These "rest" particles of the physical bodies await the day of resurrection to be reactivated by the Holy Spirit into immortality for the redeemed. Physical bodies simply envelop the faculties and personalities of the persons that are resident in the "soul." The purpose of the resurrection is to give the redeemed immortal bodies freedom from all previous dependency upon the blood, the tree of life, and food. Food will be purely optional and pleasurable, with our incorruptible bodies converting it into incorruption. The immortal body will be completely self-perpetuating. This immortal body will have flesh and bones like Jesus had after His resurrection (Luke 24:39).

Paradisiacal bodies will be similar to those of Adam and Eve before they sinned. Personality is never lost in transition from one body to another because it is resident in the "soul," not the physical body.

A RESERVE BODY

What if there are no particles of the deceased body remaining to be resurrected? In that case, Paul said the paradisiacal body would suffice as an alternative. At the death of those who are in the book of life, minute particles of the physical body may well become the embryo of the paradisiacal body. This would insure a resurrection of the physical body.

The body envelops the soul and spirit and is merely the habitation in which the real person resides. Paul compares it to the "house" in which one lives. Although it is only secondary in importance, it is necessary if one is to exist like unto Christ's glorious body and remain distinct from a spirit being.

"For *we* know that *if* our earthly house of this tabernacle (body) were *dissolved* (disintegrated), *we have* . . . an house not made with hands, eternal in the heavens" (2 Corinthians 5:1). This conveys the idea that the reserve body in the heavenly Paradise is similar to the bodies of our first parents both before and after they sinned. Once those in Paradise have partaken of the tree of life they will forever retain their identifications in visible form. "And the Lord . . . will *preserve me* unto his heavenly kingdom . . . " (2 Timothy 4:18). The properties of the "soul" within the body are more important than the particles of the body: "And fear not them which kill the *body*, but are not able to kill the soul . . . " (Matthew 10:28).[4] The "soul," unlike the "body," will never be replaced. Therefore, the salvation of the soul is imperative if it is to avoid the tragedy of the "second death."

To adopt a simple evolutionary metaphor: survival of the

firstfruit appearing on the tree identified not only the kind of tree but also what the future fruit would be like.

"Behold, I shew you a mystery; We shall not all sleep, but *we all shall be changed*" (I Corinthians 15:51).

Because the blood of Christ is in heaven (Hebrews 9:12,24; 12:24) and the blood of God is incorruptible (I Peter 1:25; Acts 20:28) we are given evidence that the "shroud of Turin," the so-called burial cloth of Jesus Christ which claimants say has a large blood spot, is not authentic.

[1] See page 241.

[2] Biologists have discovered that there are more than 2,500 diseases traceable to DNA (deoxyribonucleic acid), the basic blueprint for life; and that death in a normal healthy person sets in at the age of 20 years. "But we had the sentence of death in ourselves . . . " (I Corinthians 1:9, See pp. 87-88). Also, at birth all human beings have a variety of viruses that lie dormant unless and until some physical, emotional, or psychological condition makes them active to inflict the body.

[3] Biologists inform us that 3,800 disorders are caused by a single gene's abnormality. There are today more than 2,000 incurable diseases. There are estimated to be over 100,000 complex strands of genes.

[4] This verse infers that the lost will have *a resurrection of the properties of the souls* (rather than their bodies alone). Their souls, according to I Peter 4:6, will be enveloped in physical bodies for their appearance at the final tribunal. See John 5:29.

33

Paradise Within the Earth

The hand of the Almighty rested over the sleeping beauties within the earth (Paradise) during the time of the Old Testament.

The hand of the Almighty rested over the sleeping beauties within the earth during the time of the Old Testament. They were not forgotten because God's presence was there ("behold, thou art there"). There were millions of men, women, and children all resting in God's sanctorum.

"... Going *down* to the *chambers* of death" (Proverbs 7:27).

"In (God's) hand are the *deep places* (lower parts) of the earth..." (Psalm 95:4).

"If I (a believer) make my bed in hell (lower parts of the earth which include Paradise), *behold thou art there*" (Psalm 139:8).

"The dead... *go down* into silence" (Psalm 116:17).

"Our friend Lazarus *sleepeth*" (John 11:11).

"Now that (Christ) ascended, what is it but that he also *descended first* into the lower parts of the earth" (Ephesians 4:9).

Scripture confirms that there is a vast space within the earth where deceased souls are housed. Within recent months scientists have observed a mysterious "wobble" of the earth's axis by as much as sixteen to thirty-three feet in various directions. This, they suggest, is not the behavior of a solid body.

Paradise within the earth is conducive to quiescence and rest; hardly a place of activity: ". . . before I leave it (life on earth) to return no more, before I pass to darkness (a) land as midnight (with) no light but the shadow of death" (Job 10:21,22, *Moffatt*).

SOUL SURVIVAL

"So Moses the servant of the Lord died . . . and (was) buried" (Deuteronomy 34:5,6); compare: "And there appeared unto them Moses . . . talking with Jesus" (Mark 9:4). "And Samuel died . . . and (was) buried . . ." (I Samuel 25:1); compare "And Samuel said to Saul, Why hast thou disquieted me, to bring me up?" (I Samuel 28:15). The soul never ages. It is immortal to the first death which destroys the body because it is composed of properties unlike the atom particles of the physical body. Only the "second death" has power over the soul: "And fear not them which kill the body, but are *not able to kill the soul* . . ." (Matthew 10:28).

Some believe that when a person dies all of his entities—spirit, soul and body . . . cease to exist, but scriptures reveal that the "spirit" part of man goes to God (Ecclesiastes 12:7). There are no Scriptures to denote that the "spirit" entity (God-consciousness, conscience, and creativity) ever dies. God is a "Spirit Being" and He certainly is not dead. Angels are spirit beings and no death is indicated concerning them.

According to the following scriptural references, survival of the *soul* (which entails the personality and senses of living beings), is assured both in the lower and heavenly Paradises:

LOWER PARADISE:

Sleep: Those in the Old Testament economy entered lower Paradise when they died. Since they were asleep (not annihilated), there was no reason for the tree of life. In the Old Testament we often read that the godly "slept" with their fathers.[1]

Awakeness: Jesus said of Lazarus whom He raised from the dead: ". . . our friend Lazarus sleepeth: but I go, that I may *awake* him out of sleep" (John 11:11). This was prior to the resurrection of

Jesus before He went into the lower Paradise (compare Matthew 27:52). The word "sleep" means one has died physically, but is not annihilated: "Ye that love the Lord . . . he preserveth the souls of his saints . . . " (Psalm 97:10).

Comfort: " . . . shall be *comforted* in the nether parts of the earth" (Ezekiel 31:16, *Moffat*). "Lazarus (is) *comforted*" (Luke 16:25). " . . . *Behold (God) art there*" (Psalm 139:8).

Sight: "As for me, I will *behold thy face* in righteousness: I shall be satisfied, when I *awake, with thy likeness*" (Psalm 17:15). When Jesus entered lower Paradise both He and King David were present without their physical bodies. Thus David's "soul" and the "soul" of Jesus were viewing each other. This proves the tangibility of the soul.

Hearing: "By which also he (Christ) went and *preached* unto the spirits in prison (in the lower earth)" (I Peter 1:19). "For for this cause was the *gospel preached also* to them that are dead" (I Peter 4:6). These two texts include both those in the region of the lost, as well as those in Paradise, which are separated only by a great gulf" (Luke 16:26).

Walking: " . . . *led* captivity captive" (Ephesians 4:8). Christ led those in lower Paradise out to enter the heavenly Paradise.

HEAVENLY PARADISE:

Location: " . . . *the third heaven* . . . paradise" (2 Corinthians 12:2,4). "(God's Son) on the right hand of the *Majesty on high*" (Hebrews 1:1-3). "Father, I will that they . . . be with me where I am; that they may *behold my glory*" (John 17:24). " . . . To be absent from the body (is) to *be present with the Lord*" (2 Corinthians 5:8).

Stars Over Paradise: " . . . ye heavens (universes) of heavens (universes)" (Psalm 148:4). These universes surround the "Heaven of heavens" in which God's throne in the Holy City rests upon the planet Paradise at the center of all of God's creations.

Environment: " . . . present *with the Lord*" (2 Corinthians 5:8). " . . . Like the garden of the Lord . . . *gladness shall be found therein, thanksgiving* . . . " (Isaiah 51:3).

Nature: "The cedars of the garden of God . . . fir trees . . .

chestnut trees . . . *all the trees* of Eden (in) the garden of God" (Ezekiel 31:8,9)

Animals: ". . . white *horses"* (Revelation 19:14).

Music: ". . . and the *voice of melody"* (Isaiah 51:3).

Life: ". . . that mortality might be swallowed up of *life* (not death)" (2 Corinthians 5:4). Not destroyed with the body: "And fear not them which kill the body, but are *not able* to kill the soul (over which only the second death had jurisdiction)" (Matthew 10:28).

Tangible Bodies: "For we know that if our earthly (body) were dissolved, we have *a (body) in the heavens"* (2 Corinthians 5:1).

Fountain of Youth: ". . . ye waters that be *above* the heavens . . . " (Psalm 148:4. Compare Song of Solomon 4:15; 5:1; Revelation 22:1).

Speech: "How (Paul) was caught *up* into paradise, and heard unspeakable *words . . .* " (2 Corinthians 12:4).

Eating: ". . . to him (the redeemed) that overcometh will I give to *eat* of the tree of life, which is in the midst of the paradise of God" (Revelation 2:7).

Activities: ". . . in thy presence (are) *pleasures* for evermore" (Psalm 16:11). ". . . Thou shalt make me *full of joy* with thy countenance" (Acts 2:28).

Relaxation: ". . . Blessed are the dead which die in the Lord that they may *rest* from their labours" (Revelation 14:13).

Ultimate Comprehension: "For we know in part . . . But when that which is *perfect is come,* then that which is in part shall be done away" (I Corinthians 13:9,10). ". . . Then shall I *know* as also I am known" (I Corinthians 13:12).

Fourth Dimensional Time: ". . . one day is with the Lord as a thousand years, and *a thousand years as one day"* (2 Peter 3:8).

PRINCE OF LIFE

Jesus, citing the prophecy of Jeremiah, said on the day of the inauguration of His ministry: "The Spirit of the Lord is upon me (to) preach deliverance to the captives" (Luke 4:18). The captives were those in lower Paradise (Ephesians 4:8) for we have no record of

Christ ever releasing anyone from any bastille when on earth. To translocate those in lower Paradise was an important part of His ministry. The time finally came for this when He said to the converted thief on the cross: ". . . To-day shalt thou be with me in paradise" (Luke 23:43).

Jesus, the Prince of life, said on another occasion: "The hour is coming, *and now is, when the dead shall hear* the voice of the Son of God: and they that hear shall live" (John 5:25). "The dead know not anything" (Ecclesiastes 9:5,6), so they would have to be *awakened.* This time He would not single out certain ones as He did Lazarus, but He would address everyone. So He raised His Voice and aroused all of the Old Testament saints. Then He addressed them: "Who shall give an account to him that is *ready* to judge the quick (living) and the dead (at the Judgment Seat of Christ). For for this cause was the gospel preached also to them that were *dead* (in body, but not in soul), that they might be judged according to men in the *flesh* (which requires a resurrection), but *live* according to God *in the spirit"* (I Peter 4:5,6). Those in lower Paradise were living in unconscious sleep. However, before they will be judged, they will be raised to obtain a resurrected immortal body like flesh (Daniel 12:1,2). This shows that the souls of the Old Testament saints "slept" in insensibility in lower Paradise.

Christ gave the message to those in the other part of hell ("the lower parts of the earth") whose names were not in the book of life. However, at their (White Throne) judgment, "conscience" will play an important role in the light of their earthly deeds which were registered in the "books" (Revelation 20:11-15; I Peter 3:19-21).

Angels will be judged in "spirit" form, but human beings will be judged "in the flesh." In fact, the resurrected redeemed will some day judge angels (I Corinthians 6:3).

Why did Christ preach to those in lower Paradise? He is to bring them the good news of the "gospel" that they would have their spirit beings clothed upon with garments of righteousness (Revelation 19:8; 2 Corinthians 5:1-4). The redeemed are not to remain naked spirits after they are deceased, but clothed upon: ". . . so (as not to) be found naked." They will get their bodies of immortality at the time of the resurrection (I Corinthians 15:35-58). These must be distinguished from the paradisiacal bodies.

Did Christ preach only to the redeemed within the earth? No! He also went "into the lower parts of the earth" (Ephesians 4:9) and "preached" to them which were "disobedient" (I Peter 3:19,20) so that they, too, must have been awakened in order to hear what Christ had to say to them.

The souls in the animal kingdom, we are told, do not go into the lower parts of the earth but simply: ". . . goeth downward *to* (not within) the earth" (Ecclesiastes 3:21). The Hebrew word for soul is "nepesh," the word also used for lower life forms.

Descendants from the immoral relationship between fallen angels and the Antediluvian women were destroyed according to Deuteronomy 2:20,21: ". . . the Lord destroyed them. . . ." They probably were incarcerated in the Bottomless Pit. This hybrid race known as "giants" (from the Hebrew word "nephilim") has an uncertain etymology. The women involved in this entanglement were apparently innocent victims and were presented the "gospel" from Christ meaning "good news."

[1] Perhaps the terms "rest mass" and "rest energy" from Einstein's special theory of relativity, are apropos in expressing the properties of *soul asleep* in the fourth dimension ". . . Thou (Daniel) shalt *rest*. . ." (Daniel 12:13. "Rest" literally means quiet, not a deceased soul.

34

Chariots Are Coming

Can you imagine the excitement of those Old Testament saints as they left old Paradise and headed toward the heavenly one?

"The chariots of God are twenty thousand, even thousands of angels: the Lord is among them, as in Sinai, in the holy place. Thou hast ascended on high, thou hast led captivity captive: thou hast received gifts for men; yea, for the rebellious also, that the LORD God might dwell among them" (Psalm 68:17,18).

"The Lord is among them, as in Sinai . . . " The Old Testament saints were familiar with Sinai where Moses received the Law. God had to have a sufficient number of chariots to take aboard those in lower Paradise and to accommodate them: "The chariots of God are twenty thousand . . . "

"Thy dead men shall live (the redeemed within the earth's lower Paradise), together with my dead (Christ's resurrected) body shall they arise. *Awake and sing,* ye that dwell in dust . . . " (Isaiah 26:19).

Before Christ left lower Paradise for His own resurrection, He selected for a subsequent resurrection many saints recently buried in the vicinity of Jerusalem: "And the graves were opened; and many bodies of the saints which slept arose, And came out of the grave *after* his resurrection, and went into the holy city, and appeared unto many. Now when the centurion, and they that were with him, watching Jesus, saw the earthquake, *and those things that were done,* they feared greatly, saying, Truly this was the Son of God" (Matthew 27:52-54). What a follow-up to the opening of the graves—the bodies of the saints arising and walking in Jerusalem!

MYSTERY OF THE LIVING DEAD

"But now is Christ risen from the dead, and become the *firstfruits* (plural) of them that slept" (I Corinthians 15:20).

". . . Christ the firstfruits . . ." (verse 23).

Christ is the *firstfruits* because other saints rose from their graves with Him. This is the only recorded instance of these saints coming out of their graves and walking in Jerusalem. What happened to them? Why is there no further mention about them?

"Jesus saith unto her (Mary), *Touch me not: for I am not yet ascended to my Father . . .*" (John 20:17). Yet only eight days later Jesus said to Thomas: ". . . Reach hither thy finger, behold my hands; and *reach hither thy hand, and thrust it into my side:* and be not faithless, but believing" (John 20:27). Why this difference? What made the difference?

There remains a mystery about those estimated eight days. Where was Christ during this time? In Ephesians 4:8-10 we see that between the two events Christ ascended to His Father with great pomp and excitement. In a preview of the coming Rapture of the saints, those in lower Paradise ascended with Christ, *including those* who suddenly disappeared after walking the streets in Jerusalem.

PATHWAY THROUGH THE STARS

Vast throngs of angels must have formed endless lines to provide a corridor through which the 20,000 chariots (Psalm 68:17), loaded with ecstatic Old Testament saints, proceeded enroute to the heavenly Paradise. Can you imagine the jubilance of the saints as they left their place of rest in lower Paradise for activity in the heavenly Paradise? Can you comprehend their expressions of awe and wonderment as they fingered and touched their magnificent new garments tailored by heavenly seamstresses?

One can envision Christ in the leading chariot marshalling this parade of happy believers, including those whose names were in the book of life, such as little children. Perhaps Enoch and Elijah were with Christ in the leading chariot. As they entered the heavenly Paradise and disembarked, Christ majestically made His personal appearance to the Father while Enoch and Elijah led the great caravan into Paradise. Christ was now to occupy a new role as High Priest in the scheme of redemption: "But Christ being come an High Priest of good things to come, by greater and more perfect tabernacle (His body of immortality), not made with hands, that is to say, not of this building . . . by his own blood he entered in once into the holy place, having obtained *eternal* redemption for us" (Hebrews 9:11,12). He was now "A minister of the sanctuary, and of the true tabernacle, which the Lord pitched, and not man" (Hebrews 8:2. Compare Revelation 3:12).

EFFICACIOUS ATONEMENT[1]

The atoning blood of Christ, the incorruptible blood, is now at Mount Sion, the site of God's throne. Sinai and its shadow of good things to come have all passed away. Jesus has fulfilled the Law; He has met all its requirements as God's perfect Lamb on Calvary. There can be no demand for further payment. Christ is now the total replacement, having fulfilled the law: "*Blotting out* the handwriting of ordinances that was against us, which was contrary to us, and took *it out of the way, nailing it to the cross* . . . let no man therefore judge you in meat, or in drink, or in respect of an holy day, or of the new moon, or of the sabbath days: which are a *shadow* of things to come; but the *body* is of Christ" (Colossians 1:18).

After the deliverance of saints from lower Paradise to the heavenly one, and the completion of His role as God's new and eternal High Priest, Christ had yet to accomplish one final role on earth. He had to instruct and prepare His disciples to carry forth the work of evangelizing the world with the glorious Gospel, so He returned to earth to assure them of the need for endowment of the power of the Holy Spirit. His first reappearance on earth is recorded in John 21:3-25, where He performed His last miracle of the great catch of fishes exclusively for His disciples.

The final words of Jesus on earth were these: "But ye shall receive power, after that the Holy (Spirit) is come upon you; and ye shall be witnesses unto me both in Jerusalem, and in all Judaea, and in Samaria, and unto the uttermost part of the earth" (Acts 1:8).

These last words would naturally carry a very high priority for they were the most indelible ones. When the final words, ". . . the uttermost part of the earth," came from his lips, an unseen but gentle force lifted the Son of God bodily from the earth. As He ascended with increasing rapidity, a cloud floated beneath His feet and He—vanished!

BAPTIZING AND PRAYERS FOR THE DEAD?

"Else what shall they do which are baptized for the dead, if the dead rise not at all? Why are they then baptized for the dead?" (I Corinthians 15:29).

Prayers to the dead or baptisms for the dead have no Scriptural foundation for Christians and have no redeeming value or benefits. Paul did not say, "What shall *we* do?" or "What shall the *dead* do?" He said, "*they,*" for this was a heathen custom, neither endorsed by Paul nor practiced by the Early Church. Paul wrote to strengthen the Christian believers in this pagan city with regard to their position concerning the resurrection. He therefore penned a challenge to the pagans who opposed and persecuted the Christians saying, "And why stand we in jeopardy every hour?" It was the heathen, not the Christians, who were baptizing for the dead.

[1] If the blood of Christ were just human, he would have died only as a martyr and not as a Redeemer.

35

Heavenly Paradise[1]

Paradise is not a retirement home for the idle and infirm, not a cemetery of souls awaiting the day of resurrection.

"The way of life is *above* to the wise (heavenly Paradise), that he may depart from hell beneath (lower Paradise)" (Proverbs 15:24).

"I knew a man in Christ above fourteen years ago . . . such an one caught *up* to the third heaven . . . he was caught *up* into paradise (the Garden of God, Ezekiel 28:13), and heard unspeakable words, which is not lawful for a man to utter" (2 Corinthians 12:2-4). Paul the Apostle was the one who had this remarkable experience while lying under a barrage of stones heaped upon him outside the walls of the city of Lystra (Acts 14:19-21). He was uncertain at the time whether he was alive, but the experience was vivid.

MADAME BUTTERFLY

The deceased now in the book of life do not go "*down* into silence" to lower Paradise, but *up* into the heavenly Paradise where ". . . in thy (God's) presence is *fulness of joy* . . ." for ". . . thou (God) shalt make me *full of joy* with thy countenance" (Psalm 16:11; Acts 2:28). Paul said, ". . . we have . . . an house (body) not made with hands, eternal in the heavens" (2 Corinthians 5:1). "The dead praise not the

LORD, neither any that *go down* into *silence*" (Psalm 116:17).

The deceased body in the earth will remain there until its resurrection. In the meantime, those whose names are in the book of life will rise in the Garden (Paradise) of God in a different body than the one *planted* (I Corinthians 15:42-44) *in the earth.* It will be a translocation *from* earth *to* Paradise.

This body *raised in the Garden of God* (Paradise) will be exactly like the one planted in the earth except that it will be in a state of perfection. It will be newly created, and will be like the metamorphosis of a caterpillar when it leaves its loathsome chrysalis for its new, attractive body as a beautiful butterfly in order that it may enjoy its newfound freedom in a different dimension.

IVORY (WARDROBE) PALACES[2]

A fascinating truth in God's Word concerns the outfitting of the redeemed who will become the Royal Redeemed upon their marriage to the Lamb of God after they enter Paradise, the Garden of God. They not only will arrive in this Garden with paradisiacal bodies, but also they will be clothed upon with royal vestures which were first worn by the Lord Jesus Christ whom God chose as King to reign forever over all His creations (Psalm 45:15; I Timothy 1:17; 6:15; Hebrews 1:6). The royal *garments* will come from the ivory palaces, the wardrobes within the Holy City, which are located near the "king's palace." This is where Jesus Christ resided before He was seated beside His Heavenly Father on Mount Sion where he now makes intercession to the Father for the redeemed (Romans 8:26; Colossians 3:1; Hebrews 10:12; I Peter 3:22; Mark 16:19).

Jesus, in His prayer to the Father said, "And the glory which thou gavest me I have given them (the redeemed)" (John 17:22). Paul, in 2 Corinthians 5:1-4, reveals to us that the mortal body will be replaced with another body ". . . not made with hands, eternal in the heavens."

The body of the redeemed must be clothed. The white *robes* of righteousness will be received and worn before the royal wedding day of the Lamb of God (Revelation 6:11; 7:13,14) and the

"Bridegroom" and the "Bride" will be adorned with jewels (Isaiah 61:10). The wearers of these royal garments can be divested of them at the Judgment Seat of Christ (Revelation 16:15; 2 Corinthians 5:10) with the result being an embarassing and costly situation (Matthew 22:8-14).

Observe that the word "robe" is in the singular, showing that all of the Royal Redeemed are as *one* wife. As the Bride is wedded to the Bridegroom, Jesus Christ, she will receive a new name to identify her more closely with her Bridegroom: ". . . and thou shalt be called by a new name, which the mouth of the LORD shall name" (Isaiah 62:2). ". . . To him that overcometh (Royal Redeemed only) will I give . . . a new name . . . " (Revelation 2:17).

This white wedding robe (Revelation 19:7,8) not only will be worn at the wedding ceremony, but also it will be worn during the descent with Christ Jesus to the earth, as described in Joel 2:16: ". . . let the bridegroom go forth of his chamber, and the bride out of her closet." After this the Bride has complete access to the palatial ivory wardrobes because she is in an immortal body and can ascend and descend at will into the Holy City. There she also will be able to adorn herself with the rich collection of jewels (Isaiah 61:10): ". . . our Lord Jesus Christ . . . was rich and became poor, that ye . . . might be rich" (2 Corinthians 8:9). The Bride, as the Wife of Christ, will be vested in a royal robe to signify her status as she begins her eternal reign with the King of kings. The palatial ivory wardrobes from within the Holy City, which are accessible only to the Royal Redeemed, will contain vestures for different occasions.

WARDROBES WITHIN PARADISE

"My soul shall be . . . clothed . . . with the garments of salvation . . . " (Isaiah 61:10). "If so being clothed we shall not be found naked" (2 Corinthians 5:3). These garments will certainly have a kaleidoscope of styles and colors. The fantastic fabric will be of fourth-dimensional ultra-properties which remain permanently clean and wear everlastingly. One can visualize children garbed like cherubs and the adults wearing a variety of garments and adornments from the paradisiacal wardrobes.

All who enter Paradise unredeemed will be clothed upon with the garments of salvation because their names are still in the book of life since *salvation* remains the highest priority in God's plan of redemption. Those entering Paradise, the Garden of God, are described as follows: ". . . as the earth bringeth forth her bud, and the garden causeth the things that are sown in it to spring forth . . ." (Isaiah 61:11). What a joy it will be not only to meet friends long ago deceased, but also to possess a paradisiacal body and be vested in gorgeous garments: ". . . put on thy beautiful garments" (Isaiah 52:1).

CHRIST'S SOLILOQUY AND PRAYER

Before His incarnation, according to the Song of Solomon, the Son of God is recorded as giving a lovely prophetic soliloquy while strolling in the Garden of God (Paradise). Soon after His death and resurrection on earth He made it possible for millions upon millions, whose names are registered and retained in the book of life, to enter into this lovely garden. From among them He will choose His Bride and future composite Wife: "Ye have not chosen me, but I have chosen you . . ." (John 15:16).

God said in Isaiah 54:5 of Israel: "For thy Maker is thine *husband,*" but because of their sin God also said, ". . . by their whoredom (seeking after other gods) . . . they have defiled my holy name . . ." (Ezekiel 43:7,8). Therefore, according to God's own moral code, the *death* of either Israel or God was essential (Romans 7:3; Leviticus 21:7). In order to redeem Israel, God chose the latter course by the death of Himself through His Son Jesus Christ in order that the power of the law be severed. Thus, through the death of Jesus Christ, Israel could be redeemed (Romans 1:16) and brought in with the Gentiles (Galatians 3:13,14; 4:4; Romans 7:4; Ephesians 2:13-18). Now God, through Christ, could take upon Himself a new wife and would avoid the moral sin of adultery.

Eight times Christ Jesus makes reference in this soliloquy to "my garden" and twice to "his garden." Does this not imply it was first created by the Heavenly Father for His own begotten Son whom

He first crowned as "King" (I Timothy 1:17)? Within the Holy City is the "king's palace" (Psalm 45:15) with "ivory palaces (as royal wardrobes)" (Psalm 45:8. Compare John 12:15).

In this prophetic reflection we have Christ Jesus saying, "I am come into my garden (Paradise). . . " (Song of Solomon 5:1). Then follows this prophetic invitation: ". . . all manner of pleasant fruit . . . I have laid up for thee, O my beloved" (7:13). Jesus said to His disciples: "Henceforth I call you not servants . . . but I have called you friends . . . " (John 15:15). Christ in His prayer to His Father said, "And the glory which thou hast given me; I have given them" (John 17:22). "Father, I will that they *also,* whom thou hast given me, *be with me where I am;* that they may *behold my glory,* which thou hast given me . . . " "But this man (Jesus Christ), after he had offered one sacrifice for sins for ever, sat down at the right hand of God; *From henceforth expecting* . . . " (Hebrews 10:12,13). " . . . He is the Lord of lords, and King of kings: and they that are with him are called, and chosen, and faithful" (Revelation 17:14).

GOD'S LOVE FOR GARDENS

God has a fondness for gardens as evidenced by the following: The Throne on Mount Sion, within the Holy City, overlooks the Garden of God (Paradise) which completely surrounds the Holy City and extends over the entire palatial planet. Jesus, on Calvary's hill, observed the garden of Gethsemane where He loved to retreat, to meditate, and to pray: "Now in the place where he (Jesus) was crucified there was a garden . . . " (John 19:41). It was in a garden (Eden) that God placed Adam and Eve. It was a place of relaxation, reflection, and communication with nature. The heavenly Paradise is a place where the rippling streams from the river of life flow amid bountiful beauty. Here one can pluck delicious fruits and picnic amid singing birds perched in the branches of trees (Ezekiel 31:8). What a delightful place to come after being on an earth of turmoil, unrest, unhappiness, troubles, struggles, and uncertainties. Here one is free from sickness, pain, sorrow, and corruption. Yet it is but a transitory stopping place enroute to the greater glories awaiting in the New Heavens. Paradise eventually will be emptied and exist no longer. It will be replaced by the New Earth (the redeemed planet Earth) upon which the Holy City will descend. No longer will the rivers of life flow throughout its terrain. It will be abandoned because it will have served its purpose.

When Jesus Christ returns to earth again, this time with His Royal Redeemed, there will be a period of transformation to make the earth into a replica of the Garden of the Lord God: "For the Lord . . . will make her (earth's) wilderness like Eden, and her desert like the garden of the Lord; joy and gladness shall be found therein, thanksgiving and the voice of melody" (Isaiah 51:3). This planet Earth will become the redeemed New Earth and it will replace the heavenly Paradise when it reascends to its original place as the footstool in the third (or new) heaven.

NO ISLANDS

Paul was given an eagle's eye view of the heavenly Paradise and felt totally inadequate in describing what he saw and heard. Small wonder! Here were millions whose names were in the book of life from every clime, race, culture, and language on earth, but all were speaking and understanding a common language which was foreign to Paul. Diversity of tongues was no longer a barrier, so Paul assumed the language to be that of angels (I Corinthians 13:1). ". . . Tongues they shall cease . . . knowledge, it shall vanish away" (I Corinthians 13:8). With tens of thousands pouring into Paradise who were unfamiliar with each other's language and culture, it would have been confusing, indeed, to live cordoned into the separate cultural, lingual, and social backgrounds which had been in effect.[3] But here Paul saw a tightly-knit equalitarian society.

Language is a corrupt form of communication which was created for our physical third-dimensional senses. In our new environment, in the heavenly Paradise, our senses will be more acute as will our comprehension; therefore, a new language will be needed to adequately express ourselves beyond the third dimension.

Presumably Paradise resembles a huge planet upon which sits

the Holy City and Throne of God. It is possible that it may be much larger than our earth, which is described as the footstool of God, because it must accommodate a vast influx of immigrants. The perimeter of the heavenly Paradise surrounding the Holy City is a minimum of six thousand miles (Revelation 21:16). The river of life and its tributaries flow out from each of the twelve gates of the City's wall, so it is a well-watered place. To accommodate the 20,000 chariots known to exist, and the millions of royal white horses, there have to be massive ranchstyle enclosures (Revelation 19:11-14).

There probably are bridle paths for the equestrians and trails for the charioteers. Since the paradisiacal body will not have been resurrected or glorified as yet, outside modes of transportation within the Paradise Wonderland will be required.

A GLOWING PLANET[4]

". . . The earth shined with his glory" (Ezekiel 43:27).

Since the light of the glory of God which illuminates the heavenly Paradise has no photon particles, and therefore is not impeded by atomic particles, it appears scientifically feasible that the entire heavenly planet is consumed by a constant iridescent glow. It may be similar to a crystal ball, illuminating like "stones of fire" (Ezekiel 38:14), and therefore comparable to the Primordial Earth and the New Earth. It is conceivable that from a distance Paradise appears as a beautiful silvery shining planet, with a steady broadway of light upon which millions of individuals will be swept along at incredible speed. Those who are in the book of life will be in the escort of holy angels enroute to the heavenly Paradise.

LANGUAGE IN PARADISE

"For we know in part . . . but when that which is *perfect is come, then that which is in part shall be done away* . . . for now we see through a glass darkly; but then face to face: now I know in part; but then shall I know even as also I am known" (I Corinthians 13:9-12). When Moses and Elijah appeared on the Mount of Transfiguration, the Lord's disciples, Peter, James, and John, recognized them and communicated with them in a common language even though centuries separated them and they had no portraits of these prophets of old.

The Holy Spirit is clearly the Author of the many languages which were spoken by those filled with the Spirit. When God created Adam and Eve they instantly were able to communicate and relate.[5] At the tower of Babel when thousands of new languages were introduced by God, those of similar tongues could at once understand one another. On the day of Pentecost everyone was unfamiliar with the languages, yet they were clearly understood by the audience. Therefore, there will be great accord in the heavenly Paradise because everyone will be speaking and comprehending a common tongue.

It appears from Zephaniah 3:9 that when the Royal Redeemed return to earth with Christ they will introduce this language (of angels, I Corinthians 13:1, because it is an untainted tongue) for all to speak during the Millennium: "For then will I turn to the people *a pure* ("clean," *Moffatt*) *language*, that they may *all* call upon the name of the Lord, to serve him with one consent." All present languages will disappear as did the original language before the time of Babel. In the heavenly Paradise the same thing will happen.

NO APARTHEID PROTECTORATE

With the disintegration of the mortal body, the color difference in the pigment of the skin will be nonexistent. We shall then be "like him (Jesus)" (I John 3:2). "Who shall change our vile body, that it may be fashioned like unto his glorious body . . . " (Philippians 3:21). "But we all, with open face beholding as in a glass the glory of the Lord, are changed into the *same image* from glory (paradisiacal bodies, 2 Corinthians 3:18) to glory (immortal bodies, I Corinthians 15:53), even as by the Spirit of the Lord" (2 Corinthians 3:18). "As we have borne the image of the earthly; we shall also bear the image of the heavenly" (I Corinthians 15:49).

PARADISE: A PLACE OF OPPORTUNITY

Nothing could be more boring and devastating than to possess abounding energy, superior intelligence and abilities, and have no objectives or outlets for fulfillment. The heavenly Paradise will have a profusion of interests and activities to satisfy the most discriminating desires. There will be the opportunity to express one's self, to fulfill one's long ambition of a hoped-for-dream, and to experience the prolific explosion of one's talents. Perhaps these newly discovered talents can be expressed to the delight of an appreciative audience and the satisfaction of the performers. New stars will be born amid this society.

MENU IN PARADISE

The medicinal leaves from the tree of life must be taken shortly upon arrival as a one-time requirement for the purpose of preventing illness and death because all residents have bodies of paradisiacal flesh. Thus they can grow and develop into maturity just as did the bodies of Adam and Eve in Eden's Paradise. Genesis 3:22-24 reveals these thoughts concerning the tree of life: 1) Adam and Eve had not eaten from it; 2) their one time participation would have been adequate; 3) they chose the tree of death in preference to the tree of life. There are no Scriptural evidences that any tree of death exists in the heavenly Paradise, therefore, no choice will be required; there will be no threat of expulsion because salvation is not decided there; 4) No guardian angel will be stationed to prevent anyone from having access to this tree; 5) Genesis 2:9 and Revelation 2:7 show that the tree of life is easily accessible.

THE RIVER OF LIFE

". . . Ye waters that be *above* the heavens" (Psalm 148:1). The luxurious landscapes of Paradise will be watered by the river of life which flows from the summit of Mount Sion where God's throne is located. Scientists inform us there are many complicated molecules in the universe, so the unique properties of the river of life cannot be discounted. For instance, a child may fall into the river of life but to his amazement there will be no frightening or dangerous consequences.

VANISHING CHILDISH THOUGHTS

"But when that which is perfect is come, then that which is in part shall be done away. When I was a child, I spake as a child, I understood as a child, I thought as a child, but when I became a man, I put away childish things. For *now* we see through a glass darkly, then face to face: *now* I know in part: but *then* shall I know even as I am known" (I Corinthians 13:10-12).

In our earthly journey we are as little children with our immature questions, trying to understand Paradise. When we arrive there, childish thoughts will give way to a higher dimension with more mature contemplation, evaluation, and understanding.

Abounding enjoyment of Paradise cannot be jeopardized by negative thoughts or views about the past life on earth: ". . . *thou shalt make me full of joy* with thy countenance" (Acts 2:28). This is another reason why children of the raptured redeemed will not be left orphans to enter the Tribulation of the last days.

THE SEXES BUT NOT SEXUALITY

"And Jesus answering said unto them, The children of *this world marry,* and are given in marriage: But they which shall be accounted worthy to obtain *that world* (heavenly Paradise), *and* the resurrection from the dead (after leaving Paradise), neither marry, nor are given in marriage" (Luke 20:34,35).

"Be fruitful and multiply, and replenish the *earth"* is not applicable to the heavenly Paradise, but to the earthly paradise. However, it appears that Adam and Eve were not involved in sexuality until *after* they sinned (Genesis 4:1). Their descendants therefore, were sinners (Romans 5:19). If Adam and Eve had taken of the tree of life instead of the tree of knowledge of good and evil (the tree of death), their posterity would have proliferated until eventually righteousness would have encompassed the entire earth, and the Paradise of Eden would have circumposed the whole earth. In order to have accommodated that vast population growth, the tree of life in the Garden of Eden would have proliferated accordingly, just as the trees of life located at each of the twelve gates of the Holy City in the heavenly Paradise accelerated their production for the multitudinous inhabitants. We read that it was "east of the garden of Eden (that cherubim kept or guarded) the way of the tree of life" (Genesis 3:24. Compare Revelation 21:22). There appeared to be some kind of imperceptible restrainer surrounding Eden on earth (Genesis 3:24); likewise, surrounding the Holy City on Paradise to keep the inhabitants from entering the City for the present (Revelation 21:12-15).

EQUALITY OF THE SEXES

The sexes are on equal footing. All of the redeemed are called "*sons* of God;" all are the *wife* of Christ; Israel is spoken of as "*daughters* of Jerusalem;" converts of the Royal Redeemed are referred to as "kings' *daughters;*" those who prophesy are termed *prophets* and *prophetesses* (Acts 21:9); and all of the Royal Redeemed will reign with Christ as *kings, lords, priests, and judges.*

The creation of two sexes are essential for the propagation of the human species to enable as great a number of redeemed as possible to be in the family of God before the transition from mortality into immortality. Although this transfer will not alter the sexes, it will enable them to become fully equated and their relationship will be purely platonic.

NO NIGHTS BETWEEN DAYS

"But, beloved, be not ignorant of this one thing, that one day is *with* the Lord as a thousand years, and a thousand years as one day" (2 Peter 3:8).

The glory of God that illuminates is eternal, so there are no nights of darkness. Thus, a day in the heavenly Paradise is one thousand years of our solar time as marked by the paradisiacal calendar, meaning that only two days have elapsed since the birth of Christ in Bethlehem. Therefore, to leave our solar time and step into this utopian environment is to step backward in time.

"THEY SHALL WALK WITH ME"

"Yea, thou I walk through the valley of the shadow of death . . . *thou art with me* . . ." (Psalm 23:4),

". . . And *they shall walk with me* in white . . . for they are worthy" (Revelation 3:4).

From the final moments on earth to the first moment in Paradise, the Lord walks with His own. The fact that Christ is enthroned does not confine Him to His royal seat. The Throne only represents position, authority, and power; and like other monarchs who traverse about their realms, so Christ Jesus will walk, ride, mingle, and converse with those in the heavenly Paradise: ". . . thou (Lord) didst *ride* upon thine horses and thy chariots . . . " (Habakkuk 3:8). Those in paradisiacal bodies will require outside modes of transportation: first, because they do not yet have their immortal bodies; second, because God had to take the creatures unto Adam to be named by him when he was in his paradisiacal body (Genesis 2:19).

What a thrill to listen to Jesus Christ; to make inquiries; to become knowledgable of God's plans; to share one's personal concerns with One who really cares; to relax with Him on the banks of the river of life; to perambulate with Him down a beautiful lane; to ride astride one of the royal horses; and to join Him in His royal chariot led by a span of majestic horses racing down a bridle path through the glorious vistas of Paradise. How exhilarating and consoling!

A WORLD OF PEACE, HEALTH, AND PROGRESS

The earth would have been like the heavenly Paradise had Adam and Eve partaken of the tree of life instead of the tree of knowledge (the tree of death). While the word "overcometh" in Revelation 2:7 is an assurance to the redeemed, it does *not exclude* others who go there, otherwise the word "only" would have been inserted. Undoubtedly, the results of the progress made in the heavenly Paradise will be transferred to the Millennium, the New Earth, and the New Heaven. Such knowledge will not be lost (Isaiah 30:20).

PLANET PARADISE

The heavenly Paradise is a planet with the Holy City located upon it as shown in the scriptural texts:

1) ". . . Christ . . . is even at the right hand of God . . ." (Romans 8:34). "Father, I will that they . . . be with me where I am; that they may behold my glory . . ." (John 17:24). ". . . They shall walk with me . . ." (Revelation 3:4).

2) "He that *sitteth* (in the Holy City on His throne) in the heavens" (Psalm 2:4). "(God) *bowed* the heavens (plural)" (Psalm 18:9).

3) Abraham ". . . looked for a city which hath *foundations,* whose builder and maker is God" (Hebrews 11:10).

4) Paul said that at death ". . . to be absent from the body (is) to be present with the Lord" (2 Corinthians 5:6,8).

5) God said of the redeemed who die in the "faith" that He "hath prepared for them a city" (verse 16).

6) David said: ". . . I foresaw the Lord before my face . . . because thou wilt not leave my soul in hell (lower paradise)" (Acts 2:25,27).

7) The tree of life is associated with the Holy City (Revelation 2:7; 22:1-4).

8) Paradise is described also as a "country" (Hebrews 11:15,16).

KNOWLEDGE SHALL VANISH AWAY

What will occur when two or more hostile or antagonistic individuals who felt mutual hostility or antagonism on earth meet one another in Paradise? In I Corinthians 13:8 we read: ". . . *tongues* (languages), they shall cease (Greek root: to stop, quit, come to an end) . . . *knowledge,* it shall vanish away (Greek root: to delay or linger)." The distinction is clear. Earthly languages will come to an abrupt *end;* but knowledge shall be *delayed* until the day of judgment. "For there is nothing covered, that shall not be revealed; neither hid, that shall not be known" (Luke 12:2). It will be "hid" for awhile in Paradise for obvious reasons, but shall be manifested at the Judgment Seat of Christ: "But why dost thou judge thy brother? or why dost thou set at nought thy brother? For we shall *all* stand before the judgment seat of Christ. For it is written, As I live, saith the Lord, every knee shall bow to me, and every tongue *shall confess to God.* Let us not therefore judge one another any more . . ." (Romans 14:10-13). "For every man's work shall be made manifest: for the *day* (in judgment) shall declare it, because it shall be revealed . . ." (I Corinthians 3:13). "And now, little children, abide in him; that, when he shall appear, we may have confidence, *and not be ashamed before him* . . ." (I John 2:28).

There will be recognition (I Corinthians 13:12), but any bad relationships on earth, will be hidden by God until the day of revelation. This can be possible only within the frame of the paradisiacal body, when ". . . that which is *perfect is come,* then that which is *in part* shall be done away" (I Corinthians 13:10).

HIDDEN BEAUTY

"For *now* we see through a glass, darkly (Greek root: position in place, time, or state); but *then* face to face . . ." (I Corinthians 13:12).

"It is sown in dishonour (Greek root: disgrace, vile); it is raised in glory (Greek root: glorious, dignity) . . ." (I Corinthians 15:43).

Phillips translates I Corinthians 13:12 as "looking at puzzling reflections in a mirror" and I Corinthians 15:43 as being "raised in splendor." The New English Bible translates "sown in dishonour" as "sown in humiliation." This indicates that all persons existing on earth have an imperceptible physical beauty which will become visible only when they obtain their new paradisiacal and immortal bodies. This involves more than just going backward in time to the beauty of the days of one's youth.

Adam and Eve were each created with both third-dimensional ("image") and fourth-dimensional ("likeness") body (Genesis 1:26,27). "There is a *natural* body, and there is a *spirit*(ual) body . . . the *image* of the heavenly" (I Corinthians 15:44,49). This fourth-dimensional individualized beauty presumably is the *glorious form* of the hidden "spirit" entity. The angels possess this form (Hebrews 1:14), as well as God Himself (John 4:24).[6] Moreover, have we not been given the promise to be like God (I John 3:2) and to be able to exist in

His presence (I Timothy 6:16; Matthew 5:8).

Since this fourth dimensional entity will be separated forever from lost souls, they will never view this unexcelled hidden beauty (Ecclesiastes 12:7). Only those whose names are retained in the book of life (Revelation 3:5; Psalm 125:1) will have this hidden beauty revealed in their paradisiacal and immortal bodies. The "glass darkened" mortal bodies no longer will hide these beautiful spirit entities now clothed in the tangibility of God's creative and redemptive program.

[1] In Psalm 16:11 the word "pleasure" means gratification, delight, satisfaction. In Acts 2:28 the word "joy" means gladness, enjoyment, happiness, a good frame of mind. These are descriptions that reveal the state of euphoria in the heavenly Paradise.

[2] See page 276.

[3] Many Bible scholars believe the language in the Garden of Eden and in the heavenly Paradise to be the same. The language in the Garden of Eden terminated at the time of Babel with the creation of new languages.

[4] Why this glory has no particles: 1) Is original, because God is Light and is not created; 2) No decay, for God is Eternal; 3) No shadows (I John 1:5,6); 4) Faster than photons, because God is "Spirit."

[5] When the prophets of ancient times imparted God's message to the people, the Holy Spirit registered the words in their minds for later transcription on papyrus for inclusion in the Holy Writ.

[6] God also has a *soul* (Leviticus 26:11,30; Hebrews 10:38).

36

Perverted Justice

Full justice demands that the rights denied a person be restored.

If names are placed in the book of life only at the time persons are redeemed, then according to Revelation 20:11,15, all little children will be cast into the lake of fire! "And I saw the dead, *small* (literally, little in size) and *great* (literally, big in years) stand before God . . . and whosoever was *not* found written in the book of life was cast into the lake of fire."

If this is the case, then God completely bypasses the individual's right to choose and the innocent are pathetic victims through no fault of their own. The "will" is cruelly and deliberately disregarded, and the innocent receive no consideration from God. It declares a person *guilty even before* he has an opportunity of making a decision. This is an erroneous and unjust depiction of God: "Doth God pervert (distort, twist) *judgment?* or doth the Almighty pervert *justice?*" (Job 8:3; 34:12). "Touching the Almighty . . . he is excellent in power, and in judgment, and in *plenty of justice* . . . " (Job 37:23). "Great are thy *tender mercies,* O LORD" (Psalm 119:156). " . . . *Righteous judgments*" (Psalm 119:160).

God discloses how He wants us to interact with Him in the

conduct of our affairs: "I will praise thee with uprightness of heart, when I have learned thy righteous judgments" (Psalm 119:7).

REGISTRAR OF THE UNBORN

Names are not inscribed in the book of life when one is conceived, is born, or becomes spiritually reborn. Names were placed there *before* the foundation of the world on the basis of God's perfect foreknowledge. Therefore, to abort unborn human beings is to challenge and counter the will of God. Furthermore, the unborn, even though aborted, are still retained in the book of life. Abortion does not erase their names. The guilty must therefore face the unborn ghosts at the day of judgment concerning the denial of their God-given rights to decide their eternal destinies for themselves. Think it over! "*Woe* unto him that saith unto his father, *what* begettest thou? or to the woman, *what* hast thou brought forth?" (Isaiah 45:10). No human being is ever a "what," but a being of immense value.

A WHOLE PERSON

"And the very God of peace sanctify you *wholly;* and I pray God your *whole* spirit and soul and body be preserved blameless unto the coming of our Lord Jesus Christ" (I Thessalonians 5:23).

It is important to understand that the embryo is more than a body, or even a soul. It is a spirit person linked to the Almighty God Who is also a Spirit Being—the fountainhead of all spirit beings. The three—spirit, soul, and body—are knitted, meshed, united, and wedded together to form a single unit in the image and likeness of God.

SPIRIT BEINGS

God is a Spirit Being (John 4:24) and has within His structure all of the senses and attributes that identify personality. Therefore, the embryo in its original formation has the kernel of life, the spirit, which makes one akin to His Creator. It is, therefore, not only an insult to attempt to delete that part of the Creator closest akin to Him, but the spirit cannot be destroyed any more than one can destroy the existence of God: ". . . the spirit shall return unto God who gave it" (Ecclesiastes 12:7). "Who knoweth the spirit of man that goeth upward, and the spirit of the beast of the field that goeth downward" (3:21). "But there is a spirit in man: and the inspiration of the Almighty giveth them understanding" (Job 32:8). As *light is constant* in the universe, so the *spirit is also constant* in the entity of man.

DIVINE RIGHTS OF THE UNBORN

The controversies on the rights of the unborn child hinge on whether the unborn is really a baby, a human being, until it is born. It is said to be merely a fetus.

In Luke 1:41,44 the *unborn* John the Baptist is called a "babe." This is the same Greek word (*brephos*) as used in Chapter 2, verse 12, after the birth of Jesus Christ: "Ye shall find the babe (brephos) wrapped in swaddling clothes, lying in a manger." And in verse 16, the shepherds saw the babe (*brephos*) lying in a manger."

The *unborn* Saviour, the Lord Jesus Christ, was called a "child" *before* He was born in Bethlehem (Matthew 1:18,23).

It is interesting that the word "fetus" appears nowhere in the Scripture. Is it because God looks differently upon the unborn than do some people? It appears so.

In Luke 1:15 it is prophesied that the Holy Spirit would fill the *unborn* John the Baptist *before* he was born: ". . . and he shall be filled with the Holy Ghost, even *before* his birth" (Luke 1:41;44, *The Living Bible*).

A prediction was made about King Cyrus of Persia about 170 years before his birth: "(God) saith of Cyrus, He is my shepherd, and shall perform all *my* pleasure: even saying in Jerusalem, Thou shalt be built, and to the temple, the foundation shall be laid" (Isaiah 44:28). Even his name was foreknown by God.

Therefore in aborting the unborn, what about the divine purpose of God and the divine rights of that one? Do we have the right to make the determination as to whether or not that unborn should live in the light of God's perfect foreknowledge and *His will?*

It is known that children are instrumental in leading parents to know the Lord. They frequently make beneficial contribution to society and their forebears upon their maturity. There are, of course, circumstances in which an abortion is advisable and necessary, but not for selfish reasons. That poses another matter.

BEFORE I WAS BORN

Paul, the apostle, had this to say regarding his own birth: "For even *before I was born* God had chosen me to be his, and called me—what kindness and grace—to reveal his Son within me so that I could go to the Gentiles and show the Good News about Jesus" (Galatians, *The Living Bible*). What a tragedy if he had been aborted!

Are the aborted to be denied their rights to enjoy eternity, which would have been the case had they accepted Christ later in life? Is the eternity of their souls determined by the will of criminals or as a result of their own decisions? Be assured the guilty will not be exonerated. It is prenatal child abuse with dire consequences.

Nothing so infuriates the Almighty as the destruction of a life which was created in His own image and likeness. It is an insult to God, and a blow to the wisdom of the Almighty who has imparted a means of procreation and a modicum of Himself in every human being (Genesis 2:7; Ecclesiastes 3:21; 12:7).

If there is to be justice, there must be an accounting of the guilty, as well as a balancing of the scales to compensate the innocent. To simply pass sentence upon the accused does not adequately make up for the losses incurred by the victims. What about the normal life that was denied them? What about the potential of those lives that were cancelled by someone else's will? Full justice (because God is a "just" God) demands that the rights denied a person be restored. These rights will be granted in the New Heaven, the expansion of the Kingdom of Christ.

COMEDY OF ERRORS

In the theological position taken by some, one would have to conclude that justice and fairness have been discarded in favor of the guilty. This is submitting the entire program of God to the wheel of chance; the place it stops determines the outcome. It discredits the wisdom of God and makes the Almighty appear insensitive to the plight of the innocent. It is a ludicrous comedy of errors. The ledger must be equitably balanced because human will cannot supplant the will of God, or substitute as being divinely providential.

MY LITTLE CHILDREN

God looks upon the entire human race as children. He speaks of the "children" of Israel. The Apostle John addresses the redeemed as "my little children" (I John 2:1). The Church is labeled "the elect lady" and her constituents are called "her children." Age is irrelevant with God. He is willing that none should perish, regardless of age.

Innocence does not stop with little children. All of those within the Kingdom of heaven (or God) are called "children of the kingdom" except those who have hardened their wills against God and who have aligned themselves with Satan. They are called "children of the wicked one" (John 8:44).

Therefore, there are those beside little children who would have accepted Christ into their lives had they not been cut off from life prematurely. Only God, because of His perfect foreknowledge, knew the hearts of His creatures and their decisions even before the foundation of the world. However, their names were inscribed even in view of their rejection of God's plan of redemption, for how could they be blotted out if they never were inscribed in the first place?

LIVING TOO LONG OR DYING TOO SOON

In light of this, we cannot and must not exclude the fact of repentance and acceptance of Christ as Saviour by all human beings. There are no exceptions! This is one of the prime reasons for the New Heaven in the Lord's expanded kingdom.

Those who argue: "And as it is appointed unto men once to die, but after this judgment" (Hebrews 9:27), fail to understand that judgment does not imply an automatic indictment. An appearance at the judgment does not preclude guilt. If so, what chance does anyone have at the Judgment Seat of Christ where the redeemed are to appear? There would be no survivors even among the redeemed.

Why is it that some of us who are secure in our redemption are so ready to condemn others, especially those who never have had access to God's mercies? Few of us could survive if men were to sit in judgment over other men's souls. That is why Paul said, "Who shall lay anything to the charge of God's elect? It is *God* that justifieth. Who is he that condemneth . . . ?" (Romans 8:33,34). "Who art thou that judgest another man's servant? to his own master he standeth or falleth. Yea, he shall be holden up: for God is able to make him stand . . . So then everyone of us shall give account of *himself to God"* (Romans 14:4,12). It seems we have sufficient responsibilities to shoulder without burdening ourselves with the condemnation of others. "But them that are *without* God judgeth" (1 Corinthians 5:13); " . . . mercy rejoiceth against judgment" (James 2:13). If we err in judgment, let us err on the side of mercy.

37

Illegitimate Parents

Each mother's womb is God's Shangri-La where He sees the embryo being formed into a fetus capable of becoming a potential temple of the Holy Spirit.

All children, including those conceived illicitly, have their names registered in the book of life. God's foreknowledge not only embraces those who will be redeemed but also those whose names will have to be eradicated: "And the Lord said unto Moses, Whosoever hath sinned against me, him will I blot out of my book" (Exodus 32:32). How could names be blotted out if they were never there in the first place? One's name can be erased only by wicked unconscionable deeds or by absolute rejection of Jesus Christ as Saviour when done with clear understanding.

In the Apocrypha (Wisdom 4:6), children conceived out of wedlock will "witness . . . the wrong their parents did" (New Jerusalem Bible). Those carnal actions cannot be charged against the innocent offspring who are victims of circumstances. The parents are illegitimate, not the children.

THE HIDDEN CONNECTION

The few cells that make up the protoplasmic structure (the fetus) after conception contain the entire individual's spirit, soul, and body in minute form. These groups of cells in the embryo, called *organizers,* determine a person's personality or individuality. Actually, these triune parts of potential human beings are present in the primordial single cell. Their rapid separation and increase into other cells, called the "chemical clock," are under the direction of the DNA (the code of life) in the original first cell.[1]

All that the Spirit of God requires at the time of the resurrection is the constituent of the primordial cell which contains the DNA. Even if the atoms that make up the DNA and the first cell become "rest" particles, God has the power to reactivate them and bring the complete body back to life. The original cell might be classified as part of the invisible world—invisible to us maybe, but not to God.[2]

SHANGRI-LA OF THE HUMAN RACE

"For thou hast possessed my reins (souls with its attributes): thou hast covered me in my mother's womb . . . My substance (body) was not hid from thee, when I was made in *secret,* and curiously wrought in the lowest parts of the earth (atoms). Thine eyes did see my substance, yet being unperfect (unformed mass); and in thy book all my members *were written,* which in continuance (growth) were fashioned, when as yet there was none of them. How precious also are thy thoughts unto me, O God! how great is the sum of them!" (Psalm 139:13-17). The New Jerusalem Bible states: "You (God) . . . watched my bones take shape when I was being formed in *secret,* knitted together in the *womb* . . . all were recorded in your book, before my days." The miracle of the fetus' development is partially appreciated when one realizes the brain alone grows by an average of 250,000 cells a minute.

Each mother's womb is God's Shangri-La where He sees an embryo being formed into a fetus capable of becoming a potential temple of the Holy Spirit (I Corinthians 6:19). God, under perfect foreknowledge, has great plans for each person before he is ever conceived in the womb, the results of which are contingent upon the individual's will.

Abortions certainly will not be condoned in the Lord's manifested kingdom on earth, according to Isaiah 11:8,9; 65:25: "They shall not hurt or destroy in all my holy mountain: for the *earth* shall be full of the knowledge of the Lord. . . ." If the Lord did not approve of abortions then, why does anyone think He will approve of them now?

DEMOLISHING THE HOLY SPIRIT'S TEMPLE

The demolition of the Holy Spirit's potential temple through abortion usually is done by uninformed parents. First, the baby is not a part of the mother's body. He is an individual human being who is separated from the body and living in his own amniotic sac. He produces the umbilical cord and the placenta from his own cells; these are his life sustainers. Therefore, when a baby's life begins, the woman's prerogative regarding her body terminates. Society does not allow persons to have unlimited control of their bodies as evidenced by the fact that drug abuse, self-mutilation, prostitution, and suicide are prohibited by law.

REAL LITTLE PEOPLE

Psychologists indicate that babies know much more than is acknowledged. They are real people with mature spirits in baby bodies. For a brief time they have communication without language because they become distracted by it. They can distinguish sounds, night from day, and have the capacity to learn. Psychologists further believe that feelings of guilt start with many people when the pregnancy itself is discovered. Although a baby's brain is tiny, the mind can store experiences and understand their meaning. It really comes down to the difference between mind and brain.

The size of the brain does not necessarily determine the capacity of the mind. Entomologists say that insects with comma-sized brains are capable of complicated behavior. Fireflies often speak in a visual Morse code. The larva's tiny brain, which can slip through the eye of a needle, has the same major anatomical subdivisions as our own: cerebellum, encephalon, midbrain, cerebrum, and medulla. Within those miniscule neuroanatomical entities lie the range of complex behaviors.

WHEN TWO LIVES HANG IN THE BALANCE

The question is posed regarding priority of life between the mother and the unborn in cases of necessity. The Apostle Paul says: "But I speak this by permission, and not by commandment" (I Corinthians 7:6). Following this advice, the writer believes the mother should be given precedence for these reasons: 1) the family; 2) her husband; 3) the fact that the baby's destiny is favorably secured in the book of life, in contrast with the mother's spiritual state, which may be in question.

Under the justice of God all persons who are cut off prematurely have their names retained in the book of life. This assures that one will have the opportunity of making the decision regarding his destiny when he arrives in the New Heaven. The word "whosoever" in Revelation 20:15 confirms this fact.

AN ADAMIC INHERITANCE

A morphogenetic hereditary field (called "morphic resonance") is created in the embryo by its genealogical past. It exerts a crucial influence in the biochemical activity in the nucleus of every cell. Therefore, the fallen Adamic nature which all of us inherit, needs the cleansing and controlling influences of the Holy Spirit and the Word of God very early in life: ". . . he saved us, by washing of regeneration, and renewing of the Holy (Spirit)" (Titus 2:5). "That he might sanctify and cleanse it with the washing of water by the word" (Ephesians 5:26).

[1] Physicists refer to this multiplication of cells as "chemical communication."

[2] Without some particle remains of the deceased body, it would *not be a resurrection but a creation.*

38

Kinds of Abortions

God had no difficulty inscribing names in the book of life before the foundation of the world because His foresight is as perfect as His hindsight.

A knowledge of how abortions are performed should be understood. There are several methods: 1) Suction; 2) Dilation and Curettage, or D and C; 3) Prostaglandin; 4) Saline or salt poisoning; 5) Hysterectomy.

Under the suction procedure, a tube is inserted into the uterus and is attached to a powerful vacuum apparatus which is so strong that it sucks the baby into a container by pieces—head, arms, and legs.

In the D and C method, a hook-shaped knife which is placed into the uterus cuts the baby into small pieces. Then the baby's head is crushed with forceps so he can be removed, after which the womb is thoroughly scraped. Occasionally the consequence is profuse bleeding.

The prostaglandin method involves injecting a hormone drug into the amniotic sac, thus inducing labor and premature birth. The baby generally is born alive and has a heartbeat. He is left to die.

Saline abortion is done after sixteen weeks when a strong saline poisoning solution is injected into the amniotic sac. It is done through the abdomen and the defenseless baby is compelled to swallow this fatal solution. His kicking and jerking become violent as a result of the intolerable pain. His skin peels off as he is being burned alive. The mother delivers a dead baby.

In a hysterectomy, the abdomen is slit open and the baby lifted out, frequently crying, as he is squirming and struggling for breath. His heart is beating but the baby is set aside to die.

INFANTICIDE

In the United States, between 1937 and 1981, these horrendous abortions have averaged over 1,300,000 each year. These little ones had a right to live under the foreknowledge of God and to exercise their own wills in behalf of redemption. Someone else acted contrary to God's plan for the unborn which was a presumption on the will and purpose of God.

That is why those who are so cruelly denied God's original purpose for them must and will be granted that privilege in the New Heaven. This first chance must be accorded to them, but be assured that these unborn will rise like ghosts to haunt their murderers before God at the day of judgment.

When God placed names in the book of life *before* the foundation of the world, He ascribed a higher priority upon the souls of human beings than upon the physical structure of the earth (Luke 9:25).

DIVINE FORESIGHT

God had no difficulty inscribing names in the book of life before the foundation of the world because His foresight is as perfect as His hindsight: ". . . declaring the end from the beginning," ". . . God hath from the beginning chosen you to salvation" (2 Thessalonians 2:13). Therefore Paul could say, "According as he hath chosen us in him *before* the foundation of the world" (Ephesians 1:4). Peter said, "Elect according to the foreknowledge of God the Father" (I Peter 1:2).

Abortions will probably be done in the near future with

relatively safe drugs in the privacy of one's own home. This, however, does not alter the fact that it is a contradiction of God's Word. There is still the right of life for the unborn and a prerogative for the destiny of the unborn to be fulfilled. Remember, their names have been registered in the book of life *before* the foundation of the world.

The one redeeming factor is that the guilty, after accepting Christ as their Saviour and recognizing that wrong can be forgiven, may serve as "priests" with Christ in the New Heaven. In this spiritual ministry perhaps the opportunity will be accorded to bring the aborted ones to know Christ as their personal Saviour. Let us hope so.

39
When Innocence Ends

When knowledge of the truth comes, a decision must be made which transcends conscience.

When does innocence end and personal responsibility take charge? It is at the precise moment that conscience awakens: "For when the Gentiles, which have not the law . . . are a law unto themselves . . . their conscience . . . accusing or else excusing one another; In the day when God shall judge the secrets of men by Jesus Christ according to my gospel" (Romans 2:14-16). The actions that violate this God-given sense are placed on the "books" as described in Revelation 20:12,13. There is a point at which the accumulation of these unconscionable acts can tip the scales of God's justice sufficiently to compel Him to erase a name from the book of life.

Ignorance can be a consideration when the Gospel, through unfortunate circumstances, has not been made available or understood. The Apostle Paul is a case in point: "Who was before a blasphemer, and a persecutor, and injurious: but I obtained mercy, because I did it *ignorantly in unbelief*" (I Timothy 1:13). "And Paul, earnestly beholding the council, said, Men and brethren, I have lived in all *good conscience* before God until this day" (Acts 23:11). There are other instances when lack of knowledge has been taken into account: "Then said Jesus, Father, forgive them, for they know not what they do . . . " (Luke 23:34). "And they stoned Stephen, calling upon God, and saying . . . Lord, lay not this sin to their charge . . . " (Acts 7:59,60).

KNOWLEDGE OF THE TRUTH

One of the prime purposes of the Mosaic Law was to reveal the knowledge of right and wrong: "For until the law sin was in the world: but sin is not imputed when there is no law" (Romans 5:13). " . . . For where no law is, there is no transgression" (Romans 4:15). But God by foreknowledge knew all mankind would eventually sin, so on that premise He provided redemption for all. " . . . (God) will have all men to be saved, and to come unto the knowledge of the truth" (I Timothy 2:4). However, sin cannot be charged until actual transgression has occurred, and redemption cannot be granted without recognizing that people are inherently sinners by nature and in need of a Redeemer, in view of their projected actions.

When knowledge of the truth comes, a decision must be made which transcends conscience—and that is whether to accept or reject Christ as one's personal Saviour. However, there is a progression of unbelief which when allowed to infiltrate, will destroy the faith essential to redemption. First, "they did not believe" and then, "they could not believe" (John 9:18; 12:30). "Who changed the truth of God into a *lie* . . . " (Romans 1:25). " . . . They did *not like to retain God in their knowledge,* God gave them over to a reprobate mind . . . " (verse 28). Such is the case with individuals given over to Satanic power. They are called "workers of iniquity," and Jesus Christ addressed them as follows: "I know you not" (Luke 13:25) and "I never knew you" (Matthew 7:23). The only difference between the two is the degree of their involvement, but they form the nuclei for "false Christs" and "false prophets." How could Jesus say, "I never knew you," in the light of His perfect foreknowledge? God has the remarkable capacity of erasing from His memory that which He desires to forget: " . . . I will remember their sin no more" (Jeremiah 31:34); " . . . And their sin and iniquities will I remember no more"

(Hebrews 10:17). It is because their conscience was hardened (I Corinthians 8:7; I Timothy 4:21; Hebrews 10:22).

LITTLE CHILDREN'S ESCORTS

"Take heed that ye despise not one of these little ones; for I say unto you, that *in heaven* their angels do always behold the face of my Father which is in heaven" (Matthew 18:10). This indicates that there are no guardian angels *on earth.* The period of innocence has a special interest with God because in the event of tragedies of death, these angels in heaven descend and serve as escorts for the spirits and souls of the innocent enroute to Paradise.

Every thirty minutes, in the United States alone, twenty-two children are aborted and almost 600 children are beaten, abused, or victimized by broken homes. The abuse of the innocent will not go unchallenged and unpunished: "And whoso shall receive one such little child in my name receiveth me. But whoso shall offend one of these little ones, it were better for him that a millstone were hanged about his neck, and that he were drowned in the depth of the sea" (Matthew 18:5,6).

40
The Doctrine of Infinities[1]

The Being of the Creator also can be confirmed by the mystery of life which is contained within the purposeful and meaningful existence of matter.

"For the invisible things of him from the creation of the world are clearly seen, *being understood by the things that are made,* even his eternal power and Godhead; so that they are without excuse" (Romans 1:20).

This text clearly shows that God's existence can be proven by scientific research because the invisible Creator is visible in and through His creation. God has challenged us to believe in Him.

Scientists have accepted the concepts of mathematical infinities to the absolute infinity. They have been unable to clearly define them, but they do recognize their existence. The Scriptures unfold this mystery as being five infinities in seven dimensions.[2]

The existence of the Creator is made obvious by these infinities. As the forward momentum from departure to arrival is increased, zero-time is attained, which is the gateway into the fourth dimension and beyond. This has been calculated to be at the speed of light—186,282 miles per second. At 99 percent of the speed of light time dilation takes effect, which is the rapid slowing down of time. Near perfect time in the third dimension is believed to be 9,192,631,270 (billion) vibrations per second of the celius atom, which is but one of the atomic clocks essential to prevent even a one-second error which could send a space probe thousands of miles astray. Its vibrations cannot apply to the fourth dimension and beyond because there time would be different. Scriptures reveal that a thousand years of solar time equal but one day in the fourth dimension (2 Peter 3:8). For instancc, fourth-dimensional time allows angel spirit beings to traverse with superluminal speed in all directions without difficulty.

The Being of the Creator also can be confirmed by the mystery of life which is contained within the purposeful and meaningful existence of matter. Without this, matter would be sterile and useless. Life gives to matter more than substance; it gives to it a sense of purpose and reason for existence. There is evidence of *invisible intelligence* in all of the beauty, wonders, intricate patterns, and glories of creation.

God is an invisible Spirit Being whose existence gave birth to all matter in our third-dimensional world. Matter is composed of approximately one hundred different atoms, but who is to say that there is a limitation of atoms beyond the third dimension? Moreover, is it not possible that diverse properties exist in other dimensions which function under different law of physics and chemistry?

MORAL MATHEMATICAL INFINITY

We live in a *third-dimensional* world of length, width, and height. All of life's functions are boxed into this dimension of mortality with its continual cycles of birth and death. Nothing is permanent; not even photon light. Once the barrier that locks in a third-dimensional world can be broken, the individual can take his flight into other dimensions of infinities.

Scientists refer to the third dimension as a mathematical infinity because of the illimitable matter that exists. Who can know the grains of sand, the infinity of the stars, the innumerable plants, and the multiplicity of living creatures that exist? Yet they are all mortal.

MATHEMATICAL INFINITY OF THE SOUL

The *fourth mathematical infinity* of soul beings is evident. Those who enter the heavenly Paradise, as well as those who go into the region of the lost, are divested of their physical mortal bodies (Luke 16:22). The beings in Paradise will have their souls clothed upon with paradisiacal bodies (2 Corinthians 5:1-8). The souls in the lower region of the lost will not be clothed upon with any vestment but will be exposed to the fourth-dimensional environment and elements (Luke 16:22,23). In addition, they will be stripped of their spirit beings (EcclesiasteS 3:21; 12:7).

IMMORTAL MATHEMATICAL INFINITY

". . . This mortal must put on immortality" (I Corinthians 15:55). This chapter clarifies the method by which the believers' bodies will be "changed" from mortality into immortality at their resurrection. It indicates that their atomic particles will be altered into properties which will enable them to enter the *fifth dimension* whereby they can exist and function normally. These properties will have superluminal characteristics which defy the laws of physics and chemistry in the third dimension. Their bodies of immortality will possess a metabolism challenging all barriers of matter, time, and distance. They are the perfect gift from the Creator. Nothing can slow or restrain their instant movements.

After His resurrection, Jesus Christ possessed a body that defied the third-dimensional laws (John 20:17-31). He instantly penetrated closed structures and made sudden appearances and disappearances, negating the laws of gravity and physics as we understand them. Jesus did not require wings nor will the redeemed; they belong exclusively to angels.;[3] The redeemed will not be like angels, but like Christ.

The most unusual phenomenon was Christ's ascent and descent from earth to the Father's Throne in the third heaven in less than eight recorded days. Perhaps most of that time was spent with His Father, thus reducing to zero the time involved in transit. During those eight days in the presence of His Father, Christ Jesus certified that the plan of redemption had been sealed with His own atoning blood as God's legal Testator: ". . . For where a testament is, there must also of necessity be the death of the testator. For a testament is of force *after men are dead;* otherwise it is of no strength at all while the testator liveth . . . " (Hebrews 9:14-22).

Likewise, the believers in Christ eventually will be changed into immortality and will exceed the speed of photon light.[4] "In a moment, in the twinkling of an eye" (I Corinthians 15:52). No longer will the believer in Christ be confined within the limits of third-dimensional barriers but he will soar into the expanse of a fifth dimension of infinity.

The number of those who inhabit this dimension will be legion (2 Corinthians 5:6-9; John 17:24).

MATHEMATICAL INFINITY OF SPIRIT BEINGS

The number of spirit beings that inhabit *a sixth dimension* is legion (Revelation 5:9; 7:9; Jude 14); therefore, who can comprehend the "things" that exist in this dimension of mathematical infinity? (I Corinthians 2:9; Ephesians 4:10; 2 Corinthians 4:18; Philippians 2:10; Revelation 1:19; 5:11; 21:4; Hebrews 12:22; Psalm 68:17).

God created preternatural spirit beings, who cannot be classified scientifically as consisting of atom particles. No laws of physics and chemistry that are familiar to science apply to these created beings. They operate under their own distinctive dimensional laws which supersede third-dimensional laws.

ABSOLUTE ETERNAL INFINITY[5]

The *seventh and last dimension* brings us to the Creator Himself. Deity is Absolute Infinity—the ultimate in power and activity reaching into the infinite eternity. His omnipotence is transcending. Because He is the "Spirit" Father of all creations (John 4:24), the Creator was not created nor is He restrained to any dimensional barriers. Time is His creation, but He is timeless and dwells in timelessness. God is Eternal! One cannot equate third-dimensional laws of chemistry and physics to a higher dimension. While God in the third mortal dimension is generally invisible, in the fourth eternal dimension, however, He is both *visible* and *tangible:* "Blessed are the pure in heart: for they shall *see* God" (Matthew 5:8). God, as

Spirit entity, has properties of a greater and more noble dimension; therefore, with third-dimensional beings, He reveals His presence to their fourth-dimensional entity, the "spirit." In Romans 8:16 we read: "The Spirit itself beareth witness with our spirit, that we are the children of God."

GOD IN THE ATOM

According to Romans 1:20, the invisible God is made visible in and through His works of creation. Locked within the minute seeds of His creation are mysterious propensities which provide the traits, qualities, characteristics, and reproductive power for future generations.[6] Here the invisible Creator visually demonstrates His love of beauty, as well as His generosity, purpose and sensitivity. Redemption is the momentous demonstration of His love whereby He eventually will change our bodies from mortality to immortality.

To be able to see God and observe the magnitude of His ability to create innumerable atoms and properties, and to unite them into new "things" beyond our third-dimensional world, must be the ultimate excitement of the redeemed. The believer in Christ will behold the totality of His works and they will come to appreciate more fully the cost of redemption manifested in the surrender of His only begotten Son who came from His very heart ("bosom" John 1:18) because of His fathomless love and grace. He is not willing that any should perish but that all will come to know about the plan of redemption and have the opportunity and capacity to exercise their sovereign will toward God's manifested love (I Timothy 2:4, 2 Peter 3:9). God did His utmost from the inception to the culmination of the plan of redemption (Romans 11:33-36).

THE INNER CONNECTION

The Absolute Infinite of the Creator is subscribed to in the following texts: Deuteronomy 33:27; Isaiah 57:15; Romans 1:20; I Timothy 1:17; Hebrews 9:14. God is eternal.

The chemistry of the Absolute Spirit (Almighty God) made possible the creations which emanate from Him. To a lesser degree these are seen flowing out from the human *spirit* and involve both theory and application in the sciences and arts.

HYPOSTASIS OF THE "SPIRIT"

". . . I pray God your whole *spirit* and *soul* and *body* be preserved blameless unto the coming of our Lord Jesus Christ" (I Thessalonians 5:23).

". . . Dividing asunder of *soul* and *spirit* . . ." (Hebrews 4:12).

". . . The *spirit* shall return unto God who gave it" (Ecclesiastes 12:7).

"God is a *Spirit:* and they that worship him must worship him in *spirit* and in truth" (John 4:24).

"The *eternal* God is thy refuge . . ." (Deuteronomy 33:27).

There are various properties of the spirit, soul, and body which are distinguished by 1) names, 2) characteristics, 3) purposes, 4) destinies, 5) terminations.[7] The properties of the "spirit" are eternal because this portion of man's triune being—the voice of conscience, morality, and worship—is that which is the most allied to his Creator and can be absorbed back into Him. The "spirit" is the only constant in man's triune being just as light is the only constant in the universe. When man by his own will silences the voice of the spirit or inner being, he in essence silences the voice of the Creator and becomes a law unto himself. The hardening of the heart is the result of silencing the inner self (spirit voice) which through conscience seeks to elevate man to his better self. The Spirit of God, in cooperation with this inner self of man, seeks to return him (his soul) to the Creator and Redeemer.[8] To persistently resist such efforts, when the plan of redemption is understood, can result in one's name being removed from the book of life.

When a person is deceased, his spirit returns to God as it is an imparted entity of eternal substance. The "soul" of the *redeemed* will remain knitted with his "spirit" being because of redemption being procured; however, the "soul" of the lost will be separated eternally from his spirit being, therefore, the second death is effective upon this entity.

PROPERTIES OF THE "SOUL"

What are the components and properties of the soul?

The properties of the "soul," called by the Hebrews "the breathing spirit" or "life," contain the personality and attributes which are manifested through the functions of the body. The "soul" obviously does not die with the deceased body, but falls into a deep sleep,[9] then it is awakened in the Heavenly Paradise (see "Heavenly Paradise").

The properties of the soul are immune to the first death (that affects the body) but not necessarily to the second death (Revelation 20:6; Ezekiel 18:4,20; Matthew 10:28). The sentence imposed at the White Throne Judgment (Revelation 20:11-15; John 5:22-30) will determine the longevity of the soul under the final punishment and incarceration imposed at that time. The first death ends the molecular structure or existence of the body even though some minute seed particles will be preserved and required for a final resurrection.

Soul means "life" and that life contains the conscience;[10] will, mind, personality, emotions, instincts, and senses of the living entity. The same Biblical word, "nephesh," is used to include the animal kingdom. Only the soul (life) of mankind can be contaminated by sin (because of the impartation of "spirit"). This subjects mankind to the second death unless redeemed. Under redemption, the soul, as well as the spirit and body, will be preserved eternally because of a change into immortality by a resurrection.

Since the Creator has produced approximately 100 different atoms to constitute the "things" in the third-dimensional world, it is conceivable that He would create ultra-properties to make up the soul. The characteristics of the soul entity are analogous to the characteristics of the physical body. The personality is not lost because according to Luke 16:19-31, even while the rich man's deceased body was in the grave, his soul maintained the characteristics of intelligence, speech, thirst, taste, sight, feeling, and hearing. However, the absence of his "spirit" entity, which had returned to God, was obvious because he made no attempt to repent even though fully cognizant of his own sad state of being and the plan of redemption.

THE AMORTALITY OF THE SOUL

". . . I am tormented (not consumed) in *this* (fourth-dimensional) *flame*" (Luke 16:24). It is a *black* flame: "they shall never see light" (Psalm 49:19). In the third dimension, flame requires oxygen and is combustible as the result of a chain reaction whereby a mixture of gases is released in the form of heat and light. In the fourth dimension that flame reacts differently. It is of noncombustible material because it is influenced by unique laws of physics and chemistry, which reflect the same peculiarities as the properties of the soul, not those of the physical body.

Hell, the *temporary* abode of the rich man's soul, is not the lake of fire which is *eternal.* That will come after the White Throne Judgment. The degrees of punishment and their longevity will depend on the point at which divine justice is satisfied. The "second death" is ". . . a fire not blown (that) *shall consume* . . ." (Job 20:26). ". . . Fear him which is able to *destroy both* soul and body . . ." (Matthew 10:28). The lake of fire (second death) has one characteristic that is distinctive from hell fire; it can "consume" and "destroy." It is also a black flame: ". . . who see no light to all eternity" (Psalm 49:19, *Moffatt*).

CHEMISTRY OF THE "BODY"

The mortality of the body is without question, but its particles can be changed (I Corinthians 15:51-56) at the resurrection.

The mortality of the body and soul under the second death is evident as no judgment will pass upon any person until the soul is reunited to a resurrected physical body (I Peter 3:19,20; 4:5,6; John 5:28,29).

The Apostle Paul's appeal was that all three entities—spirit, soul, and body—be preserved. This is possible only under redemption when at the resurrection the mortal properties will be changed into immortal properties, thereby keeping all three entities intact and unified. The first death may sever the body from the spirit and soul, but under redemption in the resurrection, all three will be reunited to fulfill the prayer of Jesus Christ to His Heavenly Father that the believer will become "one as we are" and "... that they may be perfect in one... *that* they may behold my glory which thou hast given me..." (John 17:11,22-24). This was the higher objective of His prayer, not the mere uniting of all Christians into a one world Church. Scripture maintains the divisional differences of theologies as long as the prime purpose of redemption is maintained: "And God hath set some in the church... governments (plural)..." (I Corinthians 12:28). "... Till we all come in the unity of the faith, and of the knowledge of the Son of God, unto a perfect man..." (Ephesians 4:13). "....But when that which is perfect is come, then that which is in part shall be done away" (I Corinthians 13:10).

[1] When physicists determine a percentage of particles in any volume, they obviously cannot take into account any fourth-dimensional invisible properties which must remain an unknown quantity.

[2] See Illustration VI, page 608.

[3] This indicates an atmosphere exists in other dimensions (Revelation 14:6).

[4] See page 305, footnote 1.

[5] God is defined as the "Word" (Greek, Logos, John, chapter 1). This is God as a totally complete Being existing in the invisible, highest dimension as the "Spirit" entity and substance. Here we see God as the "Word," being eternal and creative. Man's words are neither. His words decrease steadily during their mission as if being consumed like fuel.

[6] See page 158.

[7] Physicists speak of entities only when they are observed. See illustration VII.

[8] The Spirit of God bears witness within man's inner self, his spirit being. If the witness were to the soul, its resident senses would *see* a vision, *hear* God's voice, *feel* God's hand, to be aroused by the senses of *taste* and *smell.* Also his mind would suddenly comprehend all truth. Faith would be cancelled. The development of the soul comes from pursuing the Word of God and having a sensitivity to the Holy Spirit.

[9] Animal souls fall "... into a *dead* sleep" (Psalm 76:6).

[10] Hardened conscience sets up a barrier *between* the spirit and soul. The status of such a person is defined as a "beast" (Romans 1:18-32; 2 Peter 2:12-22; I Timothy 4:2; Revelation 13; 14:9-11; 16:8-11).

41

The Everlasting Kingdom

All natural people who enter the Millennium, which is the first phase of the manifested Kingdom of Christ, are redeemed.

"Thy kingdom is an everlasting kingdom, and thy dominion endureth throughout all generations" (Psalm 145:13).

MYSTERY OF THE KINGDOM

"And he said unto them, Unto you it is given to know the mystery of the kingdom of God: but unto them that are without, all these things are done in parables:

"That seeing they may see, and not perceive; and hearing they may hear, and not understand; lest at any time they should be converted, and their sins should be forgiven them" (Mark 4:11,12).

"How that by revelation he made known unto me the mystery . . .

"Whereby, when ye read, ye may understand my knowledge in the mystery of Christ,

"Which in other ages was not made known unto the sons of men, as it is now revealed unto his holy apostles and prophets by the Spirit;

"That the Gentiles should be fellow heirs, and of the same body, and partakers of his promise in Christ by the gospel" (Ephesians 3:3-6).

The "Kingdom of God" and the "Kingdom of heaven" are interchangeable terms. The Gospel of Matthew addresses this kingdom as of "heaven," whereas Mark and Luke refer to it as of "God." Observe these comparisons: Matthew 4:17 to Mark 1:15; Matthew 8:11 to Luke 13:28; Matthew 13:24 to Luke 13:18,19; Matthew 13:33 to Luke 13:20,21; Matthew 19:14 to Mark 10:14.

This kingdom involves all who make any claim of association with "God" and whose destiny is "heaven." The Church is the "treasure hid in the field" and the "pearl of great price" (Matthew 13:44-46). The Lord said, "... the field is the world ..." (verse 38).

The kingdom consists of a conglomerate of people who claim to be its citizens: first, the redeemed (2 Corinthians 5:9-11; I Corinthians 3:9-15) who are in the "Church" and are called "... the body of Christ" (I Corinthians 12:13,27; Colossians 1:18; Ephesians 3:1-10). The word "Church" comes from the Greek word *ekklestia* meaning " an assembly of called out ones." It appears one hundred fifteen times in the New Testament, but nowhere in the Old Testament.

Second, it includes the redeemed who are unsuccessful in qualifying to be the wife of Christ. They are permanently stripped of their wedding garments and sent to the "outer darkness" for discipline, thereby missing the marriage of the Lamb of God (I Corinthians 3:15; I Peter 4:17,18; I John 2:28; Revelation 3:4; 16:15; Matthew 22:1-14).

Third, it involves guests from Paradise (such as little children) who are clothed in garments of salvation, but they cannot be the wife of Christ because that position is restricted to the redeemed. However, their names are still retained in the book of life because their lives were cut off prematurely (or raptured) before they could make decisions for Christ. In the New Heaven they will eventually be given the opportunity to decide for Christ. When they receive their immortal bodies after death is destroyed, they will be transported with all other redeemed into the Heavenly Father's eternal Manifested Kingdom where they will obtain positions of great eminence (Revelation 20:11-

15; Matthew 10:14; 13:43,49; Luke 18:16; Ephesians 5:5; Daniel 12:3).

Fourth, there will be claimants to the kingdom who are described in the Lord's parables as evil, wicked, and children of Satan. These will be cast out as intruders, impostors, and archenemies within the kingdom. (Matthew 13:38-42, 17-50; John 8:44).

PEARL OF GREAT PRICE

Within the Kingdom of God (heaven) is hidden the pearl of great price, the "Church." As long as it is in the world it will suffer from ". . . the gates of hell," but hell will not ". . . prevail against it." In the end the Church will triumph (Matthew 16:18). The "end" of the dispensation of the Church is predicted in Matthew 13:43 and I Corinthians 15:24.

The Church *per se* exists during the "mystery" of the kingdom which is between the two comings of Christ to earth. Then comes the Manifested Kingdom, in accordance with the Lord's prayer: ". . . Thy will be done on earth as it is in heaven" (Matthew 6:10). This is the period of the ". . . manifestation of the sons of God" (Romans 8:19,20).

The Church will be limited to reaching "all nations (as) a witness" since its period of evangelization is drawing to a close (Matthew 24:14). However, during the Manifested Kingdom, "every creature" will be reached with the Gospel of Jesus Christ (Mark 16:15): ". . . all the ends of the earth shall all know me, from the least of them unto the greatest of them, saith the LORD: for I will forgive their iniquity, and I will remember their sin no more" (Jeremiah 31:34). These evangelistic efforts will be accomplished by Israel and the Royal Redeemed as "priests" (Zechariah 8:23; Revelation 1:6).

LAUNCHING OF THE CHURCH AGE

Jesus launched His Church in a baptism of divine fire at Pentecost (Acts 1:5; 2:1-4) as foretold in Luke 12:49 and John 14:20; 16:4,23,26. Notice the frequency of the words "in that day." Jesus said, ". . . I will build (future) my church . . ." (Matthew 16:18). Pentecost began the dispensation of the Church Age ("mystery kingdom") in which

Christ is enthroned within the believers. It is that treasure hid in the field which some day Christ will return to earth to retrieve, uncover, and reveal (manifest) to the whole world (Matthew 13:44). Then all true members within the Church will emerge as "kings" to reign and rule with Christ over the earth.

Jesus said, "I will not leave you comfortless, *I will come to you*. . . *At that day* (Pentecost) ye shall know that I am in the Father, and ye in me, and I in you" (John 14:18,20). Compare John 16:25,26). He also said, "Verily I say unto you, There are some now standing here, which shall *not taste of death, till they see* the Son of man coming in his kingdom (the mystery kingdom)" (Matthew 16:18). This Scripture must be distinguished from those which refer to the visible return of Christ to the earth "in a cloud with power and great glory. And he shall send his angels with a great sound of a trumpet, and they shall gather together his elect from the four winds, from one end of heaven to the other" (Matthew 24:30,31; Luke 17:24; Mark 13:27; Luke 21:26). At Pentecost there was the appearance of "fire" but not of a "cloud." The former refers to the coming "mystery" (hidden) kingdom while the latter pertains to the "*manifestation*" (revelation) of that kingdom (Romans 8:19-23) when the Church will become spotless and pure.

The final days of the Church Age are shown in Luke 21:23: "Verily I say unto you, This generation (Church Age) shall not pass away *till all be fulfilled*" (Compare Daniel 11:32-35; Revelation 6:9-11; 14:14-16; I Corinthians 15:49-54).

IMPORTANCE OF WORKS

There are two major judgments at which Jesus Christ will preside (John 5:22). The first is the Judgment Seat of Christ after the rapture of the pre-Tribulation *and* Tribulation saints (I Peter 4:17) when everyone in the kingdom of God (Heaven) must make an appearance (2 Corinthians 5:10). The second and final tribunal is the White Throne Judgment at the end of the Millennium when all who were deprived of becoming the corporate wife of Christ must make an appearance (Revelation 20:11-15).

At both judgments works play an important part: first, in determining the degrees of honor and position one will have in the Manifested Kingdom of Christ; second in determining the degree of punishment under the *second death* of those whose names have been removed from the book of life. Those whose names have been retained will have the opportunity of making their decision for Christ in the Manifested extended Kingdom of the New Heaven (Revelation 20:15; I Timothy 2:4; 2 Peter 3:9).

All natural people who enter the Millenium, which is the first phase of the Manifested Kingdom of Christ, are redeemed. They will have been converted to Christ subsequent to the final Rapture of the Great Tribulation saints. However, their posterity will be required to make decisions regarding Christ. Even though Satan is not present, the people's fallen Adamic nature will in time cause them to sin (Revelation 19:21; 21:27; 22:14,15; Isaiah 2:4; 11:1-4; 51:5; Micah 4:1-3). While there will be a perfect government, there will not be perfect people.

The second phase of Christ's Manifested Kingdom will be expanded into the New Heaven. The same principles will prevail here as in the first phase (Revelation 21:24; 22:14,15). There will be population growth and all will be required to be redeemed or their lives will be shortened (Isaiah 65:20).

GOD'S DAUGHTER-IN-LAW

It is imperative to be chosen as the Wife of Christ (Matthew 24:14) in order to qualify for positions of high honor (Revelation 19:7) such as "... the manifestation of the sons of God" (Romans 8:19).

"... Therefore every scribe which is instructed into the kingdom of heaven is like a man that is an householder, which bringeth forth out of his treasure things new and old" (Matthew 13:52).

"... They that were ready went in with him to the marriage: and the door was shut" (Matthew 25:10).

"... To an inheritance incorruptible, and undefiled, and that fadeth not away, reserved in heaven for you ... ready to be revealed in the last time" (I Peter 1:3-5).

Since the Royal Redeemed will be the Wife of the Son of God,[1] she has to be the only *daughter-in-law* of God for He has only one Son, Jesus Christ. The Church is to be "begotten" in the same sense that Jesus Christ is the first "begotten" (I Peter 1:3; James 1:18)—from mortality into immortality (Revelation 1:5; Romans 8:29; Colossians 1:18; Hebrews 12:23). "Husbands, love your wives, even as Christ also loved the church . . ." (Ephesians 5:25).

A scientist recently remarked how great it was for God to send His only Son, and how wonderful it would be for Him to send His only daughter. Hardly did the scientist realize those were prophetic words! When Jesus Christ returns to earth with His wife, the only daughter-in-law of God, the Millennial reign of peace on earth will begin!

[1] A compound or composite unity is like the Godhead, or, husband and wife as one (John 17:11,22,23; Matthew 19:5,6).

SECTION 2

THE FUTURE

42

Prologue to Bible Prophecy

Sincerity in itself proves nothing if not supported by a right premise and proper logic.

Bible prophecy is a clear and precise revelation and manifestation of the will of God. It falls into two categories—fixed and fluid. Fixed prophecies are unalterable objectives which God has established (Hebrews 10:37). Fluid prophecies make allowances for change as a result of the proper attitude toward God (Jonah 3:10; 2 Kings 20:1-6).

Contemporary events formed the basis for the prophets' predictions of both the myopic and hyperopic views of prophecies. According to students of prophecy only one-tenth of the Biblical predictions have been fulfilled. Most of these occurred during Christ's first coming.[1] The great preponderance of the remaining prophecies began to be fulfilled upon Israel's recent return to the Holy Land and they will continue to be realized into the Millennium and beyond.

The value of Biblical prophecies was clearly demonstrated during the earthly ministry of Jesus Christ. Had the people known the prophecies concerning His first arrival on earth they would not have missed their true Messiah. Jesus verbally chastised them for this unwarranted incredulity (Luke 24:25-49).

Prophetic truth is designed to light up our pathways: "We have also a more sure word of prophecy (God's Word); whereunto ye do well that ye take heed, as unto a light that shineth in a dark place, until the day dawn, and the day star arise in your hearts" (2 Peter 1:19). Tragedy awaits one who does not avail himself of proper knowledge because confusion and spiritual lethargy are like blinders.

The most reasonable logic in the world is meaningless if it is not fortified on the right premise—God's Word! And the right premise will fail miserably in arriving at a clear perception of truth if not supported by correct logic. A good approach is the use of syllogism in which two or more premises are combined with logic to form a conclusion (2 Corinthians 13:1; Acts 15:15; Luke 24:27).

Thus, as prophecies approach the time of fulfillment, they become more clearly understood, at which point anachronistic interpretations must be discarded in the light of the clearer picture. The difficulty with the theologians in Jesus' day was that in the face of truth they would not admit or alter their prejudices (John 9:41). Jesus compared it to the fixations of little children (Matthew 11:16; Luke 7:32). Prejudice can be a dangerous shackle keeping one from the real truth.

". . . Knowledge shall be increased" (Daniel 12:4). This is not a prophecy of modern inventions and advanced academic excellence. It refers exclusively to God's people having better knowledge and comprehension of prophetic Scriptures. This is borne out by the fact that last day prophecies were sealed to Daniel. That is why Daniel said, ". . . .but *I understood not:* then said I, O Lord, what shall be the end of those things? And he said, Go thy way, Daniel: for the words are closed up and *sealed till the time of the end* (verses 8,9). Daniel was then informed that at the time "the wise shall understand" (verse 10).

It would be ideal if laymen could be as knowledgable of God's Written Word as sports announcers are of baseball, and if theologians were as informed as the team managers are regarding their players.

In the *Prophetic Ensign,* Arthur E. Blomfield, the editor, wrote as follows: "The present sad state of prophetic knowledge is due largely to the confusion caused by the unscriptural teaching of the 'authorities.'

"Anyone who starts out with . . . misinformation is bound to wind up in confusion. He will be disgusted with the whole thing. He will lose interest in prophecy . . . When a person approaches a passage of Scripture having in mind an erroneous idea concerning it, he will not be able to grasp its meaning. Either you must have no preconceived notions or a sound Scriptural background if you are to advance in knowledge by reading any Bible prophecy. Any new student who approaches the prophetic Scriptures with the background furnished by present day recognized leaders will be handicapped almost to the point of failure. They all echo each other parrot-like. There has been little independent research evidenced in many years."

The prophetic Word of God was designed to comfort and to prepare its readers for oncoming events, not to fill them with apprehension (Romans 15:4). Prophecy is both foretelling and forth-telling, and is a dependable and constant illumination during dark, troubled times (2 Peter 1:19). Recorded history frequently serves as an exemplification and admonition in the prophetic picture.

Einstein stated that "the Dear Lord may well be unorthodox, but His predictions are verifiable."

The Apostle Paul advocates that everyone study the Holy Scriptures so as to " . . . rightly divide the word of truth" (2 Timothy 2:15). To casually read it as one does fiction or history is equivalent to ambulating over diamonds with one's eyes closed. God's Word is too prolific for such surface perusal. The Spirit of God, Who is the inspiration of the Written Word, will guide the earnest seeker of truth (John 14:26; 16:15). Interpretations conflicting with any portion of the Scriptures must be discarded and brought into harmony with the whole. God is the progenitor of harmony and His Word is as infinite as God is infinite. The depths of His Word can never be fully explored.

[1] Prophecies comprise at least one-third of the Bible, including the largest part of the teachings of Jesus Christ.

43

Importance of Bible Prophecy

There would be no Bible prophecies if God did not possess perfect foreknowledge.

There would be no Bible prophecies if God did not possess perfect foreknowledge. He must not only foreknow future events, but He must have the right and the power to remove and replace persons with those He knows will carry forth His divine will and purpose. If not, events would be out of God's control and the end result would be disastrous to His divine decrees.

When Bible prophecies affect individuals directly, God does not change the wills of persons. If He must alter the course of events, He simply substitutes persons in authority (Psalm 75:6,7). When this is done, prophecy is no longer just foreknowledge but it is also predestination because the course of events has been altered to fit into God's ultimate plans.

BOOK OF REMEMBRANCE

"Then they that feared the LORD spake often one to another: and the LORD hearkened, and heard it, and a book of remembrance was written before him for them that feared the LORD, and that thought upon his name.

"And they shall be mine, saith the LORD of hosts, in that day when I make up my jewels (to unite in marriage to Christ); and I will spare them, as a man spareth his own son thnt serveth him.

"Then shall ye return (to earth) and discern between the righteous and the wicked (judgment of the nations), between him that serveth God and him that serveth him not" (Malachi 3:16-18).

This prophecy refers to the time of Christ's return to earth (verses 1,2). A record is kept concerning the manner in which those in the book of life, including the Tribulation saints, have been treated. It also contains the reasons that names are removed or retained, and the information will be divulged at the time of the final tribunal when the books are opened.

Little children, adults with serious physical and mental impairments, and all others whose names are still in the book of life, will all appear at the White Throne Judgment (Revelation 20:12). In contrast, their abusers may find that their own names have been stricken from the book of life: "*Take heed that ye despise not one of these little ones;* for I say unto you, That in heaven their angels do always behold the face of my Father which is in heaven" (Matthew 18:10,16). "It were better for him that a millstone were hanged about his neck, and he cast into the sea, than that he should offend one of these little ones" (Luke 17:2). These innocent ones are counted as the Lord's "jewels" in the book of life. The angels in heaven are assigned to receive them into the Heavenly Paradise in the event of death.

GHOSTS AND THE GUILTY

Innocent ones who are physically mature, but mentally incapacitated, are retained in the book of remembrance and in the book of life even though unredeemed. When they are capable of making their own decisions for Christ, they will have that opportunity in the New Heaven. Their names remain in the book of life, unless with full comprehension, they later reject Jesus Christ as their Redeemer and negatively determine their own destiny. The book of life includes all who have the right to life, not eternal life.

Perhaps God has allowed the existence of physically and emotionally disturbed persons in order to impress us as to the "works" (actions) for which one must be accountable at the future judgment day. Once the public becomes hardened to the idea of a life being dispensable it is easier to move from abortion toward other stages, such as the elimination of the very ill, the aged, and the nonproductive persons in society.

The guilty can rectify the situation by taking a simple course of action. They can repent and change the direction of their lives. In this way they can forward their sins to the Heavenly Father Who will grant forgiveness. If they do not, these sins will trail them like ghosts to haunt them at their time of judgment: "Some men's sins are open before hand, going *before* to judgment: and some men they *follow after*" (I Timothy 5:24). "Whosoever shall fall upon the stone (Jesus Christ as Saviour) shall be broken (repentant); but on whomsoever it shall fall (when He comes as Judge), it will grind him to powder (punishment)" (Luke 20:18).

The right of the unborn to exist, which is denied them by abortion, must likewise be accounted for by those responsible. The accused must appear at either of the two tribunals, otherwise how can it be a fair judgment? The book of remembrance certainly will be present. Redemption reaches to the unborn within the womb because their names are in the book of life (See Chapter 25).

There are no such entities as *guardian angels* because this contradicts all the facts of life and appears nowhere in Scripture. In the event of tragedies, did the protective guardian angels default? Are some more protective than others? If there were guardian angels there would be no casualties befalling the innocent.[1]

AT THE CROSSROADS

"He that being often reproved hardeneth his neck, shall suddenly be destroyed, and that without remedy" (Proverbs 29:1).

"For the scripture saith unto Pharaoh, Even for this same purpose have I raised thee up (put him on the throne), that I might shew my power in thee, and that my name might be declared throughout all the earth" (Romans 9:17).

Pharaoh of Egypt was not amenable to the will of God, thus he is called "a vessel of wrath fitted for destruction" (Romans 9:17-23). Therefore, without doing violence to Pharaoh's will, God hardened his heart further so that He might by His spectacular miracles strengthen the faith of the Israelites.

The Apostle Paul, however, was a case in reverse. While his wrath was pitted against the Apostolic Christians, God knew it was misguided zeal based on ignorance (I Timothy 1:13). When he encountered God he repented, in contrast to Pharaoh who hardened his heart. Paul said, "For even *before I was born God had chosen me* to be his, and called me—what kindness and grace—to reveal his Son within me so that I could go to the Gentiles and show them the Good News about Jesus" (Galatians 1:15, *The Living Bible*).

Some theologians believe Genesis 49:27 is a prophecy of the Apostle Paul: "Benjamin (compare Romans 11:1) shall ravin as a wolf (Acts 9:1,2): in the morning (early life) he shall devour the prey, and at night (later in life) he shall divide the spoil (minister)."

"Let the wicked forsake his way, and the unrighteous man his thoughts: and let him return unto the LORD, and he will have mercy upon him; and our God, for he will abundantly pardon" (Isaiah 55:7).

PREDESTINATION AND THE INDIVIDUAL

Predestination is a Biblical word (Romans 8:29,30; Ephesians 1:5,11). It applies only to the *redeemed and their future inheritance:* "According as he hath chosen us in him before the foundation of the world, that we *should be* holy and without blame before him in love: Having predestinated *us unto the adoption* of children by Jesus Christ *to himself,* according to the good pleasure of *his will"* (Ephesians 1:4,5). In salvation it is the determination of *our will,* but in the future inheritance it is *God's will* for us.

The word "adoption" is not the same as being "born again." The word refers to the time when those already born again will receive their bodies of immortality, the first stage toward the future inheritance: "And not only they, but ourselves also, which *have* the firstfruits of the Spirit (born again), even we ourselves groan within ourselves, *waiting for the adoption,* to wit, the redemption of our body" (Romans 8:23).

Note the words, "hath *chosen* us," which go beyond just being "called" into salvation. The words "should be" and "to himself" are clearly prophetic.

PREDESTINATION AND IMMORTALITY

"For whom he did foreknow he *also* did predestinate *to be conformed* (not transformed as in the new birth) to the image of his Son, that *he might be* the firstborn among many brethren (Christ was the firstborn into immortality). Moreover, whom he did *predestinate* them he also *called* (first in the process, compare Ephesians 1:4 "chosen"); and whom he called, them he also *justified* (second in the process): and whom he *justified,* them he also *glorified* (last in the process)" (Romans 8:29,30. Compare I Corinthians 15:45; Acts 2:23).

Predestination is not the choice we make, but what we get after we make the choice. Immortality is the ultimate goal under predestination for it qualifies and brings us into our eternal inheritance. There is no predestination for the unconverted who in God's foreknowledge will not accept the new birth. An inheritance is only possible to one who exists as a child of God.

Many people worry about the future; God never does. The difference is that God knows the end. To know the end would quiet our fears. If the end is good and we know the end, there is no cause for concern.

[1]Attribute any divine protection to the Holy Spirit, the Angel of the Lord.

44

Foundation for Biblical Prophecies

Christ was crucified on the very day Israel was preparing its lambs for sacrifice as an atonement for sin.

When the prophets of old spoke as the oracles of God, they used that which was contemporary to their time and that which was familiar to them in describing prophetic events. Birds and clouds, for example, served to describe modern aircraft. In Isaiah 31:5 the prophet foretold that as "birds flying" so would Jerusalem be delivered. This occurred in World War I when the allies delivered the city from the Turks, making it possible under a British mandate for the Israelis to begin their trek back to the Holy Land.

When John received the Apocalyptic Vision, God selected seven churches in Asia to represent the dispensational periods of the Church. Each local church depicted a specific period which had certain characteristics best representing the Church of that time. Ephesus represented the Apostolic Age; Smyrna, the first three hundred years of persecution; Pergamos, Imperialism from Constantine to the Papacy; Thyatira, the Papal Period from the 6th to the 12th centuries; Sardis, the Dark Ages from the 12th to the 16th centuries; Philadelphia, the Reformation during the 16th century to the pre-Tribulation Rapture; Laodicea, the Tribulation Church under the reign of Antichrist. The Church Age will end in a final Rapture of all Tribulation saints, called the "last" trumpet. The pre-Tribulation Rapture also has a trumpet sound (I Thessalonians 4:16).

DESIGNED TO AVOID TRAGEDIES

The tragedy of not knowing the prophetic Scriptures was demonstrated vividly at the time of Christ's first coming to earth. Most Bible prophecies zero in upon two generations covering two major events—the first and the second coming of Christ to earth.

The first coming never should have been missed by Israel, especially by their own theological hierarchy, because the prophets were very precise and explicit in their predictions. All that was required was a careful examination of the declaration of the prophets. Israel was anticipating its Messiah. Why then were they so unprepared for His coming? What went wrong? These same questions probably will be raised again after the translation of the redeemed to meet Christ in the air. The generation which will witness the disappearance of the redeemed from the earth will also witness Christ's return to earth (Luke 21:32; Hebrews 9:28).

WHEN LITTLE IS BIG

When God premeditated the sending of His Son Christ Jesus to become mankind's Redeemer, He drafted His plan very meticulously so that the Saviour would be easily identified and located. The prophet Micah (5:2) predicted the exact community in which He would be born—Bethlehem! According to Matthew 2:5, the chief priests and scribes knew their Messiah would be born in this small place, yet they missed Him!

The prophet Isaiah (7:14) predicted the Saviour would be born of a virgin, therefore His coming would be a miraculous birth. A brief investigation would have confirmed this. Any birth outside of wedlock was regarded as immoral under Hebrew law so Christ conceivably could not come under this shadow.

RED LETTER DAY

Daniel, Chapter 9, verses 25 and 26, clearly identifies the generation in which the Messiah (Christ) would come to be crucified. The seventy weeks of years (490 years) could easily have been calculated as the Hebrews kept accurate records. The reckoning began with the command to rebuild the walls of Jerusalem (certainly a most important event) which was done in the twentieth year of the reign of Artaxerxes under the direction of Neheminh in 458 B.C. A simple computation of the sixty-nine weeks of years (483 years) from this point brings one to the year A.D. 25, the time when the Messiah would be crucified.

According to historians, and assuming they are correct, Christ was born circa 8 or 7 B.C., which would make Him thirty-three years of age when He was to be crucified.[1] He was identified with lambs and young bullocks used in sacrifices, which symbolized Him as a young man to be killed on the day of Atonement.

King David predicted that the Messiah would be tortured by His hands and feet being pierced (Psalm 22:16), a custom not practiced until the advent of the Roman Empire. His garments would be disposed of by casting of lots (Psalm 22:18); and His betrayal would be with thirty pieces of silver (Psalm 41:9; Zechariah 11:12). There were numerous other predictions of equal clarity; yet in spite of these, the people failed to recognize the time of His arrival; and even after its fulfillment, that he was the true Messiah.

Jesus Christ at the closing days of His life on earth chided their unjustifiable blindness: "Then he said unto them, O fools, and slow of heart to believe all that the prophets have spoken. Ought not Christ to have suffered these things, and to enter into his glory? And beginning at Moses and all the prophets, he expounded unto them in all the scriptures the things concerning himself" (Luke 24:25-27).

SHADOWS FROM THE CROSS

One reason God instructed Israel to commemorate the various feasts and offerings annually was to continue to impress upon Israel their significance. For hundreds of years the nation observed them, but when the big event which was foretold came to pass, their rehearsals accomplished practically nothing in preparing them for the fulfillments.

Christ was crucified on the very day that Israel was preparing its lambs for sacrifice as an atonement for sin.[2] The practice was to place a long pole lengthwise through the center of the lamb and another one through the side near the heart. Then as they roasted their sacrifice they saw the lamb as upon the shape of a cross. Even the resurrection of Jesus Christ failed to awaken their conscience and understanding (Matthew 27:52,53), despite the fact that significant witnesses observed many raised from their graves and walking the streets of Jerusalem.

In our generation of God's fulfillments, are we aware of the big event which is to occur? When God had prophecies recorded about any major event, including the Lord's second advent, He was very clear and precise. But why, again, will so many in our generation fail to observe, comprehend, and prepare themselves, in spite of an avalanche of prophecies being fulfilled? Look around you and you can see why.

[1] Jesus Christ was two years old at the time Herod planned the massacre of all infants two years and under (Matthew 2:16). Herod died in 4 B.C. so Christ had to be born before that date. Dionysus Exiguus in the Julian calendar made a mistake in placing the birth of Christ in the year 753 A.U.C. (from the founding of Rome). It should have been 747. See also page 274.

[2] No fewer than 250,000 lambs were slain at the passover in Egypt by Israel; yet all are spoken of as "it," representing a type of one Lamb, the Lamb of God (Exodus 12:6).

45
Escaping the Holocaust

The Israelis in Egypt will leave there the second time when Christ Jesus returns to earth.

God always provides an escape route for those who will observe what He is saying and then act accordingly. This is evident in our Lord's prediction of the tragedy that was to befall the city of Jerusalem and its temple. The disciples asked for guidance and He gave them a precise answer as to how they might avoid the coming calamity.

"And Jesus went out, and departed from the temple: and his disciples came to him for to shew him the buildings of the temple. And Jesus said unto them, See ye not all these things? verily I say unto you, There shall not be left here one stone upon another, that shall not be thrown down" (Matthew 24:1,2).

Jesus then described exactly how the city would be captured and destroyed (Luke 19:41-44). He enlightened His disciples as to how they could avoid being caught in that web of destruction (Luke 21:20-24). He assured them of His faith in their loyalty and desire to follow instructions to the letter: "But there shall not an hair of your head perish" (verse 18).

We believe that the same gracious and merciful God has given our generation equally specific and informative instructions regarding both the Rapture and His return to earth. Thus, we too, can escape the coming tragic events scheduled for this sinful generation: "...and unto them that look for him shall he appear the second time without sin unto salvation (redemption of the body)" (Hebrews 9:28).

TIMBER!

John the Baptist said, "...the axe is laid unto the root of the trees..." (Matthew 3:10).

Josephus, the renowned Jewish historian, was an eyewitness to the devastation of Jerusalem in 70 A.D. when the Roman Emperor Titus took the city and razed it. He reported that the Jews, after crucifying Jesus Christ, grew impatient for the promised Messiah. So they took it upon themselves to break into open rebellion against the Roman government in A.D. 66.

On the day of passover (again the same celebration as on the day Jesus Christ was crucified) in A.D. 70, Titus, the Roman Emperor, with his army of 30 thousand, laid siege to the city. It was crowded with 600 thousand visitors, and the Jewish army numbered 24 thousand. There was infighting among the Jews within the walls and starvation was rampant. After five months the walls were battered down, then the temple and city were burned and destroyed. Over one million people were killed; human blood flowed deeply in the streets. Josephus, one of the 95 thousand captives, said the city looked as if it never had been inhabited. According to Busebius, the Christians escaped by fleeing beforehand to Pella, as the result of a prophetic warning.

Jesus had stated that "not one stone would be left upon another."[1] This came to pass because of the search for the gold and precious wealth hidden in the walls.

THE GOLDEN ARK OF GOD[2]

With the destruction of the temple many precious records were lost, which makes it impossible for Israel to know from which tribe they originated and the details concerning the sacrificial procedures. The

Scriptures predict that during our generation sufficient records will be recovered by archaeologists to enable Israel to obtain this information (Daniel 11:21; 12:11; Revelation 7:4-8). It will signal that the pendulum is swinging back to Israel in preparation for the coming Millennium.

The destruction of the procedures for sacrifices under the Mosaic Law in 70 A.D., and the dispersion that followed, was God's way of closing the final chapter on Old Testament sacrifices. Christ's final sacrifice of Himself for redemption ended all sacrifices of animals which were only for the purpose of forecasting the coming Messiah as the Redeemer. The lost genealogical records opened the door for all Gentiles to be redeemed on an equal basis with Israel through Christ.

Historians reveal that before the Babylonian invasion of Jerusalem, a Levite named Joshua, together with several priests, removed the ark and the temple furnishings and transported them ninety-six and one-half miles to Mount Nebo where Moses died. Here everything was hidden in a cave.

The ark symbolizes God's presence and power and is a replica of the original ark in heaven (Hebrews 8:2-5; Revelation 13:6). Israel possessed the ark for 400 years but it has been lost for approximately 2,600 years. Its anticipated recovery in the last days will symbolize the second promised marvelous deliverance of Israel from the land of Egypt.

ISRAEL'S SECOND FLIGHT FROM EGYPT

"And it shall come to pass in that day, that the Lord shall set his hand again the *second time* to recover the remnant of his people, which shall be left, from Assyria, and from Egypt . . . And the Lord shall utterly destroy the tongue of the Egyptian sea; and with his mighty wind shall he shake his hand over the river, and shall smite it in the seven streams, and make men to go over dryshod" (Isaiah 11:11,15).

"And it shall come to pass in that day, that the Lord shall beat off from the channel of the rivers unto the stream of Egypt, and ye shall be gathered one by one, O ye children of Israel" (Isaiah 27:12).

Israel's flight into Egypt from their Promised land will be as a result of the persecution by Antichrist. They also will flee into other parts of the world where they will find haven. Those fleeing into Egypt will take with them the Ark of the Covenant. There they will be concentrated into five cities, just outside the reach of Antichrist, for protection from his wrath (Isaiah 19:16-25[3]; Revelation 12:14-17; Daniel 12:1; Matthew 24:15-22). Antichrist will attempt to conquer all of Egypt but because of a naval power in the Arabian Sea and the advancing armies of the kings of the East, he will abandon his plans in order to prepare his defenses better. This will culminate in the battle of Armageddon.

The Israelis in Egypt will leave there the second time when Christ Jesus returns to earth, brings an end to Armageddon, and calls for His people to return to their land of promise: "I will hiss for them, and gather them; for I have redeemed them: and they shall increase as they have increased" (Zechariah 10:8).

At this time all of Israel will come to know Jesus Christ as their Messiah and Redeemer: "Who hath heard such a thing? who hath seen such things? Shall the earth bring forth in one day? or shall a nation be born at once? . . . Rejoice ye with Jerusalem, and be glad with her . . . Behold, I will extend peace to her like a river, and the glory of the Gentiles like a flowing stream . . . And when ye shall see this, your heart shall rejoice . . . I will gather all nations and tongues; and they shall come, and see my glory . . . to the isles afar off . . . they shall declare my glory among the Gentiles . . . " (Isniah 66:8-20).

"And so all Israel shall be saved: as it is written, There shall come out of Sion the Deliverer, and shall turn away ungodliness from Jacob" (Romans 11:26). The entire nation will be born again and restored to their land (Isaiah 11:11,16; 19:18-25; Joshua 3:5,6,15-17).

[1] Some stones measured 50 x 24 x 16 feet.

[2] The Ark of the Covenant (Testimony) will be rediscovered (2 Maccabees 2:4-9; Daniel 11:28-32; Isaiah 19:19). The ark was made of acacia wood overlaid with gold both in the interior and on the exterior, which would account for its preservation. Its dimensions were 2-½ x 2-½ x 4-½ feet, and it contained the Ten Commandments, manna, and Aaron's rod that budded.

[3] Some theologians believe this verse refers to the pyramid. The words "altar" and "pillar" would discredit this view (Exodus 40:29,36,38; Joshua 4:6,7; Numbers 4:5, 17:7,8).

46

The Cradle Throne[1]

The intent of Bible prophecy is to strengthen faith.

When God premiered His Son's arrival on heaven's stage in the blue empyrean above the shepherd's field, He must have been exceedingly disappointed in the poor attendance. Only a few shepherds were present. The vast, open grass amphitheater should have been jammed with expectant and jubilant well-wishers; but what a limited assembly there was to welcome the Messiah!

Could Israel have been excused for this poor response in greeting its Messiah? Hardly! How did the shepherds respond to this important announcement given to them by the angels? After paying homage to the Christ Child, "... they made known abroad the saying which was told them concerning this child" (Luke 2:17). Why was there not a long line rushing to receive the Messiah? Because the information they received was passed off lightly as just a mere hallucination of simple shepherds. They could, at least, have made an intensive search of the Scriptures to confirm or discredit the report. After all, they were in expectation of the Messiah's arrival, and therefore not in ignorance regarding this matter.

What kind of reaction did the shepherds receive? "And all they that heard it wondered at those things, which were told them by the shepherds" (Luke 2:18). That's it. No attempts were made to pursue the matter further. The Messiah had finally come, but there was no joy or excitement regarding his arrival. Surely there must be a lesson here for us in our present generation as we approach the impending second return of Christ to earth.

A PALATIAL HOME

Two years later wise men came 500 miles from the East to locate the predicted King of Israel. Since they logically concluded that He would be born in Herod's palace, they went there and queried, "Where is he that is born King of the Jews? For we have seen his star in the east, and are come to worship him" (Matthew 2:2).

When Herod approached the Hebrew scholars, why weren't their suspicions sufficiently aroused to inquire and probe more deeply into these events? They were familiar with the prophetic scriptures. These intellectuals could have consulted with the wise men and followed them to the place where Christ now resided. Certainly the appearance of this angel-star leading the way would have had an impressive effect upon them. Instead, they regarded the matter as seemingly insignificant. For what were they waiting to convince them that the Messiah had indeed come?

As we approach the second coming of Christ to earth, a multitude of prophecies must be reappraised by our present generation if we are to discern what God is trying to communicate to us.

FOR WHOM THE BELLS TOLL

How was God going to arouse those spiritual sleepyheads from their indifference? He had to strike a blow so hard and so loud that the entire nation of Israel would reel under its reverberations. The sound of this gong came in a form which brought the entire nation into deep mourning as prophesied by Jeremiah 31:15: ". . . a voice was heard in Ramah, lamentation, and bitter weeping; Rachel weeping for her children, and refused to be comforted for her children, because they were not."

All of the Hebrew baby boys, two years old and under, were ordered slain by Herod. If King Herod was so alarmed, why wasn't there concern among the spiritual hierarchy? Even this resounding prophecy went unheeded in what must have been most trying on God's part as He attempted to impress upon them the importance of the hour.

OUR GENERATION IN REVIEW

Unfortunately we are seeing a parallel in our twentieth century generation which not only will witness the fulfillment of many prophecies, but also will behold the actual return of Christ to earth. The rapid realization of one prophecy after another should bring the Church to a high degree of expectation and readiness.

Consider these few prophecies which have already been fulfilled: Israel's return to its homeland as a sovereign nation after 2,600 years; its occupation of the ancient city of Jerusalem; the growing hostility of its neighbors; the resurgence of the ancient Roman Empire in its embryonic state, as the European Economic Community (EEC-Common Market); the possibility of a nuclear holocaust; the rise of international terrorism under radical Arabs (Psalm 83:1-18); the Gospel reaching unto all nations of the world; and many others.

Over a period of five thousand years, predictions have been made by various archaeologists that 2000 A.D. will be a critical year. Some leading scientists today are prone to agree with this prediction because of the impact of wars, population explosion, famine, radioactive wastes, global pollution, epidemics triggered by leaks from viral or bacterial weapons, and the possibility of nuclear war. The year 2000 will also see the planets Mercury, Mars, Jupiter, and Saturn (four out of five of earth's neighbors) in closer alignment with the earth than has ever been previously recorded.

THE GUIDING LIGHT

Bible prophecy was designed and given by God to enlighten and bring into focus His plan for the believer. Its intent is not to frighten or cast shadows, but to inspire hope in the midst of darkness: "We have also a more sure word of *prophecy;* whereunto ye do well that ye take heed, as unto a *light that shineth in a dark place, until the day dawn,* and the day star arise in your hearts" (2 Peter 1:19). When it is darkest, the prophetic light shines the brightest as a beacon of hope and as a way out of the dilemma.

The "more sure word of prophecy" is "Scripture." The transfiguration of Christ to which Peter alludes was but a premiere of a major prophetic event, ". . . the appearing of Jesus Christ" (I Peter 1:7). Before the arrival of that event, however, there will be periods of dark history. The brightness of the Lord's transfiguration was ephemeral and observed by only a few—Peter, James, and John. The Written Word of God is a steady light available unto all: "For the *prophecy came* not in old time by the will of man: but holy men spake as they were moved by the Holy Spirit" (2 Peter 1:21).

WHAT YOU BELIEVE MAKES A DIFFERENCE

Peter proceeded to draw attention to false heresies as being "damnable" and the teachers as having "pernicious ways. . . making merchandise" of the people by using "feigned words" to deceive. "Of whom," Peter says, "the way of truth (is) evil spoken of. . . " (2 Peter 1-3).

Are different views of prophecy of any serious consequence since they do not alter what is destined to happen? They certainly are. One's beliefs affect the individual. Faith is tied in closely with current events and any change in one's faith can affect that person's future. If faith is not rooted solidly in God's Word it can be influenced by false teachings and channeled into unscriptural actions, attitudes, and conclusions detrimental to the believer's spiritual well-being.

Faith that is poised is not easily staggered by the changing winds of doctrines. When one is firmly anchored to God's Word the enemy's vicious and poisonous teachings cannot shatter the hope that radiates from a positive faith. Faith is a sacred commodity and any vicissitudes can subject the believer to confusion and destroy hope. When faith is shaken, hope dies and turns to gloom. The intent of Bible Prophecy is to strengthen faith.

[1] Based on archaeological findings, historians and chronologists have established September 29, 8 or 7 B.C., as the birth of Christ: October 14, A.D. 21 His baptism; and April 3, A.D. 25 as the date of His crucifixion. See also page 265.

47

Cosmic Bridges

When a believer in Christ is free from the encumbrance of his fleshly exterior, he can bridge the cosmos in an instant.!

". . . This mortal must put on immortality" (I Corinthians 15:55).

". . . Neither can they die any more . . ." (Luke 20:36).

In order to exceed the speed of photon light superluminal properties are required. Scientists call these hypothetical properties tachyons (after the Greek word "tachy," meaning quick). To their knowledge, no observational evidence is available to prove their existence but they are completely consistent with the theory of relativity. The problem is that the physicists have not consulted the source which has evidence in abundance—the Holy Scriptures! They refer to nonatomic particle objects as spirit beings, paradisiacal bodies, and immortal bodies.

The Heavenly Father does not possess a body of immortality; He is "Eternal," as is the Holy Spirit. The resurrected body of the Son of God, Jesus Christ is immortal. One must first possess a mortal body before one can have an immortal body. Under the divine decree of suffering eternal punishment, angels are not permitted to possess the body of a human being. It is an exclusive right of the Holy Spirit (Acts 2:4; 4:8; 5:3; 9:17; 13:9). The angels who defy God's pronouncements are under divine indictment of the most serious nature (Jude 6; Matthew 8:28-34).

There is one distinction between the manner in which Jesus Christ in His incarnated body was filled with the Holy Spirit and the way the redeemed were filled: "...God giveth not the Spirit by measure unto him" (John 3:34). "For in him dwelleth all the fulness of the Godhead bodily" (Colossians 2:9). Jesus never baptized in water because He is the Baptizer in the Holy Spirit (Matthew 3:11).

RELEASE OF MORTAL HOSTAGES

The Apostle Paul refers to this glorious immortal body[1] as incorruptible; therefore, it is free from every disease and sickness. There are no aging processes at work in this body. Its survival is not dependent upon vital physical organs or upon blood because it has its own built-in pulsating life from the indwelling Spirit of God (Romans 8:11).

It is a glorious body: "It is sown in dishonour; it is raised in glory..." (I Corinthians 15:43). Its radiance can become as luminescent as bright stars (Daniel 12:3; Revelation 21:23).

It will be an untiring body: "...it is sown in weakness; it is raised in power" (I Corinthians 15:43). There will be no more fatigue or necessity for relaxation as it will possess superior endurance. Rest will be a time solely for reflection and socialization.

BELIEVERS' WARDROBES

"Thou hast a few names even in Sardis which have not defiled their garments; and they shall walk with me in white; for they are worthy" (Revelation 3:4).

It appears "the garments (plural) of salvation" must be distinguished from "the robe of righteousness" (Isaiah 61:10; 52:1; Revelation 3:4; 16:15). Because the fabrics of both are of unknown materials, their wearing quality and capacity to remain spotlessly clean are identified as being eternal. The garments and robes that fill the wardrobes in Paradise and in the "ivory palaces" (as we shall describe later) within the Holy City apparently are ineffably adorned with luxurious and artistic embellishments.

NEW LAWS OF PHYSICS

Immortality is gained when mortal substance is converted into a body of a higher dimension. Its composition will then equate with the spirit beings but maintain a visible appearance like flesh. It will enjoy total freedom from all third-dimensional environments that have inflicted and limited the physical body. No imperfect bodies will exist because the Holy Spirit, which is the life of the immortal body, acts as Anti-Matter destroying the very roots of death cells.

A change from mortal into immortal substance is necessary when the following scientific data is understood: third-dimensional *particles cannot* remain intact as mass *at* the speed of light but will dissolve into pure energy; fourth-dimensional *properties can* remain intact as mass even *beyond* the speed of light. Subparticles can also survive the speed of light when not a part of mass, but they cannot escape the third dimension.

Scripture shows that we will be like God (I John 5:2). In Job 41:12 we read that God is *"comely proportioned."* A description of Christ in the Song of Solomon (5:10-16) is interesting. "My beloved is white, and ruddy . . . His head is as the most fine gold, his locks are bushy, and black as a raven (not gray from age or worry for we read): "Casting all your care upon him; for he careth for you" I Peter 5:7). His eyes are as the eyes of doves . . . His cheeks are as a bed of spices, as sweet flowers; his lips like lilies . . . His hands are as gold rings set with beryl; his belly is as bright ivory overlaid with sapphires. His legs are as pillars of marble . . . His mouth is most sweet; yea, he is altogether lovely . . . ."

Because Christ Jesus at the present time is the only one possessing an immortal body (I Timothy 6:16), He is called the "firstfruit from the *dead,"* and "firstfruit of them that *slept* (were deceased)" (Romans 8:29; Colossians 1:18; I Corinthians 15:20,21). Christ is the *first* of many who later will receive bodies like unto His "glorious body" (Philippians 3:21).

THE INCORRUPTIBLE BODY

Although the physical organs of the body will be retained, they will be changed biologically into incorruption (I Corinthians 15:42,53,54), as will the perishable food they receive (Luke 24:41-43). "Change" does not mean removal but simply a biological alteration. Eating and drinking will not be a necessity, but it will be a normal, pleasurable function.

The Holy Spirit, who is the Life Sustainer dwelling in this new body, will convert perishables into imperishables as they enter the digestive tracts (Romans 8:11; I Corinthians 15:52). Even angels, as spirit beings, have been known to consume perishable food when appearing in visible form (Genesis 18:1,2,8; 19:1-5). The Royal Redeemed will eat a variety of foods, as verified by the fact that they will be participants in the "Marriage *Supper* of the Lamb" (Revelation 19:9). Since this immortal body can never be injured or destroyed, it is unlikely that blood will flow through it (Luke 24:39; Hebrews 9:12).

Another major distinction of immortal bodies is that they will be visible forever. This is in contrast to the angels that are fundamentally invisible in their mobility and total free flight, but who occasionally make instant appearances.

THE QUANTUM LEAP

All matter is made up of atomic particles subject to the third-dimensional laws of physics and chemistry, but what if there were structures of ultra-properties? They would follow laws of the fourth dimension or even higher. Physicists have alluded to this possibility.

Nuclear forces hold the smallest mass together; *electrical* forces bind the intermediary mass; and *gravitational* forces restrain the larger objects. The cosmic phenomenon, electromagnetism,[2] is the force acting between charged particles and it is responsible, among other things, for the transmission and reception of radio broadcasts. The Unified Field Theory brings together the four main forces in the universe: gravitation, electromagnetism, strong and weak nuclear forces—the ultimate in a third dimension.

Motion cannot be described in terms of the classical concepts of an object always moving continually forward from one place to another. In quantum physics, when an electron changes paths, or makes the "quantum leap," it does so in a flash as it reaches a velocity of 40,000 times the speed of light, after which it instantly slows down to less than the speed of light. Therefore, according to the electron's clock, it moves backward in time. This superluminal and independent sharp break with the past caused Einstein to make the comment: ". . . the good Lord does not play dice with the universe," unless, of course, (our words) the dice were loaded (with fourth-dimensional energy) in God's favor. Randomness does exist in the universe. It gives life distinct uniqueness in its unpredictability. Einstein, in recognizing the "quantum leap," found that it challenged his concept of "determinism," which is the law of cause and effect in the third dimension.

Quantum physics is viewed from two perspectives: one physicist sees a *particlelike* electron; another one observes a *wavelike* electron because it is practically mass-free. Do the laws of fourth-dimensional physics affect both time and speed? We would consider such an introduction to the third-dimensional world a *miracle* defying the laws of physics. Einstein proposed to use the quantum theory in search of a greater theory. Perhaps in quantum mechanics we see God operating within His creation. This lends credence to: 1) Romans 1:20, "For the *invisible things* of (God) from the creation of the world are clearly seen, being understood by the things that are made, even his eternal *power* and Godhead; so that they are without excuse;" 2) I Corinthians 15:52, "In a moment, in the *twinkling* of an eye;" 3) 2 Peter 3:8, ". . . a thousand years is *as one day,*" 4) I Corinthians 15:43, ". . . raised in *power.*"

Ouantum physics in no way threatens Einstein's equation of $E=mc^2$ (energy equals mass times the speed of light squared) which confines itself to the third-dimensional world. Here the theory is that mass cannot travel faster than light without changing into pure energy and that time is a fraction of velocity. The equation limits itself to the total structure of many particles as mass and not to a single particle or wave being accelerated near or beyond the speed of light. This is because of the pressure exerted upon the total mass which is largely space.

Physicists indicate that the physical universe is "made up of paradoxes of which we are usually unaware since we do not see both sides at once." They offer these possibilities: 1) the discontinuous jump of an electron from one orbit to another, with the particle mysteriously leaving no trace of its paths; 2) the instantaneous collapse of a wave of probabilities into a single real event; 3) the link between two separate locations, events, or ideas; that magical moment when the previously inexplainable is suddenly clarified, and a radical new theory is born. Therefore, according to the quantum leap theory, a mortal body *suddenly can be changed, faster than light,* from mortality into immortality; from corruption into incorruption; from dishonour into glory; and from weakness into strength—all into another dimension, leaving the one path of a third-dimensional world, to another path of a fourth-dimensional world.

In dimensions beyond the third and fourth, the garments of salvation and the robe of righteousness should be compatible, as should any spirit being functioning under its own laws as an incorporate entity composed of neither particles nor waves, but of an elemental constituent—supernatural properties.

The laws of physics and chemistry, as we know them, are restrained to the third dimension where particles of matter as mass are involved, but there may well be other astrolaws superior to ours which govern creatures of a different composition. While matter and force are indestructible, they can be changed from third-dimensional *particles* and *waves* to fourth-dimensional *properties* and *pure energy*.

NEW BIOLOGICAL MATTER

This body of immortality will possess a metabolism that will challenge all barriers of matter, time, and distance. It is the perfect gift. Nothing can slow or restrain its instant movements. After His resurrection Jesus Christ possessed a body that defied the third-dimensional laws (John 20:17-31).

The resurrection and translation of the redeemed at the partial return of Christ to earth (Rapture) will be "In a moment, in the twinkling of an eye . . ."[3] (I Corinthians 15:52). Time and distance will be irrelevant.

When the "spirit being" is in command, the metabolic body can no longer dictate or limit its movements, activities, or abilities. It becomes subject to the will of the person.

HYPERSPACE

Hyperspace is the passageway through black and white holes in the cosmos. It is also referred to as bridges, subways, and tunnels through the universe. Here time and the laws of physics and chemistry have been crushed out by massive gravity and heat from the death of stars.

Hyperspace is believed to be as calm as the eye of a hurricane. There is no matter and time so the moment of entrance is the moment of egress. Even light (photons) has been crushed out, so the speed of light is no barrier to velocity.

In order to pass through hyperspace one would require shielding by metabolistic material of a higher dimension. The Holy Spirit offers such protection by changing mortal substance into immortal properties. When the believer in Christ dies, his soul and spirit are clothed upon with both the garments of salvation and the robe of righteousness for transit to the heavenly Paradise. Others, such as little children, will have only the garments of salvation, not because they are redeemed, but because that opportunity is awaiting them in the future. The spirit entities of the lost will ascend to God but their souls will descend into the lower region of the lost to be reunited with their bodies at the resurrection to the White Throne Judgment.

COSMIC SUPER-EXPRESSWAYS

Space scientists theorize that the pure energy within a black hole is not held there but is expelled into "white holes," and converted back into matter.[4]

A well-known astronomer, when asked about white holes,

remarked that they must be gifts from the Creator since they appear to be part of the initial conditions of the origin of the universe. In other words, if the white holes did not exist, matter eventually would forever disappear bringing an end to all of the galaxies in the universe. Even where galaxies are born out of cosmic matter from exploding stars, black holes would, in time, consume all matter.

White holes are believed to be the reproductive systems or celestial wombs reprocessing pure energy into matter which is limited in quantity. Eventually black holes would consume all matter, therefore they have been called the "time bandits" and "scavengers" in the cosmos.

As one enters a hyperspace, hypothetically speaking, he discovers that he has appeared instantly on the other side many megaparsecs[5] away, conceivably into a different dimension where the laws of physics and chemistry could be altered. Scriptures call these stations in space "dominions" and "principalities." They are suitable only for the habitation of spirit beings. Jesus called them "dry places."

EXPANDED OPPORTUNITIES

The redeemed are royalty (I Peter 2:9) and some day they will reign as kings, priests, judges, and lords in the performance of specific services to Christ (Revelation 1:6; I Corinthians 6:2,3; I Timothy 6:15). In order to fulfill their functions in the far-flung regions of the manifested kingdom of Christ they will need immortal bodies.

The "robe of righteousness" probably will be worn only during the transaction of important assignments, and the many "garments of salvation" will become the regular apparel in their vast wardrobes.

All of the redeemed included in the "first resurrection" will receive bodies of immortality. All others who are redeemed after the "first resurrection" must wait until death is destroyed (I Corinthians 15:26) in the distant future of the New Heaven to receive their bodies of immortality. Those in the "first resurrection" are an exclusive body of believers. They will be the corporate wife of Christ with specific inheritances that are unavailable to others. How important it is that we realize the great honor and privilege that can be ours if we conduct ourselves properly, wisely, and with pure motives.

GOD'S FAMILY COMPLETED

An immortal body will be unable to reproduce itself into other immortal bodies. This is the sole function of the Holy Spirit (Romans 8:11,23; Matthew 1:20; John 6:63; I Peter 3:18). Christ, who possesses such an immortal body, will never reproduce Himself. He has been, is, and will always remain the "only begotten Son of God" (John 3:16).

The redeemed who come in after the Marriage of the Lamb of God, which is after the "first resurrection," will not be in the same classification as the Royal Redeemed. But after they receive their immortal bodies at the time death is destroyed, they will be ruled by those in the "first resurrection" (Revelation 20:6). Nowhere in Scripture are they addressed as "kings" or by other titles bestowed upon the Royal Redeemed.

Since Jesus Christ will always be the "only begotten Son of God," the Church as His corporate wife will forever be His only helpmate. She is complete. If those who are redeemed after the "first resurrection" were to be included in that corporate relationship, it would mean that Christ married a child-bride or an incomplete wife. Therefore, these two groups of redeemed will be forever differentiated throughout eternity.

NEPHILIM GIANTS

Some theologians believe that fallen angels once materialized and cohabited with the "daughters of men" thus producing abnormally massive beings of super-intelligence (Genesis 6:1-4). They contend the angels were punished severely and are being held in the bottomless pit (Jude 6; Revelation 9:1-11).

These "giants" (Hebrew, *nephilim,* meaning fallen ones), according to their theory, were the root cause of the devastating wickedness that brought on the judgment of the Flood (Genesis 6:11-13).

WHEN DEATH IS THE QUARRY

Immortality is the devourer of death, which is the result of sin; therefore, those in immortality will remain forever free from the desire and results of sin. The soul (personality) not only has been united to the "spirit" within a person, but also it will eternally receive its life of excellence from the Holy Spirit (Philippians 3:21).

Immortality means life, not death. The Giver of eternal life brought immortality into existence, thus the Royal Redeemed will remain forever immortal (Revelation 3:5). This is "eternal security" personified.

[1] Josephus calls it the "pure body."

[2] Some scientists regard electromagnetism as an unseen fourth-dimensional spacewarp within the third-dimensional world. Millennial transportation may fully utilize this source of energy (see chapter 80). The existence of higher dimensions beyond the fourth utilizes other forces, which according to physicists, are rolled up, or (in their jargon) "compactified."

[3] See page 305 footnote.

[4] See illustration V, page 607.

[5] Parsec equals 3.26 light-years, or close to 19 trillion miles.

48
The Cosmic Families

When one attempts to mentally digest the magnitude of the universe, one's imagination is stretched beyond belief.

When one attempts to mentally digest the magnitude of the universe, one's imagination is stretched beyond belief. The universe consists of billions of intergalactic groups known as Globular Groups. One of the largest of these is 60 million light-years wide and consists of enormous families called Herculean Groups. The Milky Way, which is one of a dozen galaxies bound together gravitationally, is a galactic cluster and is referred to by astronomers as our local group or a suburb of our cosmos. The Milky Way galaxy is approximately 100 thousand light-years in diameter and about 15 thousand light-years in thickness or depth and our sun is approximately 30 thousand light-years from its galactic center.

That which science has been able to convey to us concerning the immensity of our universe is incomprehensible; for example, Andromeda is 2,000,000,000 (billion) light-years away. Light travels at the incredible speed of six trillion miles per year. It takes light from Proxima Centauri, our nearest star in the system of Alpha Centauri, 4.3 light years, or 26 trillion miles in distance to reach us. It would take the Apollo-Saturn rocket 70 thousand years to reach it.

No less than 400 billion galaxies are known to exist, and astronomers believe they have reached only a little more than halfway across the universe with their most sophisticated radio telescopes. Some astronomers do believe they have reached near the edge of the universe.

Distance in our cosmos is unfathomable. To travel to the center of our Milky Way would require 30 thousand light-years. The distance between stars is analogous to a few baseballs scattered over all of North America. Stars that appear to be stacked together are actually separated by many light-years. Some stars in other galaxies are so huge they would dwarf our galaxy. Astronomers believe some comets[1] which were born 100 thousand years ago have not yet completed their first full swing into deep space. The vastness of the cosmos can also be realized when one understands that the constellation Cygnus is expanding at the rate of 20 thousand miles per hour.

TIME-DILATION EFFECT

The time-dilation effect is part of Einstein's theory of relativity. It states that the passage of time slows down drastically when one travels at velocity approaching that of light. For example, to reach the center of our galaxy, a distance of 30 thousand light-years away, under the dilation effect the traveler would have become no older than thirty-five years of age, while the passage of time on earth would have exceeded tens of thousands of years.

In taking advantage of this time-dilation effect, it still would be impossible to span the cosmos in one's lifetime even if one could survive such speed. Since it would take thousands of time-dilation-effect-years to traverse the cosmos at a speed approaching that of light, this offers no solution to spanning the universe in a reasonable time.

Taking full advantage of its maximum possibilities, there is no way under the known laws of physics to bridge the universe. There must of necessity be other modes of transportation to conquer the restrictions of time, distance, hostile environments, and the known laws of physics. God has already provided for this by introducing such entities as spirit beings, paradisiacal bodies, and immortal bodies to enable one to go from earth to glory. Furthermore, in doing so He has removed the problems of time and distance.

[1] Comets travel approximately 26 miles per second.

49

Tribulation Syndrome

Paul firmly believed that the first resurrection included a translation of the living believers.

The Apostolic times are an approximation of the predicted Tribulation day under Antichrist, so let us consider how attitudes can affect a person's faith. The Caesars ruled with an iron hand. Christians were persecuted and martyred because they were regarded as enemies of the state. From those dark shadows through which the early Church passed, Paul kept a vigil on anyone who might advocate a teaching that would smother the spiritual flame burning so intensely in the Christians' hearts.

Ultimately he heard that teachers from the church at Ephesus were spreading a strange doctrine capable of extinguishing the hope of the Church. They taught that the resurrection of believers had already passed (2 Timothy 2:18). The inherent dangers lurking behind this insidious dissemination became apparent. If the first resurrection had already passed, then important cardinal truths of Scriptures were being jeopardized. Under this doctrine, the living clearly had missed two significant events: first, the coronation of the believers by Christ when He descended with angels laden with crowns and awards for those meeting Him in the air (Revelation 22:12; Hebrews

9:28; 2 Timothy 4:8; I Peter 5:4; I Corinthians 9:25; Revelation 2:10; James 1:12; I Thessalonians 2:19). Secondly, the coronation of Jesus Christ as King of kings by the redeemed before the Father's throne (Revelation 4:4,10). They also would forfeit appearances at the Judgment Seat of Christ (2 Corinthians 5:9,10; Romans 14:10); at the Marriage of the Lamb of God (Revelation 19:7-9); and at the Millennium.

We are told: "Blessed and holy is he that hath part in *the first resurrection:* on such the second death hath no power . . . but they shall reign with him a thousand years" (Revelation 20:6). This verse states a very important hope of believers—immortality! To have missed the first resurrection meant that any hope of being "like Christ" had eluded the believers (I Timothy 6:16; I John 3:2). Their only hope was a reuniting of spirit and soul to a mortal body in a resurrection to the White Throne Judgment. Paul could not allow this insidious doctrine to go unchallenged.

INDISSOLVABLE MARRIAGE

Paul firmly believed that the first resurrection included a translation of the living believers, and realized that too much of what the believers hoped for was at stake. After all, this hope of living Christians must be preserved—that of joining Christ in company with their friends and deceased redeemed loved ones. This hope served as an important purifying element.

The Apostle Paul found it difficult to uproot this false teaching and to dissuade the Christians who already were embracing it. Paul called it an "error" that threatened to "overthrow the faith" of believers (2 Timothy 2:17,18; I Timothy 1:19,20).

So persuasive were these false teachers that Paul's attempt to discredit and dislodge this false doctrine from the minds and hearts of those being influenced met with firm resistance: "At my first answer no man stood with me, but all men forsook me." But he was encouraged to firmly pursue his efforts for he said, "Notwithstanding the Lord stood with me, and strengthened me; that by me the preaching might be fully known, and that all the Gentiles might hear . . . " (2 Timothy 4:16,17).

DEVELOPMENT OF DOCTRINAL ERRORS

What was the basis for teaching that the resurrection had already passed? The premise must have been Matthew 27:52,53 since apparently no other incident is recorded in Scripture.

Almost immediately after the resurrection of Christ many saints who were raised from their graves strolled the streets of Jerusalem. These were not resurrected into immortality, for only Christ has immortality. Their brief appearance suddenly ended with their disappearance into the heavenly Paradise. Undoubtedly, some of the living Christians were acquainted with them and so the occasion was confirmed by reliable witnesses.

Perhaps other Scriptures, plucked out of context, were used to support the false teachings (such as from Daniel 11:45; 12:1,2; John 6:25-29). The only good emerging from this record in the Word of God is that it showed the Apostolic Church did believe in a pre-Tribulation Rapture. The one major flaw in the doctrine was their failure to link the translation of the living believers to it.

In order to discredit this false doctrine, something obvious must be associated with it to expose its error. Paul firmly believed that the Rapture of the living redeemed was related to the resurrection of the believers, and that the events must occur simultaneously (I Thessalonians 4:13-18).

A THRONELESS TEMPLE

Paul pointed out in 2 Thessalonians 2:1-4 that at the time Antichrist arrives on the world scene he will sit in the temple in Jerusalem declaring himself to be God personified. None of the previous Roman rulers have done this, so it is to be one of the signs in the last days of the arrival of the predicted final Tribulation upon the Church and Israel. There were rumors in Paul's day that the Romans were about to lay siege to Jerusalem and destroy it, along with the temple, as Jesus had predicted (Matthew 24:2; Mark 13:2; Luke 21:6). Therefore, the Christians may have concluded that they were in the final Tribulation days. However, the Antichrist who will *sit* in the temple will *not destroy it,* for it is the Tribulation temple that Christ must enter when He returns to earth. The two witnesses in Revelation, Chapter 11, also will be placed as sentries before the Tribulation temple in Jerusalem.

Paul's contention, therefore, was that the Apostolic Church was not in the predicted Tribulation under Antichrist, and that the resurrection of believers had not passed. He supported this assertion by declaring that the translation of the living in Christ must be concurrent with the resurrection of believers in Christ. And it remains our safe course of action today in determining whether or not we are living in the final Tribulation days.

INSEPARABLE TWINS

In His Olivet discourse, Jesus Christ was the first to reveal that a translation of believers would precede the Tribulation under Antichrist: "And when these things *begin* to come to pass, *then* look up, and lift up your heads; for your redemption (of the body, Romans 8:11,12) draweth nigh" (Luke 21:28). This promise was followed by the admonition: "Watch ye therefore, and pray always, that ye may be accounted worthy to escape *all* these things *that shall come to pass,* and to stand before the Son of man" (Luke 21:36).

There have been times in the history of the Church when persecution raged so relentlessly against it that it was rationalized that the final Tribulation was upon it and that the current political leader was the predicted Antichrist. There was always the realization, however, that the translation of believers was the determining factor as to whether or not they were living in the predicted Tribulation days. How else could they be certain whether or not they were going through that period? This kept the believers doctrinally in balance regarding any post-Tribulation conclusion that the "first resurrection" had passed.

We are not in the final Tribulation period today because there is no Jewish temple situated in Jerusalem on the Temple Mount as will be necessary in order for Antichrist to sit there (2 Thessalonians 2:4).

The Tribulation under Antichrist cannot come until the redeemed are first removed from the earth. Those who make up the body of the Tribulation saints will later share in the post-Tribulation translation of the saints. This will occur before the final Tribulation upon Israel. All Tribulation saints will share immortality as part of the Church in the the final resurrection so that all may appear at the Judgment Seat of Christ.

ACTS, CHAPTER 29

A common misconception is that the Church Age ends at the pre-Tribulation translation; however, the Church Age will not terminate until *all* of the Tribulation saints have been harvested from the earth (Revelation 14:14-16; I Thessalonians 5:9,10; I Corinthians 15:51) just preceding the Judgment Seat of Christ and the Marriage of the Lamb of God. These final Tribulation saints also will reign with Christ during the Millennium (Revelation 20:4; 1:6; 5:10).

The line of demarcation separating the pre-Tribulation Church from the Tribulation Church under Antichrist is the translation of believers. Why do those who strongly advocate the belief of a pre-Tribulation Rapture decry the teaching of a post-Tribulation Rapture, or vice versa, when both are right? One does not cancel out the other. They only need to readjust their thinking into proper sequential order.

THE TRIBULATION TEMPLE

According to Acts 15:14-16, the Tribulation temple will not be built until *after* the pre-Tribulation Rapture: "Simeon hath declared how God at the first did visit the Gentiles, to take out of them a people for his name (Church) . . . *After that, I will return* (for the redeemed), and will build again the tabernacle (temple) of David, which is *fallen down* (in 70 A.D.); and I will *build again the ruins* thereof (this will not happen to the Tribulation temple Malachi 3:1), and I will set it up."

According to the above prophecy Christ has to come *after* 70 A.D. when the temple was put in ruins and *before* the Tribulation temple is to be built. So the words "I will return" refer to the pre-Tribulation Rapture. The Tribulation temple will not fall into ruins before Christ returns to earth for He will enter that temple and begin the period of the Millennium. The "residue" must therefore refer to the Tribulation saints who will seek the Lord after the pre-Tribulation Rapture (Acts 15:17).

The Tribulation temple *must* be built on the Temple Mount in Jerusalem beside the Temple of Omar accordinq to Revelation 11:2 and Christ is to build the Millennial Temple upon the site of the ruins of the temple of 70 A.D. Christ will then raze the Golden Temple of Omar and the Tribulation Temple as He expands the Millennial Temple to cover the Temple Mount where once Solomon's Temple stood.

How long did it take to build Herod's Temple in Jesus' day? In John 2:20 we read: "Then said the Jews, *Forty and six years* was the temple in building. . . ." In making reference to the last-day generation when all these things He mentioned must come to pass, including the rebuilding of the temple, Jesus said, ". . . This generation shall not pass away till all be fulfilled" (Luke 21:32). This is a reference to the generation which was alive when Israel set up its parliament on May 14, 1949, one year after declaring its independence as a nation. Allowing for the last day Tribulation period, we can readily understand the imminence of the hour in which we live, and how quickly the times of the Gentiles are coming to an end.

The altar of burnt offerings once stood upon the great rock on which now rests the Moslems' Dome of the Rock. This means that the altar at the time of the dedication of the Tribulation Temple will not be located on its original site.

The Temple of Omar (Golden Domed Mosque) presumably will be destroyed when Christ Jesus begins construction of the Millennial Temple. There will be no followers of Islam left to make a protest at the time (Obadiah 18).

50

Harvest Time

The Church must equate immortality with Christ in order to "be like him" (I John 3:2).

The farmer looks forward with anticipation to harvest time when he expects to be rewarded for his investments and energies. Preparing the soil, cultivating, and allowing adequate time for the crop to grow and mature under the sunshine and the rains take considerably longer than the harvesting. For almost two thousand years, Christ has been waiting to harvest His anticipated bumper crop: ". . . Behold, the husbandman waiteth for the precious fruit of the earth, and hath long patience for it . . ." (James 5:7).

SECOND SALVATION

"So Christ once offered to bear the sins of many; and unto them that look for him shall appear the *second time* without sin *unto* salvation" (Hebrews 9:29).

The word "salvation" means safety. When we accept Christ as our personal Saviour we are saved from the *power* of sin, but when He receives us unto His Glory we are saved from the *presence* of sin. The first assures the immortality of the *soul;* the second, the immortality of the *body:* ". . . the soul that sinneth it shall die" (Ezekiel

18:4,20). ". . . But we are . . . of them that believe *to the saving of the soul"* (Hebrews 10:39). When one accepts Christ as his Redeemer, the *spirit and soul* are wedded together, as it were. Because the "spirit" is eternal, the *"soul"* of the redeemed receives eternal life. It is the *redemption of the soul.*

The word "death" means separation, not cessation. At the first death the souls not in the book of life are separated from their spirits as well as their bodies. At the resurrection they will appear at the White Throne Judgment with both their bodies and souls reunited, but not with their spirit entities; therefore, the second death will determine the longevity and finality of the reunited body and soul (Revelation 20:14,15).

FIRSTBORN FROM THE DEAD

Jesus Christ is the "firstborn from the dead" (Colossians 1:18). He is the *first* to receive an immortal body (I Timothy 6:16). As a result of this honor Christ has been elevated to be "Head" of the Church which is His Body (I Corinthians 12:27; Colossians 1:18). It is inconsistent that the "Head" be immortal and the "Body" mortal, so the Church in order to be eternally joined to Him as His corporate wife must of necessity also possess immortality. The Church must equate immortality with Christ in order to "be like him" (I John 3:2).

The requirement that the Church receive immortality is borne out in Romans 8:29: "For whom he (God) did foreknow, he also did predestinate *to be* conformed (not transformed, as in being "born again") *to the image of his Son,* that *he might be* the firstborn among *many* brethren."

FIRSTFRUITS OF THE HARVEST

The resurrection of Jesus Christ into immortality is called "the firstfruits" (I Corinthians 15:20). This means His resurrection will be followed by multitudes of believers coming into that same glorious experience. This harvest will represent all ". . . that are Christ's at his coming" (verse 23).

The "time" of the harvest of the redeemed will not be accomplished in a single day (Matthew 13:30), but in a season in which the "husbandman," Jesus Christ, will gather into His heavenly barn the bountiful wheat harvest.

The dispensation of the Holy Spirit and the Church run parallel (Matthew 24:14; 2 Thessalonians 2:7). The *seed* is the Word of God (I Peter 1:23; I John 3:9) and the *watering* is the Holy Spirit (Acts 2:17; 3:19; 14:17; James 5:7). The Church will be ready for the final sheaves to be harvested at the end of the dispensation. The Holy Spirit will then step aside to allow the "wrath of God" to be poured upon a rebellious world (Revelation 14:19; 15:1,4; 16:11).

SPIRITUAL RAINS

In the Middle East two major rainfalls are essential for a good harvest—the "former" to germinate the seed (James 5:7) and the "latter" for the growth and maturity of the crop (Joel 2:23). Thus, spiritually speaking, the former rain comes "moderately" which implies that the greater rainfall, or the outpouring of the Holy Spirit, will come after the Apostolic Age during the closing days of the Church Age. Since the Holy Spirit will produce effective witnesses, it follows that the greatest witnessing and harvest will come in the last days of the Church. The entire period in which these rains come is called "the last days" (Acts 2:17-20; Joel 2:28-32; James 5:3,7-9).

The latter rain will extend deep into the final tribulation of the saints under the rise and height of the power of Antichrist. During Israel's tribulation of the last days there will be a great harvest of souls to Christ, otherwise there would be no one to enter the Millennium. The Millennium also will be a time of vast reaping of souls to Christ under the ministry of Israel and the Royal Redeemed as "priests" to reach everyone with the Gospel of Jesus Christ (Joel 3:1; 2:30-32; Acts 2:19-21).

ETHNIC HARVEST

The Church will presently succeed in reaching only the *nations* of the world before the Church Age ends (Matthew 24:14). Further evidence that not even all the *cities* will be reached with the Gospel is clear from Matthew 10:23: ". . . Ye shall *not* have gone over the *cities of Israel,* till the Son of man be come." In the time of Christ on earth the cities in Israel had been covered, and if they weren't, they could have been in a few weeks. But not all of the Israelis were living there at the time. Many of the tribes of Israel were and still are scattered over the globe, and the cities where they reside will not be reached before the return of Christ to the earth. During the Millennium all *ethnic* groups finally will be reached. There are 228 nations in the world, not counting the many thousands of ethnic groups within each country, which have their own individual cultures and dialects. India, alone, has over three thousand such groups separated from the mainstream. The United States has over one hundred. The Christian Church is hopelessly behind in reaching these tribes, as well as the lost tribes of Israel. During the Millennium, Israel and the Royal Redeemed will seek to reach the offspring of those who entered the Millennium (Jeremiah 31:34; Hebrews 8:11). This is one reason for the introduction of a one world tongue (Zephaniah 3:9). It will take such a megamovement to meet the challenge of total evangelization of the world's people.

James 5:7-9 shows that there are two spiritual rainfalls. The former rain covers the Apostolic Age and the latter rain involves the closing days of the dispensation of the Church which extends well into the final Tribulation period. The "righteous nations" which oppose the Babylonian regime of the last days presumably will have a vast majority who will accept Christ as He returns to earth. This will be simultaneous with Israel's complete conversion to Christ. These can be redeemed for they have not taken the fatal mark of the Beast.

Going beyond the Apostolic Age, James 5:8 says: "Be ye *also* patient; *establish your hearts* for the coming of *the Lord draweth nigh."* The Holy Spirit actually never leaves the earth in His mission to effect the redemption of souls (John 14:16; 16:8-11) but simply steps aside for the days of God's wrath, "the day of the Lord" (I Thessalonians 5:1-11; Amos 5:18-20; Malachi 4:5). The Holy Spirit is absolutely essential for anyone to be redeemed (I John 5:6; Galatians 4:6). Without conversions to Christ at this time, there would be no one to enter the Millennium, because all who will have taken the mark of the Beast will have been destroyed (Revelation 14:9-11; 19:20).

The Holy Spirit remains present during the Millennium and on into the New Heaven because the redemptive process will continue until all who are to be redeemed will receive bodies of immortality when the last enemy, death, is destroyed: "Then cometh the end . . ." (I Corinthians 15:24-26). No one can be redeemed apart from the aid of the Holy Spirit (John 16:7-11; Romans 8:14-16).

YOUR HONOR, THE JUDGE

The phrase, "coming of the *Lord,*" does not refer to any rapture of the redeemed but rather to the Lord's return all the way to the earth to become its Ruler.

James 5:9 points to the Judgment Seat of Christ: "Behold, the judge standeth before the door." Thus the dispensation of the Church Age will terminate immediately preceding that judgment so that all of the redeemed coming into immortality ("first resurrection") can stand before Christ.

This judgment is necessary, not only for rewarding the faithful, but also for making amends for injustices in the family of God. All of the redeemed who are to be in the corporate wife of Christ must appear at this judgment (2 Corinthians 5:10). Those whose conduct merits disciplinary action will be placed in "outer darkness" and will later emerge rehabilitated; however, they will suffer irrecoverable losses throughout the eternal future, such as not being incorporated into the wife of Christ, denial of honorable positions of great excellence, and disqualification from rewards which they might have retained (See "Mystery of the Kingdom"). Some simply will lose their rewards and special honors.

The Church Age and the outpouring of the Holy Spirit must include *all* of the final Tribulation saints coming into immortality who will be included in the "first resurrection" just before the Judgment Seat of Christ (Revelation 6:9-11; 7:9-17; 2 Corinthians 5:10). This is what Jesus meant when He said, "But many that are first (in the kingdom before the last day Tribulation saints) shall be last" (Matthew 19:30). Many in the Kingdom of God before this final Tribulation will find themselves in outer darkness, stripped of their rewards, whereas those newly redeemed in the last days will become incorporated into the wife of Christ and endowed with great honorable positions and distinctions (compare 2 Peter 2:21).

51
Space Invaders

Prophetic Scriptures make a difference between the appearance and revelation of Antichrist.

As the Christians are "caught up" and "changed in a moment,[1] in the twinkling of an eye," there will be a great thunderous sonic boom. As they meet Christ in the air, thousands of royal chariots loaded with golden crowns will descend with millions of royal white horses. Innumerable angels will place crowns on the heads of the Royal Redeemed who will be reigning as "kings" with Christ, the "King of kings." After the coronation, they will board the royal carriages, led by spans of royal steeds, and proceed to the third heaven, which is the "Heaven of heavens," to appear on the sea of glass before the Throne of God.[2] The people on earth will observe blips on the radar screens which they can only classify as Unidentified Flying Objects.

Three words are used in connection with this first harvest Rapture. Each carries "military overtones. The first is the word "shout" used in I Thessalonians 4:16 as a "shout of war," which is derived from the Greek verb "keleuo" meaning "to command" or "to order." It is a reveille call "to arise." On only two other occasions did Jesus Christ raise His voice to the level of a shout—once, in the garden before the tomb where He "cried with a loud voice, Lazarus come forth" (John 11:43); second, at His own death when He "cried with a loud voice, saying, Eli, Eli lama sabachthani" (Matthew 27:45).

The next time the Lord Jesus raises His voice with a "shout," it will be to break the power of death again. This time it will be to take microscopic organisms of the deceased disintegrated bodies and change them into immortal bodies, a process called "the redemption of the body" (Romans 8:23): "It is sown in corruption; it is raised in incorruption" (I Corinthians 15:42).

MILLIONS OF MIRACLES

Myriads of *deceased* redeemed will have their remaining minute atomic particles resurrected into immortal bodies like time-elapsed pictures which are speeded to instant finished products.

Innumerable living redeemed also will rise instantly in bodies transformed into immortality (I Thessalonians 4:13-18; I Corinthians 15:35-38; Romans 8:11, 19-25). As a result, millions of mute lips silenced by death will speak again; blind eyes will see; deaf ears will hear; and paralyzed limbs will become normal. In many cases, these faculties will have been in a state of inertia for thousands of years.

These million-fold miracles will excel any and all miracles of healings and resurrections of the past: ". . . at that time thy people shall be delivered . . . and many of them that sleep in the dust of the earth shall awake" (Daniel 12:1,2). The word "many" does not mean "all." It is limited to the redeemed.

RESCUE OPERATION

When the deceased redeemed are resurrected, the living will simultaneously join them. This is referred to as the "Rapture." The synonyms for "Rapture" are inadequate as there are few suitable expressions in Scriptures for "caught up." Paul came close when he spoke of "our gathering together unto him." The word "together" expresses both a reunion with Christ and reunion of the redeemed.

For "Rapture" Jesus used the word "redemption" (Luke 21:28), referring to the redemption of our bodies[3] (Romans 8:23; Ephesians

4:30); however, this word does not embrace all of the truth of that event.

The thought behind the words "delivered" (Daniel 12:1), "escape" (Luke 21:36), and "sudden destruction" (I Thessalonians 5:3) expresses the process taking place at the Rapture. Since so many truths are involved in this event, the word "Rapture" has been selected by Bible scholars as a convenience, and its reference is easily understood.

The word "Rapture" (rapere) is found in the Latin translation of I Thessalonians 4:17 where Paul said the redeemed will be "caught up" to meet Christ in the air at His coming. In 2 Thessalonians 2:1 we read "coming" (literally, "presence of the Lord Jesus Christ"). The word "escape" is from the Greek "ekpheugo" and has the same meaning as *deliverance*—safety, get free from, to miss danger by being out of the way. The same word is used in I Thessalonians 5:3 and Hebrews 2:3.

The word "escape" in Luke 21:36 means "to be taken away from," the same word Paul used of his escape in a basket (2 Corinthians 11:33). He was not sheltered in the midst of those seeking his life. He completely escaped their company, even as those who "are not appointed unto wrath" (literally, "to attain deliverance from it" I Thessalonians 5:9).

Paul uses for "Rapture" a peculiar, strong word, "harpaza" (I Thessalonians 4:17), meaning "to seize hastily; to draw to one's self by swift, sudden movement." Luke used the same word in Acts 23:10 to describe how Paul was torn away by Roman soldiers from a violent mob. In I Thessalonians 1:10 Paul calls the event of the Rapture "a salvation from the coming wrath." He uses a word which means "a rescue with power," the same word which he used to describe his preservation in Nero's court, his "rescue out of the jaw of the lion."

According to I Corinthians 15:51 the saints who are "caught up" to meet Christ in the air will leave no discarded bodies on earth. Their bodies will be *changed* into immortality, not cast away.

At the time of the Rapture, bodies will be caught up in the

twinkling of an eye as they break the sound barrier, cause a loud sonic boom, and are changed enroute to meet Christ in the air. The bodies will acquire a brilliant irridescence as they become glorified (Daniel 12:3; I Corinthians 15:43), thus sending a flash of brightness that will flood the skies and startle those left behind. Microscopic atoms are sufficient for the resurrection (as in conception) of those with deceased bodies: "And the Lord . . . will *preserve* me unto his heavenly kingdom . . ." (2 Timothy 4:18).

JUST IN TIME

As covered under the chapter, "United States in Bible Prophecy," the Rapture will snatch the redeemed from the earth just as the nations are about to unleash warfare on each other: "And I will send a fire (thermonuclear bombings) on Magog (Soviet Union) and among them that dwell carelessly; (having a feeling of security or safety) in the isles (nations outside the old world)" (Ezekiel 39:6). The United States has over thirty thousand thermonuclear bombs; the Soviets also over thirty thousand. Recent estimates indicate that a nuclear clash between the United States and the Soviets would kill over three billion people, or eighty percent of the world's population.[4] Six nations today have thermonuclear capability. By 1985 there will be an estimated thirty-five countries with thermonuclear weapon capability. Jesus Christ, in describing the people subsequent to the Rapture, said they would enter "the beginning of sorrows" (Matthew 24:8), which will be synchronized with "earthquakes in divers places." Jesus promised that the redeemed can expect to "escape *all* these things that shall come to pass, and to stand before the Son of man" (Luke 21:36).

The redeemed have been the force restraining the final Tribulation judgment and tragedies about to visit the world, therefore their departure will be like the removal of a huge dam which suddenly unleashes deadly, swirling, on-rushing waters that will sweep away everything in their path and thereby cause panic and terrible destruction.

FALL IN RANK

The second word, "voice," denotes the army command to "fall in rank." As the redeemed arise in glorified bodies, they will gather at a central place in the skies to join their Commander-in-Chief. Here in the corridor above the North, they will be given "crowns" and "rewards" (Revelation 22:12; Philippians 4:1; I Thessalonians 2:19; 2 Timothy 4:8; James 1:12; I Peter 5:4). Remember, they cannot lose their rewards at the Judgment Seat of Christ if they have none to lose.

This is "coronation day" for the redeemed (Revelation 22:12). Christ will descend with the holy angels in charge of the golden crowns and place them upon the heads of the redeemed in accordance with their status.

MARCH AWAY

The words "trump" and "trump of God" are God's clarion call to "march away." Under angelic escort similar to that given Christ at His ascension (Psalm 68:17,18), this Royal Party of redeemed will ascend with Christ to the Heavenly Father on His throne of Imperial Majesty.

[1] "In a moment" (I Corinthians 15:52) in the original is *en atomo* (dat. of Greek, *atomos*), literally, *in an atom.* According to the atomic clock, chanqe will occur in 1/9,192,631,770th (billionth) of a second; 50,000 times faster than light; probably at the speed of "thought."

[2] See chapter 86.

[3] Greek *apoluttrosis* is not the usual word meaning "to purchase, or buy back," but "to call for a purchased article and take it away."

[4] A thermonuclear bomb will generate 20 million degree Fahrenheit heat, with winds up to 2,000 miles per hour. The blast will destroy life and property within fifty square miles. For hundreds of miles there will be radioactive fallout creating serious burns and even death. Some scientists believe that the earth could face such darkness resulting in 50 degrees below zero (Fahrenheit), causing a year-round snowfall and eventually great destruction of plant life and worldwide starvation. An ice-covered earth is considered a strong probability.

52

First Harvest Day Rapture

Christ will rechristen each of the Royal Redeemed with names of special significance, chosen and conferred by Him (Revelation 2:17; Psalm 147:4).

The harvest represents more than one day's gleanings for the Husbandman, Jesus Christ, in bringing the fruits of His labors into His barn. "Christ . . . the *firstfruits of them that slept*" (I Corinthians 15:20) is followed by these words: "but *every* man in his *own order*" (that is, chronological); "Christ the firstfruits, *afterwards* they that are Christ's *at his coming*" (I Corinthians 15:23).

This plurality of harvesting days is attested to by a simple statement Jesus made: ". . . The day*s* (plural) will come, when ye shall desire to see *one of the days of the Son of man* (these harvest days), and ye shall not see it" (Luke 17:22). Each harvest day will be a point of no return to the former days. Each event will be followed by increasingly difficult times in the world.

Therefore, the partial return of Christ to the earth in the first of these harvest days is called "the day of *Christ*" (Philippians 1:6,10; 2 Corinthians 1:14; Zechariah 14:1-4; I Thessalonians 5:2). Christ means "the Anointed One."

At these harvest days we are promised that we will see Jesus Christ: "And if I go and prepare a place for you, I will come again, and receive you *unto myself;* that where I am, there ye may be also" (John 14:3). "Father, I will that they also, whom thou hast given me, be with me where I am; that *they may behold my glory,* which thou hast given me . . . " (John 17:24).

THE DAY OF CHRIST

The "day of *Christ"* must be distinguished from the "day of the *Lord"* (Malachi 4:5; Amos 5:18-20; Joel 1:15; 2:1,11; 3:14-17; Zechariah 14:1-4; I Thessalonians 5:2). The "day of Christ" refers to the glorious reunion with him in the air. The "day of the Lord" is associated with the closing days of the final Tribulation in which God's wrath will be sent upon the earth, up to and including the actual day of the Lord's return to earth. The title "Lord" implies reigning and ruling with authority.

There is one exception to the above in 2 Thessalonians 2:2. The words, "the day of Christ is at hand," as they appear in the Authorized (King James) Version, should be translated properly as "the day of the Lord," which has been done in other versions. The message is that the rendezvous with Christ in the air (the first harvest day Rapture) will occur *before* the Lord's return *to earth* and prior to the final wrath of God just preceeding His coming to earth. Verse 3 shows that the Tribulation under Antichrist will also precurse the Lord's return to earth.

Additional evidence that the first harvest day Rapture will come before the Tribulation is that Antichrist does not appear on the scene until after the redeemed are in heaven prior to the opening of the seven seals. The first seal that is opened shows Antichrist first coming as a "conqueror." He is not yet shown as the "Beast" in Revelation, Chapter 13.

There are no Scriptures showing the harvest Rapture of the redeemed as a one-time event. There will be a Rapture preceding the last day Tribulation of the saints, and another immediately prior to the Tribulation of Israel. This distinction becomes obvious by comparing I Corinthians 15:52 with I Thessalonians 4:16. They are two separate events. We must recognize that the "first resurrection" encompasses *all* of the redeemed in both harvest day Raptures, including the Tribulation saints, who are to be resurrected or raptured in bodies of immortality (Revelation 20:6). If there is a "last" trumpet sound heralding the Rapture of the Redeemed, then there must be a preceding trumpet sound.

GOD'S STARS ON STAGE

In Revelation 5:9 the twenty-four elders are seen worshipping Christ in song with these lyrics: ". . . Thou art worthy . . . and has *redeemed us* to God by thy blood out of every *kindred,* and *tongue,* and *people,* and *nation.* And hast made us unto our God kings and priests, and we shall reign on the earth" (See also Revelation 20:4,6; 22:5).

The twenty-four elders are actual characters selected from among the redeemed in Paradise to participate in this live drama revealed to John on Patmos. They apparently are chosen from among the Old Testament saints and prophets (Compare Revelation 17:1 with 19:10; 21:9 with 22:8,9).

In this vision which came to John about 80 A.D., the elders had to be *representative* of the vast multitude of redeemed in the first harvest Rapture because obviously there are more than twenty-four families, languages, individuals, and nations in the world from which shall emerge the untold numbers of redeemed who will be in the first harvest day Rapture. Because the Apocalypse is a prophetic book (Revelation 22:7,10) extending beyond John's time, his vision on Patmos included innumerable unborn persons who would be in the Church and the pre-Tribulation Rapture.

KINGS WITHOUT CROWNS

After the redeemed are Crowned in mid-air by Christ (Revelation 22:12), they will be seated before the Heavenly Father's Throne: ". . . sitting, clothed in white raiment: and they had on their heads crowns of gold" (Revelation 4:4). We know from this scene that the first harvest day Rapture had taken place or the redeemed would not be in heaven witnessing the unfolding drama of the opening of the seven-sealed book. The fifth and sixth seals therein relate to the final

Tribulation saints who are yet to arrive, so chronologically, there has to be a pre-Tribulation Rapture and Resurrection.

Visualize the setting before the Heavenly Father's Throne with the uncrowned Christ standing in the midst of the redeemed whom He has invested with royal crowns. These redeemed, symbolized by the twenty-four elders, see the scars from the wounds[1] that were inflicted upon Christ's body, while their immortal bodies have no signs of physical defects. In the glorious portrayal of Christ in Revelation 1:13-16, there is not even a crown upon His head. The price of redemption is evident.

The scene is extremely graphic, overpowering, and even embarassing. Without any cue, all simultaneously and voluntarily remove their crowns and cast them before the feet of Jesus Christ; then they fall down to worship Him. When they arise they see upon His head five crowns which are representative of all the crowns to be awarded the redeemed (Revelation 19:12; I Corinthians 9:25; James 1:12; I Peter 5:4; I Thessalonians 2:19; 2 Timothy 4:8). Christ is truly King of kings and Lord of lords. He is King over the redeemed and Lord over all created beings, even the angels. After the redeemed receive their bodies of immortality, they are crowned, honored, commissioned, and enthroned.

HONOR GREATER THAN CROWNS

The redeemed before the Father's Throne witness the unsealing of the book of seven seals containing the events of the last day Tribulation. They are now "kings" without crowns, but they will have something far more invaluable—their new relationship as Christ's royal wife. They, as the Royal Redeemed, will reign with Him as "Queen" over all creations. They shall "inherit all things" (Revelation 21:7).

This new relationship of the Royal Redeemed with Christ, as His corporate uncrowned wife, will involve the royal distinction of being kings, priests, judges, and lords (Revelation 1:6; I Corinthians 6:2,3). The various degrees of rank will be determined at the Judgment Seat of Christ (Luke 19:11-27; Matthew 25:14-30). The Royal

Redeemed will have special access to Christ, even to the consternation of the angels, because as the Royal Wife of Christ they will share with Him the honor of sitting on His powerful throne (I Peter 1:12; Psalm 8:4; Hebrews 2:6; I Corinthians 6:2,3; Revelation 3:21).

The last Tribulation saints, who are resurrected and raptured, will join the other saints to also reign as "kings" without crowns (Revelation 20:4). Christ already will have been crowned "King of kings" by both the resurrected and raptured pre-Tribulation saints (Revelation 3:21; 14:14-16). This is the reason that the last Tribulation saints will receive "palms" in lieu of "crowns" at their resurrection and rapture. When all of the Royal Redeemed (including the Tribulation saints) in the "first resurrection" (Revelation 20:6) return with the crowned Christ (Revelation 19:12-14), they are without crowns, yet they are called "kings" under the "King of kings" (Revelation 19:16).

In this book we will be distinguishing those in the first resurrection as the "Royal Redeemed" while all others who receive Christ later will be referred to as the "redeemed." Those not included in the "first resurrection" will appear at the White Throne Judgment in bodies of mortality.

Jesus Christ is "King of kings" over the Royal Redeemed, but to earthly rulers He is the "*prince* of the kings of the earth" (Revelation 1:5). Christ does not have that same close relationship with them as He has with the Royal Redeemed.

ROYAL CHRISTENING

Names are important to God and to individuals. They give a certain distinctive identity, therefore Christ will select names of special significance and rechristen each of the Royal Redeemed (Revelation 2:17; Psalm 147:4).

MIDNIGHT RIDERS

Scripture refers to the Tribulation of the last days as the time of "darkness, not light" (Amos 5:18,20). The apocalyptic riders in Revelation, chapter 6, gallop during the darkest hours of mankind's political and religious history.

There is a clear distinction between Antichrist and Christ riding on white horses. Antichrist has a bow; Christ has a sword (Revelation 19:11-16). Antichrist has a crown; Christ has many crowns. Finally, Christ is opening the seals so He could not possibly be part of its contents.

A MILLION YEARS OF MINISTRY

Who are the four creatures (beasts) in John's vision that are selected from Paradise for this live drama in the company of the twenty-four redeemed elders? They are the ones who represent the redeemed in their final roles as priests. The redeemed are to reign with Christ as both *kings* and *priests* (Revelation 1:6) as Christ is not only "King" of kings (Revelation 17:14) but also the High Priest forever (Hebrews 3:1; 7:17). The elders have golden crowns upon their heads, so obviously they are the kings. By the process of elimination that leaves the four living ones symbolizing the future priesthood of the redeemed.

The priesthood of the Royal Redeemed falls into four categories: first, those who proclaim Christ, the Lion of the tribe of Judah (Revelation 5:5); second, those who present Him as the crucified One, who has made a sacrificial offering for sin, as depicted by the young bullock or calf; third, those who restore men to God because of all of God's created creatures they are the only ones to have been provided redemption; fourth, those who grant hope to all believers in Christ that they will possess immortality where all bondages and restraints upon a physical mortal body will be removed. Time, distance, speed, and all other barriers will become passe, as symbolized by the flying eagle. These four categories comprise the redemptive plan of God to mankind which will extend for one million years into the New Heaven (Psalm 22:30,31; 102:18; 105:8; Deuteronomy 7:9; I Chronicles 16:15; 2 Peter 3:8).

PRIESTHOOD APPRENTICESHIP

The Royal Redeemed are also a "Royal Priesthood" (I Peter 2:9) and must be distinguished from others who will be redeemed later. The ministry of the Royal Redeemed will begin with the Millennium (Revelation 5:10; Romans 8:19-22). As "priests," *all* of the Royal Redeemed will *minister* in their relationship with Christ rather than being ministered *unto* by others. They no longer will be "children of God," but mature persons in the corporate "Wife" of Christ (Ephesians 4:13). "For even the Son of man came not to be ministered *unto* but to *minister* . . ." (Mark 10:45). *Faithful ministers* on earth have this wonderful promise of Christ's special nearness in both Christ's and the Father's future Kingdoms: ". . . a faithful priest (minister) . . . shall walk before mine anointed (Jesus Christ) for ever" (I Samuel 2:35).

This priesthood begins only *after* one is a "king" (Revelation 1:6). It cannot commence until all of the redeemed, including the last Tribulation saints, have participated in the "first resurrection" (Revelation 20:4,5). The White Throne Judgment will include everyone redeemed *after* the "first resurrection" as well as the unredeemed still in the book of life (John 5:29; Daniel 12:1,2; Revelation 20:11,15).

THE INNOCENT AND THE RAPTURE

One of the most sensitive areas in the matter of the Rapture concerns small children and the mentally retarded who are all incapable of accepting Jesus Christ into their lives. Will Christian parents abandon all hope of a rapture for them and expect them to go into the final Tribulation period? Is the Rapture exclusively for the redeemed who have accepted Christ as their Saviour and have become the corporate Wife of Christ? If so, then the children of unredeemed parents will fare better because their families will remain intact.

Since the innocent go to Paradise upon their death, why shouldn't they go there at the time of the Rapture as well? "For by *one* man's disobedience MANY (not all) were made sinners, so by the obedience of *one* shall MANY (not all) be made righteous" (Romans 5:19). Fortunately, Scriptures do not state at what age innocence and the state of mental incapacity end or when personal responsibility begins. Since the innocent cannot be charged with guilt (Romans 4:15; 5:13), they will pass into Paradise when they die, as King David indicated regarding his young son (2 Samuel 12:23). It is the "enemies" and "foes" of Christ that enter into the region of the lost (Matthew 22:44; Acts 2:35). They obviously made their premeditated decision to reject Christ, so all others will enter Paradise to later make their decision for Him in the New Heaven after the White Throne Judgment (Acts 17:30; Luke 13:3,5; 2 Peter 3:9).

Luke 21:28 states: ". . . your redemption draweth nigh . . ." which is a reference to the redemption of the body and not the soul (Romans 8:23). Will these innocent ones be able to partake of the tree of life in Paradise? Certainly. Adam and Eve in their innocence had access to the tree of life.

WHO NEVER HEARD THE GOSPEL

What about the living whose names still remain in the book of life but who have not as yet accepted Christ as their Saviour at the time of the Rapture? They obviously will be the prime candidates as final Tribulation saints (Revelation 13:8).

Many of the unredeemed whose names still remain in the book of life will have their opportunity to receive the Gospel of Jesus Christ during the final Tribulation. They will be adequately warned about not taking the mark of the Antichrist (Beast) which brings eternal condemnation upon souls (Revelation 14:9-11). There is no reason why innocent children and mentally incapacitated adults will not be raptured into Paradise. Their disappearance certainly would serve as a "sign" throughout the world that something very momentous had taken place and would trigger an intensive investigation: ". . . and upon the earth distress of *nations, with perplexity* . . . *For as a snare shall it come* on the face of the whole earth" (Luke 21:25,35). Can you envision a world without a single innocent child left on it?

There also will be multitudes of "stained-glass" Christians who never have had a spiritual rebirth but who have attended church services on a fairly regular basis. The Rapture will serve as an "eyesalve" (Revelation 3:18) to open their eyes and large numbers of them will make up the first contingent of Tribulation saints.

Since the Royal Redeemed will reign *forever* as "kings" with Christ, they must of necessity have subjects in the manifested Lord's kingdom over whom they will rule. These subjects will consist primarily of those who are the natural children—descendants of the original redeemed millennialists (Revelation 21:27; 22:14,15; Isaiah 65:20).

[1]There are five kinds of wounds. Christ suffered all five: 1) contusion, 2) laceration, 3) penetration, 4) perforation and 5) incision.

53

Catalyst of the Tribulation

Individuals, churches, and religious groups would be wise to stockpile Bibles in various locations to alleviate the difficulties of Tribulation saints in obtaining the Word of God.

"And when he had opened the fifth seal, I saw under the altar the souls of them that were slain for the word of God, and for the testimony which they held" (Revelation 6:9).

The first assault upon those left on earth after the pre-Tribulation Rapture will come from pseudo-Christs (Matthew 24:4,5). People will be left in a state of shock, confusion, loneliness, and fear; implacability will reign because of panic and hysteria. People will be able to relate the event only as having some connection with Christ. Those who have had some association with fundamental Christianity and with Christians knowledgeable about prophetic events will have a general consensus of the reason for the disappearance of millions of Christians from the earth. But to most people in the world the event will be one of mere curiosity; hundreds of millions will seek to diagnose what happened. Where will they turn for solution? Deceivers with false answers will have a ready audience throughout the world. Of course, the multitudes who have at least a fragmentary knowledge of God's Word will know the truth, turn to repentance, and get right with God. They will be among the first

Tribulation saints and will be the catalyst for a great spiritual awakening.

GODS OF CONFUSION

Those with little or no knowledge of Scriptures can be deceived easily and destroyed spiritually by the dangerous Tribulation mine fields planted by deceivers: "And take heed to yourselves, lest any time your hearts be overcharged with surfeiting, and drunkenness, and cares of this life, and so that day come upon you unawares. For as a snare shall it come on all them that dwell on the face of the whole earth" (Luke 21:34). A broad concept of God's Word will expose pseudo-Christs and false scriptures, as well as individuals with a shoddy comprehension of the Holy Scriptures. It is characteristic of false religions to claim they are the sole repositories of truth, so these impostors will declare they have the answers. The seemingly reasonable explanations of impostors will prepare their adherents to embrace Antichrist when he arrives on the scene to solve the world's many problems as its messiah. In Luke 21:8 the word "Christ" is in italics, showing that it was not in the original text but that it was supplied by the translators. It should read, "For many shall come in my name, saying, I am; and the time draweth near. . . ."

HOW THE RAPTURE WILL ALTER WORLD EVENTS

Why will the Rapture or translation of born-again Christians trigger the rise of the European conqueror? Why will it hasten the invasion of the land of Israel by the Soviet Union? The United States and its allies are basically Christian and have the largest number of Christian believers, therefore, in Christian nations there will be major dislocations and disruptions. It is ironic that they will suffer the major disorganization and immobilization, even though it is temporary, while the Soviets and their supporters will remain practically unimpaired. As a result, NATO and Israel will have a greater vulnerability because their allies will be unable to rush to their defense. The opportune moment for the European conqueror and the Soviet Union to simultaneously strike will be while their adversaries are momentarily paralyzed and ineffective. As we know from Scriptures, the Soviets will be soundly vanquished. Since they and their allies are anti-Christians, it is understandable why God will bring about their great defeat. The European conqueror, however, will remain untouched in his military advances. This will confuse the world as it considers the mystery surrounding the man. Is he the coming world Messiah who is unmolested by God as he continues to swallow up nation after nation, and now with the defeat of the Soviets, advances into the eastern portion of the ancient Roman Empire to consume many of these nations? No, it is simply God's way of bringing about the reconstituted Roman Empire into prophetic alignment and fulfillment as the final Gentile power upon the earth.

SYRIAN CONFEDERACY[1]

Immediately after the defeat of the U.S.S.R. and its allies, the European conqueror will rapidly emerge to become a serious threat to the Middle East nations, Asia, and Northern Africa. As a result, two separate confederacies will be formed. One will consist of the Islamic nations in the Middle East and Asia; the other will be made up of Egypt and the countries of Northern Africa. When this is done, many Tribulation saints in the Islamic confederacy will become martyrs of the Tribulation (Revelation 6:9-11). Islam will credit the victory over the Soviets to God, not to Jesus Christ, and Israel will be feared because God is considered as being in its defense.

STRANGE FAMINE

"Behold, the days come, saith the Lord GOD, that I will send a famine in the land, not a famine of bread, nor of thirst for water, but of hearing the words of the LORD; And they shall wander from sea to sea, and from the north even to the east, they shall run to and fro (in a state of panic) to seek the word of the LORD, and shall not find it (because of the mass destruction and disappearance of the written Word of God—the Holy Scriptures)" (Amos 8:11,12).

The pseudo-Christs will capitalize on the general ignorance of the people in spiritual matters, especially concerning the Word of

God. People will be needing spiritual therapy as they seek for the truth. Distraught, bewildered, desperate, and near panic they will go anywhere and to whomever they believe will offer a reasonable answer. False Christs will expound their claims and deceptive theologies to many followers. They will have instant erroneous answers to the people's dilemma.

UNSUNG HEROES

The spiritual anointing and boldness of the first segment of Tribulation saints of the last days is beautifully described in Daniel 11:32-35, and we must do something to help their ministry at this juncture in fulfilled prophecy as we near the pre-Tribulation Rapture.

Individuals, churches, and religious groups would be wise to stockpile Bibles in various locations to alleviate the difficulties of the Tribulation saints in obtaining the Word of God. What an investment this repertory of stockpiled Bibles would be in the harvesting of souls, especially when one realizes that Bibles may be impossible to procure from retail outlets and churches where they would either be sold out or confiscated.

WHY THESE FRIGHTFUL APOCALYPTIC EVENTS

How merciful God is in permitting these tragic events to occur, and in allowing us to see them loom up on the horizon! What if the pre-Tribulation Rapture ended it all and no one could be saved from their sins after that event? There would be no Tribulation saints. Think of the multitudes who would be lost who otherwise would have accepted Christ (Revelation 7:9-17).

[1] See chart D.

54

Mysterious Weapon of Antichrist

This prophecy of Habakkuk is an avionic sword of Damocles hanging over the territories of the ancient Roman Empire today.

In the Apocalypse (6:2) we see Antichrist as the white horse rider with a mysterious weapon in his conquests—an *arrowless bow* which indicates no fatalities, as opposed to arrows which symbolize casualties. In this age, when nations are well armed with the destructive power of conventional and nuclear weapons, how can Antichrist make conquests without plunging the world into a nuclear holocaust which would obliterate over seventy percent of the human population?

His method of conquest will be unlike that of any in the history of wars. He will seize control of three heartland countries which are bordered on the north by the Soviet Union. He actually will scoff at the big powers and paralyze them into making decisions as to how to cope with his unusual strategy.[1]

DUST BOMBS

The prophet Habakkuk outlines the strategy of this conqueror so beautifully in chapter 1, verse 5: "Behold ye among the (Gentiles), and regard, and wonder marvelously: for I will work a work in your days, which ye will not believe, though it be told you."[2]

This prophecy was directed for the last days: "And the Lord answered me, and said, Write the vision, and make it plain upon the tables, that he may run that readeth it. For the vision is *yet for an appointed time,* but in the *end* it shall speak, and not lie: though it tarry, wait for it; because it will surely come, it will not tarry" (Habakkuk 2:3).

This prophecy in Habakkuk is like the avionic sword of Damocles, as it were, hanging over the territories of the ancient Roman Empire today. The words, "... that he may run that readeth it," implies that those possessing this knowledge in the last days must proclaim the Gospel message quickly for that time will be rapidly terminated once this truth is revealed.

How did Habakkuk describe the rise of this last day conqueror? "... They shall fly as the eagle ... they shall come all for violence ... they shall gather the captivity as the sand. And they shall scoff at the kings ... they shall deride every stronghold; and they shall heap *dust* and take it" (Habakkuk 1:8-10).[3]

RADIOACTIVE DUST

Four top scientists, namely., Dr. Hans Bethe, physicist at Cornell University; Dr. Frederick Seitz, physicist of the University of Illinois; Dr. Leo Sziland, biophysicist; and Dr. Harrison Brown, chemist of the University of Chicago, made a joint statement for the press in 1950, which showed that radioactive *dust* mixed with cobalt and oil could be more deadly in the next war than an atomic or hydrogen bomb.

The way in which such a weapon could be applied effectively by Antichrist and his gang of international terrorists was explained in the July, 1950, issue of *Coronet* under the title, "War Without Death," by David B. Parker.

The article described planes releasing clusters of small crescent-shaped bombs fluttering down silently spreading the black radioactive *dust* on every street and structure. Once this has been achieved, an announcement could be made to the populace that planes have dropped weakened radioactive *dust* on the city, the effects of which would not be experienced until after five days, more or less. After that, its radiation of deadly gamma rays, which cannot be heard or felt, would make a person ill. If the people were to stay in the city, death would be inevitable.

It would be imperative that government authorities call a meeting to negotiate peace terms with the adversary. The black *dust* would make the production of planes, tanks, or other war materials impossible. Air bases, submarine pens, factories, and homes would be abandoned. There would be nothing left to use militarily and no one left to fight.

A NEW RADIOLOGICAL WEAPON[4]

Black *dust* has been called a "new radiological weapon." It is inexpensive and easy to produce in a very small area such as a basement, attic, shop, garage, or home. If these radioactive materials were encased in bombs, the explosion would scatter the poisonous *dust* over a large area.

As a radioactive agent, atoms shoot out at high speed invisible radiations of several forms which vary in speed and penetrating power. All are harmful to living cells and impossible to see, hear, feel, or smell.

The radioactivity does not harm inanimate objects. It simply coats them with a thin layer of disintegrating atoms whose lethal rays injure only living things.

Scientists say radioactive agents can be produced with various life spans—one day to several days. An enormous quantity can be stored in a very small area. When released they adhere like invisible glue upon whatever they contact. The benefits of this type of radiological warfare are that buildings are not destroyed and that the entire contaminated area can be captured intact by the enemy after the danger of contamination has passed.

ROMANCE OF TERRORISM

We are now at that point in time of prophetic fulfillment where international terrorists are rapidly on the increase and speak of their activities as "romance." The potential allies of Antichrist may possibly be assembling the black *dust* radioactive weapons at this very moment in clandestine locations.

Scripture suggests that the idea for this devious weapon will be implanted in the mind of Antichrist by Satan: "And his power shall be mighty, but *not* by his own power: and he shall destroy wonderfully" (Daniel 8:24).

This was a burdensome prophecy for Habakkuk to see: "... spoiling and violence are before me: and there (is) strife and contention" (Habakkuk 1:1-4).

This method to be employed by Antichrist is so shocking, so violent, and so cruel that it is labeled: "... that bitter and hasty nation (to) possess dwellingplaces that are not theirs. They are terrible and dreadful ..." (Habakkuk 1:6,7).

[1] See chart C.

[2] "For, lo, I raise up the Chaldeans.... There are no Chaldeans now. Those ancient people have disappeared. It seems to be common practice for the prophets to call a future nation by the name of a past one. The name is applied to people of similar characteristics, rather than to people in the same place.

[3] Verse 11 changes "they" to "he" showing that Antichrist directs the operation.

[4] A proposed Super-Radiation Bomb (SRB) would leave no survivors.

55

Deadly Shadows

During the final Tribulation none of the plagues of God's judgment will come upon the redeemed.

"Blow ye the trumpet in Zion, and sound an alarm in my holy mountain: let all the inhabitants of the land tremble: for *the day of the Lord cometh,* for it is *nigh at hand;* A day of darkness and of gloominess, a day of clouds and of thick darkness . . ." (Joel 2:1,2).

The disappearance of millions of believers in Christ will be a reminder that the Millennial reign of Christ is just ahead. False christs (religions) will rise like a plague of locusts announcing the future arrival of their messiah.

Islam, which does not accept Jesus as Christ (Messiah), will declare God is about to come to earth to establish worldwide peace in accordance with its teachings in the *Koran.* Immediately, Syria, according to prophecy, will spearhead a Middle East confederacy of all Moslem nations to unite and prepare themselves for the arrival of their God—Allah!

People who are undeceived and accept Jesus Christ as their personal Saviour will become the first of the Tribulation saints. They will find themselves facing the hostility of the bigoted and intolerant religious fanatics. Many will be assaulted, persecuted, and even martyred, especially where Islam is strong within the Confederacy (Revelation 6:9-11; Obadiah 7; Psalm 83:5).

CASCADING OF EVENTS

Immediately after the pre-Tribulation Rapture there will be a cascade of events utterly changing the course of world history. First, there will be the impact of the Rapture itself which will convince millions to accept Christ as their personal Saviour; second, there will be the dynamic ministries of dedicated Bible teachers who failed to make the Rapture. Their powerful ministries, accompanied by miracles, will encourage Tribulation saints (Daniel 11:32-35); third, there will be the rise of a European conqueror who will instill in people an intense fear of impending thermonuclear holocaust (Ezekiel 39:6). In Luke 21:26 we read: "MEN'S HEARTS FAILING THEM FOR FEAR, AND FOR LOOKING AFTER THOSE THINGS WHICH ARE COMING ON THE EARTH: FOR THE POWERS OF *HEAVEN* (Greek "ouranos," the same word from which we have derived the word *uranium*) SHALL BE *SHAKEN* (or fissioned)."

Today's thermonuclear bombs have one million times more explosive power than those which wiped out Hiroshima. They weigh less than one-fifth the amount (500 pounds), and equal one million tons of conventional explosives (TNT). Each blast can shatter a four hundred mile diameter. The United States estimates there are in excess of sixty thousand thermonuclear warheads—over three tons for every man, woman, and child on earth. A single missile can carry ten or more such warheads to strike many different targets and fifty percent of the shots fired will fall where they are aimed. We have now run out of new targets for the additional warheads being manufactured.

Scientists have a growing concern about the mysterious agent called EMP (electromagnetic pulse)[1] which would be triggered in the event of a thermonuclear strike by a hydrogen bomb. It would travel halfway around the world at the speed of light if released 50 to 500 miles above the earth. The EMP, alone, would injure and kill hundreds of thousands of people. For anyone to touch or be near any metal object would result in severe burns, even death. The United States is spending $20 billion dollars annually to find a defense.

Finally, there will be the advance of Soviet forces into the Middle East because the European conqueror will serve as one of the prophetic "hooks" in the jaws of the Soviets to bring them down into that area (Ezekiel 38:4). It will be a case of taking the Middle East nations before the European conqueror does.

Prophetic Scriptures make a distinction between the *appearance* and *revelation* of Antichrist (Revelation 6:2; Daniel 7:8; 2 Thessalonians 2:8). First, he will appear as the European conqueror; second, he will be revealed as the "Beast" of Revelation 13. The "beasts" in Daniel's prophecies depict *territories,* while in the Apocalypse they represent *tyrants,* as depicted here.

SOVIET UNION DEFEATED IN ONE DAY[2]

Scriptures show that before the Soviets advance into the Middle East, they will effectively bomb the United States and its allies, perhaps even from bases which they have established in the Western Hemisphere. The United States, in turn, will seek to pulverize the Soviet Union (Ezekiel 39:6); nevertheless, the Soviet Union will continue to advance because of its first-strike effectiveness. This will be followed by an encounter with God Himself, Who will use nature to prevent the Soviet forces from deeply penetrating into the Middle East, thus finally completing Russia's defeat. Whereas the Nazi war machine was obstructed by cold and snow in Russia, the Soviet war machine will be immobilized by mud, swollen streams, landslides from earthquakes, devastating lightning bolts, and massive hailstones (Ezekiel 38:22). It will be a sequel to what is recorded in Joshua 10:11: "... more which died with hailstones than (slain) with the sword."

This defeat of the Soviet forces and its allies in "one day" is not only imperative, but it also will have a worldwide spiritual benefit: "Thus will I (God) magnify myself, and sanctify myself; and I will be known in the eyes of many nations, and they shall know that I am the LORD" (Ezekiel 38:23). This will have a tremendous effect upon

Israel: "So will I make my holy name known in the midst of my people; and I will not let them pollute my name any more; and the (Gentiles) shall know I am the LORD, the Holy One in Israel" (Ezekiel 39:7)

After the devastating defeat of the Soviets and their allies, no one will dare to challenge Israel militarily until Antichrist invades the land seven years later. This will be after Israel has buried the Soviet casualties and their military hardware, just three and a half years before the Lord's return to the earth (Ezekiel 39:9-14). Syria, however, after the defeat of the Soviet forces and their allies, will confederate the thirty-two Islamic nations of approximately 700 million Moslems. Tribulation Christians then coming to Christ will be well aware of the hostility against them from Islam and will begin to see the handwriting on the wall.

The Christians trapped within the confederated Islamic lands will become the first Tribulation saints to be imprisoned[5] and martyred. Infiltration into non-Islamic countries by Islamic terrorists will become a source of deep apprehension. Nevertheless, according to Revelation 6:9-11, Christians will be very bold in their testimony, undoubtedly encouraged by the realization that these are the last days: "And when he had opened the fifth seal, I saw under the altar the souls of them that were slain for the word of God, and for the testimony which they held."

TRIBULATION MIDNIGHT[3]

Are all of these frightful prophecies necessary from God's point of view? Certainly. They will account for the greatest harvest of souls for Christ in history and will prove to be a blessing in disguise. What a tragedy if no advance stormclouds were seen over the horizon! These signs serve as a warning to be prepared: "And when these things begin to come to pass, then look up, and lift up your heads; for your redemption draweth nigh . . . Watch ye therefore, and pray always, that ye may be accounted worthy to escape all these things that shall come to pass . . ." (Luke 21:28,36).

How unfortunate if God suddenly closed the door of redemption and there were no Tribulation saints. Millions would be lost who otherwise would make their way into the Christian fold. It is better to come under these stressful conditions than to have no opportunity to be redeemed.

No doubt when the Soviet Union and the United States, together with their allies, launch retaliatory thermonuclear(?) bombings on each other, there will be tens of thousands killed who will have accepted Christ immediately after the Rapture. Many Christian survivors also will be martyred shortly thereafter. Those who survive will make the final last trumpet Rapture before the intense judgments of God are poured upon the earth, wiping out hundreds of millions. This is known as "the day of the Lord."

Thank God that multitudes, who otherwise would have been swept up into the maelstrom of God's judgments in "the day of wrath," will be able to make their decision for Christ under these dire circumstances. In describing the latecomers into redemption, Jesus said, ". . . the days will come, when ye shall desire to see one of the days of the Son of man (before this final Tribulation on the earth), and ye shall not see it" (Luke 17:22). It is the point of no return to the more favorable days of accepting Christ and serving Him. Now the twilight hours of the "times of the Gentiles" are bringing the chill of the Tribulation winter. The Millennial summer looms just ahead beyond the Tribulation: ". . . Behold the fig tree, and all of the trees; when they now shoot forth, ye see and know of your own selves that *summer* is now nigh at hand. So likewise ye, when ye see these things *come to pass* (already arrive), know ye that the kingdom of God is nigh at hand (the manifested kingdom of God on earth under the reign of Jesus Christ)" (Luke 21:29-31).

Those who will not accept Christ during the final Tribulation will face one of the greatest deceptions of all times (2 Thessalonians 2:10-12; Romans 1:21-32). They will be the potential candidates for the mark of the Beast with its serious eternal consequences to their souls: "And all that dwell upon the earth shall worship him (Antichrist, the Beast) *whose names are not written in the book of life of the Lamb* slain from the foundation of the world" (Revelation 13:8).

This Scripture unfolds the fact that these worshippers by their demeanor already have justified God's eradication of their names from the book of life. They would have been lost souls even before their act of pagan worship; it is only that their viciousness and malfeasance are now unmasked.

BLACK SMOKE FROM THE ANTARCTIC

"And he opened the bottomless pit; and there arose a smoke out of the pit, as the smoke of a great furnace; and the sun and the air were darkened by reason of the smoke of the pit . . . And in those days shall men seek death, and shall not find it . . ." (Revelation 9:2,6).

The earth is located near the apogee of the cosmos at the point closest to the bottomless pit, which lies just outside and below the cosmos, as observed from God's throne. Thus the smoke that will bellow forth from the pit will reach the earth at the Antarctic and then envelop the earth. The heavy overcast with its sulphur-like odor will darken the sky and fill the air with suffocating effects.

Once the air begins to clear, a mysterious malady will come upon the people striking every part of their anatomy with painful, scorpion-like stings which result in horrible red blotches. The medical and pharmaceutical professions will be unable to produce an effective antidote or immunization. Business no longer can be conducted as usual.

It is soon apparent that the Christians appear immune. Like Stephen of old, who chastised his audience for their sin and rebelliousness, and consequently became a martyr at their hands, so will these Tribulation saints similarly suffer as martyrs: "Neither repented they of their *murders* . . ." (Revelation 9:21).

We see from Revelation 6:10,11 that these Tribulation Christians become martyrs at the hands of the ungodly: " . . . How long, O Lord, holy and true, dost thou not judge and *avenge our blood on them that dwell on the earth* . . . and it was said unto them, that they should rest yet for a little season, until their fellowservants also and their brethren, that should *be killed as they were,* should be fulfilled."

Some theologians believe that those who are protected from this plague and who are marked by God consist of only the 144,000 sealed Israelites, but this cannot be proven. It is more likely that they represent all of the last day Tribulation saints for these reasons: 1) there is no indication that any of the Tribulation saints are stricken by the plagues; 2) instead they appear protected: "... Come out of her *my people,* that ye be not partakers of her sins (Babylonian world-wide regime), *and that ye receive not of her plagues"* (Revelation 18:4); 3) "... *after that ye believed ye were sealed* with that holy Spirit of promise" (Ephesians 1:13); 4) "Because thou hast kept the word of my *patience* (compare Revelation 13:10; 14:12), I also *will keep thee* from the hour of temptation (Tribulation of the last days), *which shall come upon all the world,* to try them that dwell upon the earth" (Revelation 3:10).

COSMIC KILLERS ON THE LOOSE

The 200 million mentioned in Revelation 9:16 are not a reference to Armageddon nor to the kings of the East, but to evil angels in the form of insects who are possessed with an instinct to kill. These fallen angels in insect bodies will be released upon the earth from principalities in the cosmos by four powerful holy angels who have restrained them for five months from joining their wicked comrades from the bottomless pit (Revelation 9:9,14).

These fallen angels from the cosmos will be led by "the prince of Persia," who is also called "the kings of Persia" (Daniel 10:13,20). They will possess scorpion-like stings enabling them to torture their victims by causing red blotches, itchings, and burning fevers. The prey will be one-third of the human race whose names are *not* in the book of life. Hundreds of millions will die from this dreadful plague for which there is no available antidote. Yet we read of those who manage to recover: "... repented not of their *murders* ("upon the Tribulation saints" Revelation 9:21). All of this occurs under the first two of the three woes (Revelation 9:12). The final "woe" will be the seven last plagues (Revelation 11:14,18; 15:1): "... Woe, woe, woe, to the inhabiters of the earth ..." (Revelation 8:13).[4]

PURPOSE OF THE FINAL TRIBULATION

"... Come out of her, my people (Tribulation saints), that ye be not partakers of her (Babylonian) sins, and *that ye receive not of her plagues*" (Revelation 18:4).

"Seek ye the LORD, all ye meek of the earth... it may be ye shall be *hid* in the day of the LORD's anger" (Zephaniah 2:3).

The purpose of the final Tribulation is to bring into the plan of redemption all whose names are in the book of life, and to bring God's judgments upon those not registered. Between the pre-Tribulation Rapture of the saints and the return of Christ to earth, there will be a constant flow of those being redeemed: first, those who are converted to Christ during the first three and a half years and who are martyred; second, those who are redeemed during the next three and one-half years of "great tribulation" of the *saints.* Many will be martyred but many will survive to go into the post-Tribulation Rapture; third, those who are redeemed during the Great Tribulation of *Israel,* including both Israelis and Gentiles. This is confirmed by the fact that there will be a judgment of the nations with the "righteous" entering the Millennium. These redeemed ones, because they come in after the "first resurrection," do not have immortal bodies, so they will go into the Millennium in their natural bodies and repopulate the earth. The Royal Redeemed who were in the "first resurrection" will reign with Christ over them and their descendants.

None of the plagues of God's judgments during the final Tribulation will come upon the redeemed Tribulation saints. They will be protected by God (Ephesians 1:13; Revelation 18:4; 7:3; 9:4). Their deaths will come at the hands of the ungodly, not at the hands of God (Revelation 3:7,10; 9:21; 11:7; 12:17; Daniel 11:33). The judgment of God will not be poured out indiscriminately. God did not send the Flood without providing an ark of safety for the righteous. He did not send fire on Sodom without first providing an avenue of escape for Lot and his family; nor did he allow the destruction of Jerusalem in 70 A.D. without adequate warnings to Christians so they could avoid being caught in the holocaust. God faithfully follows this precept (Ezekiel 14:14,16; Revelation 14:9-11).

The judgments of God also are designed to diminish the wrath of man in committing genocide of Christians and Israelis, but it will not stop them, as evidenced in Revelation 9:20,21. The immunity God supplies to the final Tribulation saints will provide them with an advantage over the ungodly under His judgments.

[1] It would knock out delicate silicon chips in computers and affect banks, the Federal Reserve system, the stock market, the military, the IRS, federal, state, and local criminal justice agencies, credit companies, and privately owned companies, as well as radio, television, and telephone services. It would affect the nation's electric power systems by popping open circuit breakers. Street lights would go black, burglar alarms would start ringing, and many other critical areas in normal life would be disrupted.

[2] Ezekiel 38:14,19; 39:8,11,13,22.

[3] See chart K.

[4] See chart O.

[5] Imprisonment proves some will not survive for the Tribulation Rapture: "And *some* of them (not all) of understanding shall fall (be martyred)" (Daniel 11:35).

56
Coronation Day of Antichrist

The "little horn" of Daniel, chapter 8, must be distinguished from "another little horn" in chapter 7.

". . . And behold a white horse: and he that sat on him had a bow; and a crown was given unto him . . ." (Revelation 6:2).

"And a mighty king shall stand up, that shall rule with great dominion, and do according to his will" (Daniel 11:3).

First, the conqueror with a "bow" symbolizes conquests; next, he is given a "crown" representing rulership over his kingdom.

AUSTRIA: THE CONTINENTAL BRIDGE

The new kingdom will involve the reconstruction of classical Europe in its entirety which is nothing other than a new European alignment along the limits of the Old Roman Empire. Austria, which is a neutral nation on the eastern flank of non-Communist Europe, acts as a bridge between Eastern and Western Europe. The Soviets permitted it to have complete freedom in 1955, provided that it would become a permanent neutral state unattached to NATO.

Vienna, Austria, along the banks of the Danube River, was the seat of authority for the Holy Roman Empire for hundreds of years, and again it will become the focus of attention in the reconstituted

Roman Empire. Therefore, is it possible that the United Nations organization, which is now located in New York City, may suddenly be transferred to the U.N. City section of Vienna?[1] Will it give its full support to this conqueror who suddenly vanquishes these lands and assists in the enthronement of the coming European conqueror, the future Antichrist?

Kept on display in the Schatzkamer or Royal Treasury in the Old Imperial Palace (Holburg) in Vienna is the crown of the Holy Roman Empire dating back to the time of Otto the Great in the tenth century. It seems to be waiting for one final coronation.

Scripture shows that at this future time the Soviets no longer will be a threat to world peace as they will have been defeated by God (Ezekiel, chapters 38 and 39). The "Comecon" nations (East Germany, Poland, Czechoslovakia, Hungary, Rumania, Albania, Bulgaria, Mongolia, Vietnam, and Cuba), the counterpart to the EEC, will have fallen into the hands of the European conqueror. NATO also will be nonexistent.

ASSASSINATION OF A KING

". . . But within few days he shall be destroyed, neither in anger, nor in battle" (Daniel 11:20).

". . . Deadly wound . . . which had the wound by the sword . . ." (Revelation 13:12,14).

Chronologically, the "little horn" in Daniel 8 must follow "another little horn" in Daniel 7, just as the False Prophet follows Antichrist, the first Beast in Revelation, chapter 13. The "little horn" must be distinguished from "another little horn" in Daniel's prophecy.

RAM WITH TWO HORNS[2]

"And I beheld another beast coming up out of the earth: and he had two horns like a lamb (a young man), and spake as a dragon" (Revelation 12:11).

The Persian Empire, which had conquered Babylon, was made up of two countries—Media, with its ancient people; and Persia, with its more modern caste. It was called the Medo-Persian Empire, thus fulfilling the prophecy of "a ram with two horns" (Daniel 8:3,20).

Daniel 8:22 predicted that the Grecian Empire would break up into four divisions after conquering the Medo-Persian Empire. The Persian Empire lasted about two hundred years; then Alexander the Great of Greece invaded it and became ruler of the known world at the time. After his death, his generals, who were unable to keep the empire intact, saw it separated into four divisions—Asia, Africa, Syria, and Europe. Only the European division became another great empire and that was the Roman Empire.

The False Prophet described as "... two horns like a lamb" will therefore arise from this area once occupied by Medo-Persia, which is depicted as "...a ram with two horns." But from which nation?

THE LITTLE HORN

This territory which is predominantly Islamic or Moslem, is anti-Christian and anti-Israel. Out of this area a confederacy of Islamic nations will be formed and headed by Syria from which will emerge a religious figure known as the False Prophet (Obadiah 7 and Psalm 83:5). He will unite with Antichrist in persecuting both Christians and Israelis.

"And out of one of them came forth a little horn, which waxed exceeding great, toward the south, and toward the east, and toward the pleasant (Holy) land" (Daniel 8:9). Syria is the only nation from one of the four divisions of the Grecian Empire which was once a part of the Medo-Persian Empire. It is located in the right place to fulfill this prophecy.

According to Daniel 8:8-10, the *"little horn,"* the smallest of the four divisions, is the one country among the forty-two Islamic nations which will form a confederacy that will be separate from the southern one headed by Egypt in Africa. This is stated clearly in Scriptures: "For they have consulted together with one consent: they are *confederate* against thee (Israel): the tabernacles (mosques) of Edom (Arabs)..." (Psalm 83:5,6). Obadiah 7 states: "All the men of thy *confederacy* have brought thee even to the border (of Israel): the men that were at peace with thee have deceived thee, and prevailed against thee (Israel): the tabernacle (mosques) of Edom (Arabs) . . . " (Psalm 83:5,6).

Because this confederacy will make a concerted attack upon Israel, God has decreed that not one Islamic worshipper will survive to enter the Millennium: "For thy violence against thy brother Jacob's shame shall cover thee, and thou shalt be cut off forever" (Obadiah 10). " . . . There shall not be any remaining of the house of Esau (Arabs); for the Lord hath spoken it" (Obadiah 18).

JEHOVAH DISPLACED

A major reason for God's indictment upon Islam is the substitution of Allah for JEHOVAH! "Let them be confounded and troubled for ever; yea, let them be put to shame, and perish: That men may know that thou, whose name alone is JEHOVAH, art the most high over all the earth (Psalm 83:17,18).

Islam also has set up its own kind of clergy and God predicts concerning them: "Shall I not in that day, saith the LORD, even destroy the *wise men* out of Edom . . . " (Obadiah 8).

In addition, Islam has substituted the *Koran* for the Holy Scriptures, which means that they must bear the burden of judgment for this violation. The Book of Revelation, the book that completes the prophetic Scriptures, the canon of God's Word, has this warning: "For I (Jesus Christ) testify unto every man that heareth the words of the prophecy of this book (the entire Bible is prophetic), If any man shall add unto these things, God shall add unto him the plagues that are written in this book: And if any man shall take away from the words of the book of this prophecy, God shall take away his part out of the book of life, and out of the holy city, and from the things which are written in this book" (Revelation 22:18,19).

MORALITY ACCORDING TO CAIN

Islam conquers by the "sword," the symbol of its many conquests and power; its hatred of Israel is unmasked. The founder, Mohammed (or Mohammad), has a history of violence and cruelty.[3] His savagery towards the Jews at Medina is described as follows:

In the year 627 A.D., Mohammed attacked one of the Jewish tribes that resided in Medina. The Jews held out for twenty-five days, but when all hope vanished they surrendered. The prophet Mohammed condemned the males to death and the defenseless women and children were placed in slavery. Between six hundred and nine hundred men were beheaded in the central marketplace of Medina and their bodies were thrown into huge open trenches. There is no tolerance against nonbelievers under the banner of Islam. We have seen demonstrated recently the violence of which Islam is capable under its spiritual and political leaders.

The manner in which the European conqueror is assassinated is prophetic of the violence of Islam: ". . . which had the wound of a sword . . ." (Revelation 13:14). The manner in which Tribulation saints will be martyred is predicted: ". . . with the sword . . ." (Revelation 13:10).

Scripture says this European conqueror is slain by a sword: ". . . neither in anger, nor in battle" (Daniel 11:20; Revelation 13:14). It all happens according to Daniel's prophecy "in a few days" after his enthronement.

We know this takes place in Rome for it was here the "Beast" rose. ". . . Up out of *the* (Mediterranean) *sea* . . ." (Revelation 13:1). Rome juts out near the middle into the Mediterranean Sea.

BLINDED WITH RAGE

The Islamic confederacy led by Syria ("the little horn") comes into formation about the same time that the European conqueror is making his conquests. It is conceivable, therefore, that this may begin prior to the pre-Tribulation Rapture. The threat of a Soviet invasion into the Middle East and the rapid gains of the European conqueror could account for the forty-two Islamic nations considering a federation as the course of self-protection. The tragedy is that this confederacy will be used primarily to attack Israel when the Antichrist breaks his covenant of peace with that nation.

RED HORSE RIDER OF THE APOCALYPSE

"And there went out another horse that was *red:* and power was given to him that sat thereon to take peace from the earth, and that they should kill one another: and there was given unto him *a great sword"* (Revelation 6:4).

This will be an Islamic sword wielded by Moslems against Christians during the Great Tribulation. The rider will be the False Prophet who is blinded by rage, seeing "red" as it were, and finally removing ". . . peace from the earth. . . ."

So immediately after the pre-Tribulation Rapture, two mighty forces will come upon the earth—one will be political under the European conqueror; the other will be religious under a confederacy led by Syria, out of which will emerge the False Prophet.

FROM THE "KING" TO THE "BEAST"

When the European conqueror recovers, his wrath will be directed against *God* because of what happens by the sword of Islam. He nearly loses his kingdom as a result: "And I saw one of the heads as it were wounded to death . . . " (Revelation 13:3). His kingdom is the eighth in the scenario of Revelation 17:10,11. The third verse shows how the wounding of the conqueror and his kingdom occur simultaneously.

It is interesting that when the conqueror recovers and becomes the first Beast, he does not blame Christians. Rather, his insults and blasphemies are against God. Islam does not recognize Jesus Christ as the Son of God; furthermore, it recognizes Allah (or Jehovah) as God (Psalm 83:18).

The threatened loss of his kingdom is indicated: ". . . to whom they shall not give the honour of the kingdom" (Daniel 11:21). This is because he now is a different person and is incarnated by Satanic power. In using his charm and flattery to regain the kingdom, he makes two important concessions to appease the people: first, a conciliatory dialogue with the Vatican; second, a seven-year peace covenant with the Zionists (Daniel 9:27). These agreements, however, will be short-lived.

HEALING OF THE REVIVED ROMAN EMPIRE

In order for this "Beast" to retain his position as ruler over the reconstituted kingdom, the power and influence of the Roman Catholic Church is employed in an attempt to show that he is not anti-Christian. The Vatican's latest peace overtures to the Palestinian Liberation Organization (PLO) could enable the False Prophet, when he arises as the scarlet-colored beast (Revelation 17:3), to temporarily tolerate the Holy See. Also, the Vatican opposes Israel's claim to the "eternal and undivided" capital of Israel, as well as her Biblical claim to the West Bank. This places the Vatican in more sympathy with Islam than with the Israelis. Attitudes toward Israel's rights may possibly divide the "great whore" (Roman Catholic Church) from the last day "great tribulation" Christians.

In Revelation 17:3-9 we see the Roman Catholic Church symbolized as the "whore" that sits upon the scarlet-colored beast. It soon becomes apparent that she is a liability to the coalition for she will not embrace Antichrist as "Allah" (or God) incarnated, so she will be discarded shortly. According to Revelation 17:6: ". . . the ten horns (ruling hierarchy in the kingdom of the False Prophet) . . . shall hate the whore (Roman Catholicism as the symbol of Christianity), and shall make her desolate and naked, and shall eat her flesh, and burn her with fire."

When the Middle East confederacy initiated by Syria ("little horn") unites with the kingdom of Antichrist, we have the predicted "Babylon" consisting of both the political and ecclesiastical systems. (See map on "Babylon.")

TWO BEASTLY KINGDOMS CONFEDERATE

When Antichrist is enthroned at Vienna, Austria, with the crown last worn by Otto the Great, last king of the Holy Roman Empire, he will have completed his conquests of both Western and Eastern continental Europe. The apportionment of the kingdom will be into seven units (districts or states) under a body of ten rulers. The crowns are on the horns, not the heads (Revelation 13:1; compare Daniel 7:24).

These will be the strongholds under the Antichrist. When Antichrist moves to Rome he will form a coalition with the False Prophet and establish headquarters in Rome. Presumably when the Pope and his hierarchy are removed from the Vatican, both Catholic and Protestant Christians will face genocide under the new alliance (Revelation 17:3,9,16; 13:7; Daniel 11:37,38; 7:21). At this time, Rome will become the center of the Babylonian coalition, with the False Prophet probably occupying the Vatican, and Antichrist possessing the palace in the city of Rome. Islam (Moslems) sees the Pope as a symbol of Christianity; therefore, his assassination will be their objective. Under the Islamic False Prophet, in conjunction with Antichrist, they will successfully murder the ruling Pontiff of that period.

THE INTERNATIONAL PAC-MAN[4]

Antichrist's agreement with the Vatican will be broken quickly after he holds a "one-hour" conference with the False Prophet for the purpose of reconciling differences and making an accommodation. The False Prophet will emerge from this conference declaring that Antichrist is "God" (Allah) incarnate, and will place a vestment upon him. He will then unite his ecclesiastical kingdom of Islamic followers with those of other religions (Revelation 17:15) in support of Antichrist: "And the ten horns (ruling Islamic clergy) which thou sawest upon the beast (False Prophet), these shall hate the whore (Catholicism, identified with Christianity), and shall make her desolate and naked, and shall eat her flesh and burn her with fire" (Revelation 17:16). Observe that in the Apocalypse "heads" means nations, kingdoms, or divisional units, and "horns" (upon which are crowns) refer to the powerful ruling body.

The Antichrist is now just what his name indicates—*anti*-Christ—and he will tolerate no affront to his new position as "God" (Allah) incarnate. Christianity in any form, whether Catholic, Protestant, or independent, will face genocide. Martyrs' blood will flow and the concentration camps under the Nazis will be considered mild by comparison. The tragedy is that so many nominal Christians will be martyred who are not, by divine experience, born-again. The blood of martyrs has no atoning value over sin; only the blood of Jesus Christ atones for sin.

A RADIOACTIVE CITY[5]

The future destruction of Rome as the Babylon of prophecy will be by a nuclear bomb followed by a tidal wave: "The merchants of these things, which were made rich by her (Rome, Babylonian Capital), shall stand *afar off for fear of her torment* (dangerous radioactivity) . . . they saw the smoke of her burning . . . " (Revelation 18:15,18). "And they cast *dust* on their heads, and cried, weeping and wailing, saying, Alas, alas, that great city . . . for in *one hour* is she made desolate" (Revelation 18:19). Soviets will have been wiped out in *one day*.

The past destruction of ancient Babylon was not by a flood; it simply fell into ruins. Therefore, the following text is prophetic of the last day Babylon (Rome), as well as other Mediterranean cities: "*The sea* is come up upon Babylon: she is covered with the multitude of the *waves* thereof. Her *cities* are a desolation, *a dry land,* and a wilderness, a land wherein no man dwelleth . . . " (Jeremiah 51:42,43). Prophecy declared that ancient Babylon never would rise again: "And thou shalt say, Thus shall Babylon sink, and *shall not rise* . . . " (Jeremiah 51:64). In verse 26, " . . . thou shalt be *desolate for ever,* saith the LORD."

This devastation of Rome (Matthew 25:31-46) will bring about the destruction of the composite kingdom of Antichrist and the False Prophet because the "kings of the East" will move upon the kingdom to initiate the battle of Armageddon.

Tyrus was a prophetic name for the last day Babylon and carries this prophecy: "What city is like Tyrus, like the destroyed in the midst of *the sea?*" " . . . In that time when thou shalt be broken by the seas in the depth of the waters: Thy merchandise and all thy company in the midst of thee shall fall" (Ezekiel 27:32,34).

"And the Lord GOD of hosts is he that toucheth the land, and it shall *melt* (thermonuclear blast and the heat from the plague), and all that dwell therein shall *mourn,* and it shall rise up wholly like a flood: and shall be drowned, as by the flood of Egypt.

"It is he that buildeth his stories in the heaven (heavenly universes), and hath founded his troop on the earth; and he that calleth for the waters of *the sea,* and poureth them out upon the face of the earth. The LORD is his name" (Amos 9:5,6). In Amos 8:11 we see this was prophetic: "Behold, the days come, saith the Lord GOD . . ."

[1] The U.N. had a training camp for PLOs in Beirut. It is conceivable it will support and even enthrone the conqueror who will be the Antichrist.

[2] The language of the so-called nondescript beasts in Bible prophecies is clarified by the interpretations that follow.

[3] The *Koran* clearly states in the words of Mohammed that the Jews will never return to Palestine. Hence for Moslems, there can be only two possible conclusions: either the Jews must be driven out of Palestine or the *Koran* and Mohammed have lied.

[4] A video game in which Pac-Man is the devourer.

[5] An international committee of scientists in 1983 reported through the World Health Organization that half of the world's population will suffer casualties in an all-out nuclear war.

57
A King's Nightmare

Conditions are now ripe for a massive end-time reshaping of the world's political systems.

"And as the toes of the feet were part of iron, and part of clay, so the kingdom shall be partly strong, and partly broken.

"And whereas thou sawest iron mixed with miry clay, they shall mingle themselves with the seed of men: but they shall not cleave one to another, even as iron is not mixed with clay" (Daniel 2:42,43).

Revelation 17:10,11 traces the major kingdoms as follows: Egyptian, Assyrian, Grecian, Medo-Persian, Roman, and the revived Roman Empire. Daniel's prophecies record the rise and fall of such great empires as the Babylonian, Medo-Persian, Grecian, and Roman. When the Roman Empire exists again in the last days, it will be a reconstituted empire which must begin and end at Rome. It will consist of a land area covering one-third of the earth, in contrast to the ancient Roman Empire which covered two and one-half million square miles, or about five-sevenths the size of the United States.

Today we see numerous nations within the boundaries of the ancient Roman Empire; some are strong like iron, and others are weak like clay. These nations include Italy, Spain, Portugal, France, Belgium, Netherlands, West Germany, Switzerland, Austria, Luxembourg, Bulgaria, Hungary, Rumania, Yugoslavia, Albania, Greece, Asia Minor (Turkey), Syria, Lebanon, Israel, Jordan, Egypt, Tunisia, Libya, Algeria, and Morocco, as well the Mediterranean islands.

FALL OF THE ANCIENT ROMAN EMPIRE

Nebuchadnezzar beheld a great image in his dream. In Daniel's interpretation of the meaning of the image, he described the pair of legs as the two divisions of the Roman Empire in its demise—the Eastern division in Byzantium (later Constantinople; presently Istanbul), and the Western division centered in Rome. With the fall of the Western Empire in A.D. 476, a transitional period followed with the Empire occupied and governed by kings of the Germanic race. The revival of the Roman Empire will be a non-Aryan Empire.

The ancient Roman Empire began its final demise politically in 1918: first, it started with the deposed czar (meaning, "Caesar"), Nicholas II of the Imperial Russian Government, who was executed on July 16, 1918; second, it continued with Kaiser (meaning "Caesar") Wilhelm II's merger of the demised nations of the Hapsburg Empire (Austria, Bohemia, and Hungary) into the greater Third Reich in 1938. The ecclesiastical phase remained at Rome.

With the fall of the Russian czar, and the approaching collapse of the Soviet Union predicted in God's Word, it is highly probable that Antichrist not only will take both Western and Eastern Europe, but also much of Russia. In addition, China will conquer a great portion of the Eastern territory of the U.S.S.R. which penetrates deeply into Siberia. It is from the Far East that the "kings of the East" will assemble their military might in preparation for the invasion of the last day Babylonian regime under Antichrist and the False Prophet. They will enter from both the east and the north across the present Soviet territory (Revelation 16:12; Daniel 11:44). This invasion will culminate in the battle of Armageddon (Revelation 16:16).

AMERICA AND EUROPE AT THE CROSSROADS

The nations of continental Western Europe are almost totally dependent upon the Mideastern and North African oil. Energy-starved

Western Europe will begin to turn more to the Soviets for its source of energy as the present Middle East oil supplies deplete. In fact, in 1980 the Soviets were supplying thirty percent of the natural gas consumed in West Germany. In addition, Soviet oil reserves are now estimated to be as high as 4.55 trillion barrels, which is many times greater than the combined reserves of the Middle East oil fields. Western Europe's dependence upon the Eastern bloc for energy sources is growing increasingly.

After World War II, war-torn Europe looked to the United States, not only for its military protection, but for a societal model as well. America was the "Promised Land." In recent years, however, a wave of isolationism has been sweeping across Western Europe, causing the United States and Western Europe to steadily drift apart. Europeans now are fearful of American military equipment coming into their territory. They feel it may offend the Soviets, and give them justification to discontinue Europe's energy supplies; also, they fear it may bring about the U.S.S.R.'s interference in their lands.

Conditions are now ripe for a massive end-time reshaping of the world's political systems. Therefore, it is mandatory that the Common Market nations maintain a peaceful coalition with the Soviets once the gas pipeline is completed. Any indication of a confrontation with the Soviet Union could trigger the predicted European conqueror to move swiftly to neutralize the Soviet threat of an invasion. It also would establish the conqueror's objective of being a world power.

A blitzkrieg seizure of mainland Europe by the European conqueror for the purpose of achieving independent solidarity could force the Soviets to make their military drive into the Middle East which will end their power.

It is likely that Antichrist will conquer all of the territory into which the Grecian empire extended—Asia, the Middle East, Northern Africa, and Europe (a Grecian name): ". . . the four winds of heaven . . ." (Daniel 8:8. See verses 21-24; also, Habakkuk 1:9; 2:2,3). Because Germany under Kaiser Wilhelm II was the *last political remnant* of the Roman Empire to *fall,* in all probability, Germany will be the *first* nation to *fall* to the European conqueror (the future Antichrist) in his attempt to reconstitute the Roman Empire.

THE TREATY OF ROME

The Roman Empire is in its embryonic state of revival, and since Rome is the place where the *religious* power of the ancient Empire ended, it is also the place from where it must begin its revival. On March 25, 1957, under "The Treaty of Rome," the European Economic Community (EEC or Common Market) was formed. The original signatories to this document were Italy, West Germany, France, Belgium, Netherlands, and Luxembourg. The former three are the heartland countries of the EEC; the latter three, known as the Benelux nations, were created as a customs union later in 1957. On January 1, 1973, Denmark, Ireland, and the United Kingdom were added to the EEC. On January 1, 1981, Greece officially became its tenth partner. Spain, Portugal, and Austria have made applications for membership. On May 20, 1982, King Carlos of Spain was awarded the Charlemagne Prize at Aachen, West Germany (Charlemagne's old capital), as the figure most advancing the cause of Europe's unity. Could this conceivably mean that when Spain becomes the tenth and final partner of the EEC *within* the boundaries of the ancient Roman Empire that the final crisis of the Apocalypse will begin?

The Eastern leg of Nebuchadnezzar's image is composed partially of the Comecon nations (Council for Mutual Economic Assistance), the Soviet's counterpart of Western Europe's Economic Community, the EEC.

HEARTLAND COUNTRIES OF EUROPE

The heartland countries of the EEC have a unique place in Bible prophecy. They are scheduled to be the first to fall prey to the European conqueror of Revelation 6:2.

At the "heart" of these nations is a diminutive country which prophecy refers to as "another little horn" (Daniel 7:8,24). This little nation will be the base of operations from which the conqueror ("in this horn were the eyes of a man") and his band of international terrorists will subdue the heartland nations in a planned solidarity of all Europe. It will begin in the reconstituted Roman Empire, predicted for the last days, which will envelop all of the nations within the ancient empire: "I considered the horns, and behold, there came up *among them another* little horn (the 11th and it is a small nation), before whom there were *three of the first horns* (original signatory nations of the EEC) plucked up by the roots . . . And the ten horns out of the kingdom (all ten territorial units of one or more nations must be within the boundaries of the ancient Roman Empire before this event will take place) are ten horns that shall arise *after them* (not before); and he (the conqueror to dominate the scene) shall be diverse from the first (the ancient Roman Empire will be smaller in territory than the revived empire whose final influence will be worldwide and referred to as "Babylonian" world power), and he shall subdue *three* kings" (Daniel 7:8,24).

The present lineup of nations in the EEC does not yet total ten *within* the boundaries of the ancient Roman Empire. Denmark and Ireland are outside of it. We can, therefore, expect to see other nations coming into the EEC *before* this European conqueror will make his appearance to conquer the heartland nations.[3]

ANOTHER LITTLE HORN

Sandwiched among the only three original signatory nations to "The Treaty of Rome" is the little nation—Switzerland! It is right in the center at the right place at the right time. In this tiny, unpretentious country, many world conclaves are held intermittently.

Switzerland offers all of the potential elements of fulfilling the prophecies relating to the rise of this conqueror with a "bow" (Revelation 6:2)[1] It has been a neutral nation since the 16th century. Swiss citizens are forbidden to participate in war anywhere. Although it is host to many world organizations within the United Nations, it has not joined that body. It does not have an army; it is an army. Switzerland has an obsession for building fallout shelters which are capable of accommodating fifty percent of its population. It expects it will be able to provide protection for all of its people from the effects of a nuclear war by the year 2000.

The Swiss banks, with reinforced concrete underground vaults or bunkers, are hidden deep in the Alps. They are bulging with gold, silver, precious valuables, and gems, as well as deposits by deposed and current rulers, wealthy individuals, corporations, and industries. Everything is held under a *number system* known only to its depositors.

These deposits are regarded as the safest in the world. However, here is ready cash for the "conqueror" in his bid for a revival of the ancient Roman Empire. Antichrist will incorporate a world number system in his Babylonian rise to world power. Here in diminutive Switzerland, plans already may be in the making by this conqueror and his international terrorist henchmen.

LAST DAY BABYLON

The European Economic Community (EEC) under Antichrist will become part of the revived Roman Empire. Nations will be divided into seven units ("heads"), to be controlled by ten governors ("horns"). This kingdom of Antichrist must be distinguished from the kingdom under the False Prophet. When these two kingdoms combine they will cover one-third of all of the land surface of the earth, and will be referred to as "Babylon" (in prophecy).[2] This is the reason so many of the judgments are zeroed in upon one-third of the earth.

More than one hundred nations now maintain accredited representatives at the Commission's headquarters in Brussels. Babylon, under Antichrist and the False Prophet, will have great influence over the world's economy, and will seek to dominate the world scene militarily, politically, and religiously. The system will institute the predicted 666 code which is referred to in Scripture as the "mark of the Beast." Because of its identification with Satanic power backing it, it will face the wrath of God.

THE EUROPEAN TRIANGLE

". . . and a crown was given unto him . . ." (Revelation 6:2).

". . . and upon his horns ten crowns . . ." (Revelation 13:1).

Italy, Switzerland, and Austria will form the European triangle in the initial stages of a reconstituted Roman Empire. Besides the coronation of the European conqueror, there will be the enthronement of the ten surrogate rulers in his egalitarian kingdom.

From Italy (Revelation 13:1) the European conqueror will move to Switzerland to begin his conquests (Daniel 7:8, 24). Once these have been achieved he will be enthroned in Vienna, Austria, where the crown of the last ruler of the Holy Roman Empire is kept (Revelation 6:2). From here the crowned European conqueror will return to Italy where his kingdom will combine with the Syrian kingdom (confederacy) to form Babylon. Through control of the world's economy and religions, the Antichrist and False Prophet will seek to conquer the rest of the world; they will not succeed, however. The "kings of the East" will begin a military move to overthrow this Babylonian dynasty which will result in the most calamitous destruction in all history, the Battle of Armageddon!

[1] See chapter 54.

[2] Babylon is the apostasy to which Paul refers in 2 Thessalonians 2:3,4 concerning the falling away from the faith when Antichrist is revealed.

[3] Ten in Scripture is an indefinite number (Revelation 2:10; Jude 14; Luke 15:8). Therefore, the EEC could be more than ten nations before the rise of Antichrist.

58

Antichrist Profiled

There can be no simultaneous world conquerors. One will be eliminated, which will be the Soviet Union.

The word "Antichrist" appears only four times in Scripture—I John 2:18,22; 4:3; 2 John 7). He is described as a conqueror, beast, vile person, man of sin, son of perdition, proud man, abominable branch, raiser of taxes, king of fierce countenance, eyes of a man (in "another little horn").

The first prophecies of Antichrist are in Genesis 3:15: "And I will put enmity between thee (Satan), and the woman (symbol of the human race), and between thy seed (Antichrist) and her seed (Christ. Compare Galatians 3:16); it shall bruise thy head (destroy Satan's power), and thou shalt bruise (persecute) his heel (Church: the Body of Christ which is on earth).

The second prediction is found in Genesis 12:3: "And I will bless *them* (Gentiles, plural) that bless *thee* (Israel), and curse him (Antichrist, singular) that curseth thee, and in thee (Israel) shall all families of the earth be blessed (during the Millennium and in the New Heaven)."

SIMILARITIES

Christ was born in Bethlehem, a small place; Antichrist will rise from a small nation. Christ will have a kingdom; so will Antichrist. Christ is called a prince; so is Antichrist.

CONTRASTS

Christ will bring peace; Antichrist will produce violence. Christ came as an innocent babe; Antichrist will arrive as a wicked adult. Christ was surrounded by simple shepherds; Antichrist will be flanked by terrorists. The shepherds had staffs; Antichrist's supporters will be armed with deadly weapons. Christ was persecuted and killed; Antichrist will be the persecutor and killer. Christ was filled with the Holy Spirit; Antichrist will be filled with an evil spirit. Christ elevated women; Antichrist will pervert them. Christ honored marriage; Antichrist will disdain marriage. Christ is the Lamb of God; Antichrist will be the incarnated Beast of Satan. Christ is voluntarily worshipped; Antichrist will demand worship. Christ was "raised up a horn of salvation"; Antichrist is "a horn of destruction." Christ was crucified and resurrected to teach truth and life; Antichrist will be assassinated with a deadly wound but will recover to spread his evil and violence. Christ will have the inscription "Holiness unto the Lord" on all products intended for commerce during the Millennium; Antichrist will have a mark placed within the flesh of persons for the purposes of trade during his reign.

GOD HOOKS THE "JAWS" OF THE SOVIETS

The Soviet Union cannot make a move into the Middle East to attack Israel until God's time as denoted: "And I will . . . put hooks into thy jaws, and I will bring thee forth, and *all* thy army . . . even a great company . . . . " (Ezekiel 38:4). Before the Soviets make their invasion into the Middle East they must first take Persia (Iran) and Turkey (Ezekiel 38:5,6). The military move of the Soviets will coincide with the sudden rise of the European conqueror because the U.S.S.R. is distrustful of his goals. The ambition of the European aggressor is to conquer the world; the Soviets have the same goal. Since there will be no simultaneous world conquerors, one will be eliminated, and that one is the Soviet Union.

Immediately preceding its invasion into the Middle East, the U.S.S.R. will rain bombs on the United States and its allies, who in turn, will retaliate: "And I will send *fire* on Magog (Soviet Union) and among them that dwell carelessly in the *isles* (United States and her allies) . . . " (Ezekiel 39:6). The word"carelessly" subtly suggests this will occur *after* the pre-Tribulation Rapture (Luke 21:28,36). Those who are prepared, and not putting their trust in the military strength of their nation, will be raptured to meet Christ.

GOD'S CANNONADE[1]

" . . . I will rain upon him, and upon his bands, and upon the many people that are with him, an overflowing rain . . . " (Ezekiel 38:22). This will create mudslides which will immobilize their vehicles by burying them deeply and hopelessly in the mire. The defeat of the Soviets will be so complete that only one-sixth of their military forces will be intact. The bedraggled remnants will flee into Northern Siberia near the Bering Sea (Ezekiel 39:1-5). There will remain such an abundance of carcasses and military debris that it will take Israel seven years to cleanse the land (Ezekiel 39:8-12). It will take seven months just to bury the dead.

KINGDOM OF SEVEN PRINCIPALITIES

When the iron and clay nations finally translocate from the bottom of the image to the horns on the head of the Beast, there will be a realignment of national boundaries into seven divisional units or states within the revived Roman Empire. They will be under ten rulers who are surrogates of the Antichrist. The Bible predicts the existence of this world-encompassing economic and political bloc at the close of the age. It will be combined with the ecclesiastical bloc of nations which is also divided into seven divisional states under a governing body of ten clergymen. This is known as Babylon.

The Bible describes the expansion of the restored Roman

Empire, Babylon, as the embryonic stage of a mass attempt at genocide of Christians first, and then Jews. Antichrist's conquest will be swift: "And with the arms of a flood shall they be overflown from before him, and shall be broken . . . " (Daniel 11:22). Therefore, the cry arises from his admirers and worshippers: " . . . Who is like unto the beast? who is able to make *war* with him?" (Revelation 13:4). He seems invincible.

The first Beast (Antichrist) knows the source of his power and gives credit to the Devil by including him in the worship: "And they worshipped the dragon which gave power unto the beast . . . " (Revelation 13:4). Certainly if there had been any previous doubt as to the identity of Antichrist, it is now crystal clear: "And then shall that wicked (Satanic power incarnated in Antichrist) be *revealed,* whom the Lord shall consume with the spirit of his mouth, and shall destroy with the brightness of his coming" (2 Thessalonians 2:8).

This is not the Battle of Armageddon involving all of the nations of the world. That lies in the future. As severe as this war will be, comparatively it is but a skirmish. It is in the wrong place and at the wrong time (Ezekiel 39:11. Compare 12:11).

COSMIC MONSTER INVADES THE EARTH

Satan is highly organized in the planetary heavens. When he is dislodged and evicted to the earth, he will quickly regroup his fallen comrades into a formidable power. In order to accomplish his aims of survival and possession of a kingdom on earth, he must incarnate certain promising personages and leaders.

Satan desires to establish a powerful political and religious kingdom on earth that will prevent Christ from ruling the earth with His holy angels and Royal Wife. Antichrist's form of government will be similar to that of Satan's in his heavenly dominions. He will take all of the clay nations within the boundaries of the reconstituted Roman Empire and make them into iron nations by dividing them into seven sections (principalities, states, regions, or provinces). These will be under a controlling body of ten persons called "kings," in imitation of the Royal Redeemed. Observe the crowns are upon the horns, not upon the heads (Revelation 13:11).

One of these seven heads (regions) conceivably could be the Vatican state at Rome because of its strategic location and influence. This ecclesiastical state, which is a symbol of Christianity in the eyes of the world, will be wounded when the pope is assassinated. The Rapture of the redeemed also will have a devastating effect upon the Vatican State and threaten its very existence. Its survivors will be hard pressed to justify why they were not translated with other Christians. However, Antichrist will temporarily give his support (Revelation 17:16,17) since this is his Mother Church (Daniel 11:37).

When Satanic powers enter man, they turn him into a *bitter beast,* but when the Holy Spirit enters, He makes him a *better man.* The ancient Roman Empire was originally pagan, and it will end the same way when it is fully reconstituted and united with the kingdom of the False Prophet in the last days. Both Denmark and Ireland lie outside of the ancient Roman Empire and cannot be included as the ten nations to reconstitute its revival in the last days. Therefore, two more nations that are within its boundary must yet come into the EEC (Common Market) to fulfill the prophecy of the ten horns in Daniel 7:24. The ten horns will be in place within the boundary of the ancient Roman Empire ("out of this kingdom") before the rise of the European conqueror, the future Antichrist, who subjugates the first three signatory nations to "The Treaty of Rome."

[1] See page 457.

59

Covenant With Death

God has always provided an escape for His people.

An erroneous conception is that the number 666 will be the actual mark of the Beast. If this were the situation, all that would be necessary to decode the facade is *knowledge;* however, we read that *wisdom* will be required.

The number is clearly that of a "man." It is not a machine, such as the giant computer in Brussels, which has been termed by some as "the Beast." Neither is it a card system since it will be placed directly in the forehead or right hand (Revelation 13:16).

The number is not designed as trickery to entice Christians to accept it. First of all, this will be an anti-Christian regime and it will not want them in its program. Moreover, names of the believers are in the book of life and cannot be placed on a register of death with the Beast of Revelation 13:6. Christians will not be rounded up like cattle for branding against their will. Any such attempts would certainly be unsuccessful: ". . . *that if it were possible* (which it is not) they shall deceive the very elect" (Matthew 24:24). Only those whose names are *not* in the book of life and who take the mark of the Beast will receive maximum penalty (Revelation 13:8; 14:9-12).

THE ESCAPE CLAUSE

God has already placed His brand upon the foreheads of the Tribulation saints (Revelation 9:4). This will be clearly understood by Satanic forces as a "hands-off" policy. If there is no escape clause in the code 666, how can there be any "righteous nations" to enter the Millennium? (Matthew 25:31-46). How can Christ and the redeemed reign if there are no subjects?

The code 666 offers three options: 1) a mark on the right hand or forehead; 2) the number of his name; 3) the name of the beast (Revelation 13:16,17). Only the first two of these are associated with worship of the Beast or his image (Revelation 14:9,11; 15:2; 19:20; 20:4). The last option under the name of the Beast enables nations to engage in commerce with the revived Roman Empire without the requirement of worship. God has provided this way of escaping extermination (Revelation 14:9-11; 19:20).

The Beast is not just a number. He has a name. He is a person, a human being, who is incarnated by Satanic power. One possessing divine wisdom can avoid the curse of the mark and number of his name and continue to survive. Worship of the Beast or the image is never associated with knowing the "name" of the Beast.

Naturally, people who do not worship the Beast or his image (especially those living within the revived Roman Empire) will be viewed by his followers with askance and as enemies of the State, but an international code for righteous nations will still make it possible to participate in business with the revived Roman Empire. The curse of God is not on the code which is essential for international trade and business transactions. It is the *worship* of the Beast and his image that violates God's code.

It is clear from Scriptures that the code 666, which is associated with the "name" of the Beast, not only is essential to transact general business, but also is necessary to allow survivors of the Tribulation to enter the Millennium. However, it can be a frightening experience to many people when they see this international code on products which they need to purchase, but because of lack of wisdom they refrain from it and perhaps suffer and famish needlessly.

God always has provided an escape for His people. He did it for the early Apostolic Church when Jerusalem was destroyed in A.D. 70. There were those who were rescued when the fire fell on Sodom, Gomorrah, and the cities round about. When the Great Flood came in Noah's day, there was deliverance. Since He is the same God, He will again make a way of escape. However, neglecting the way of salvation through Jesus Christ can close forever all routes into an eternity with God: "How shall we escape, if we neglect so great salvation . . ." (Hebrews 2:3).

[1] The word "in" could indicate some kind of insertion under the skin.

60

Middle East Aflame

After the Soviets are defeated, Israel will require a period of seven years of peace to fulfill the prophecy in Ezekiel 39:9.

Psalm 83 shows a clear picture of the recent hostility of Islam toward Israel as a result of the activities by the Palestinian Liberation Organization. Let us scrutinize some interesting passages from the Scriptures.

The PLO is described as a "confederacy" against Israel (verse 5). It has worship centers, called mosques, which the Psalmist identifies as "tabernacles" and "houses of God," and it makes a "tumult" by stirring up hostility toward Israel and by taking "crafty counsel against" it. Perhaps this is best summed up in verse 4: "They have said, Come, and let us cut them (Israel) off from being a nation; that the name of Israel may be no more in remembrance."[1]

The Psalm concludes with God's anger at these Islamic disciples for changing the name of God from JEHOVAH (to ALLAH): "Let them be confounded and troubled for ever; yea, let them be put to shame, and perish: That men may know that thou, whose *name alone is* JEHOVAH, art the most high over the earth."

LEGAL OWNER OF THE HOLY LAND

God calls the Holy Land "my land" (Ezekiel 38:16), and of Israel He said, ". . . I shall place you in your own land . . . " (Ezekiel 37:14). "Moreover I will make a covenant of peace with them (Israel): it shall be an *everlasting covenant with them:* and I will place them, and multiply them, and will set my sanctuary in the midst of them for evermore. My tabernacle also shall be with them; yea, I will be their God, and they shall be my people. And the (Gentiles) shall know that I the LORD do sanctify Israel, *when my sanctuary shall be in the midst of them* for evermore" (Ezekiel 37:26-28). Expect this temple to be erected soon on Mount Moriah, the Temple Mount.

There is no way Islam or the Arabs ever will be able to break God's prophetic Word because there will be no Millennial peace on earth until Jerusalem becomes a city of peace (Psalm 122:6). God, the owner of the Holy Land, can give it to whomsoever He pleases, and it pleases Him to give it to Israel.

ISRAEL AND THE PROMISED LAND

More than three million Palestinian Arabs and about the same number of Israelis live in Israel. Half of these PLO's reside in Israel on the West Bank; the others in the Gaza Strip. The majority live in Jordan. When Lord Allenby led the British forces into Jerusalem in December of 1917, Palestine had a population of about 600 thousand which consisted of ten percent Israelis with the rest being mostly Moslem Arabs. Previous to the British occupation the country had been under the Islamic rule of the Arabs and under the dominion of the Turkish Ottoman Empire for thirteen centuries. A month before Allenby entered Jerusalem, British Foreign Secretary Arthur Balfour had made his famous declaration that "the British government looked with favor on the establishment of a national home for the Jewish people in Palestine."

An advantage to the Jews at that time was the fact that Zionism was already an influential force internationally. The Ottoman Empire was falling apart and its various nationalities, such as the Arabs,

Armenians, Kurds, and Jews were pressing their claims to the right of self-determination for all people as promoted by President Woodrow Wilson.

Under the League of Nations mandate, which formalized British rule in Palestine, the Jewish right to migrate and settle in the land was recognized. By 1939, the Jewish population formed almost one-third of a total of 1.5 million. The first uprising was by Arabs opposing British rule and Jewish settlements. The arguments over whether there should be a partition or a unified state were inconclusive. The British, unable to reach an agreement or to impose a solution, laid down the mandate and placed Palestine to the judgment of the United Nations. As a result, Palestine was divided into no less than eight parts in that peace body: "... they ... parted my land. And they have cast lots for my people..." (Joel 3:2). The Scriptures reveal in this verse that God is diametrically opposed to the promised land being partitioned in the last days, and some theologians believe this is the reason for the dismemberment of the British Empire upon which the sun had never set.

The Zionists accepted the plan of partition but the Arabs rejected it. The result was war and Israel's proclamation of independence. The Gaza Strip and the bulge of the West Bank were held by the Arabs, but the Jewish forces drove them out.

Israel became a state with international recognition and powerful support, notably from the United States. Two more wars, in 1967 and 1973, eventually strengthened Israel militarily.

As of this writing, the Middle East has been an endless source of trouble in a strategic area which produces most of the oil consumed by Western Europe and the United States. The PLO, which was trained in Libya and backed by the Soviet Union, has increased its record of terrorism. The Camp David agreement in 1978 calmed the waters between Israel and Egypt, but it has increased the anger of the Palestinians and the Arab sympathizers.

The PLO has refused to recognize Israel's right to existence as an independent state. The EEC (European Economic Community or Common Market) was asked to intervene for new security arrangements and the right of self-determination for the Palestinians, but Israel has flatly refused such a suggestion and the United States has remained cool to the idea. "No sane Israeli will accept or countenance a scheme of things which is tantamount to the extermination of Israel," said the Israeli Ambassador to Britain.

ISLAM AND BIBLE PROPHECIES

Psalm 83, as well as Obadiah, depicts the demise of the Islamic religion by the time of the Millennium: "Let them be confounded and troubled for ever: yea, let them be put to shame, and perish: That men may know that thou, whose name is JEHOVAH, art the most high over all the earth" (verses 17,18).

ALLAH is not the same as JEHOVAH. Moslems believe that ALLAH is one God and scorn the concept of a trinity. Therefore, they have in essence tragically blasphemed the Holy Spirit (Matthew 12:31,32). They reject Jesus Christ as the Son of God and as man's Redeemer and have reduced Him to the level of a prophet like Mohammed. This is why they can join so easily with Antichrist against Christians who believe in Jesus Christ as God and as mankind's only Redeemer.

BAHAIS

The Bahais (Mohameddan sect) believe that Mohammed Husayn Ali (1817-1892), called Bah'u'llah (Glory to God), is the latest prophet in the prophetical line. This has divided the Moslem faith, and the Bahais are being persecuted and killed by the Moslems in many cruel ways.

ISRAEL'S SEVEN YEARS OF PEACE

The Soviets and their Islamic allies ultimately will invade Israel in an unsuccessful move to exterminate it. By this time Israel will have recovered the golden Ark of the Covenant (Testimony) and probably will have started, or even completed, the construction of the Temple. The Ark of the Covenant apparently will play an important part in the defeat of its enemies just as it did in ancient times (Ezekiel 38:21-23; 39:1-5).

Syria, *after* being defeated together with the Soviet forces, will form a confederacy of Islamic nations in the Middle East and Asia for protection against the advancing European conqueror.

After the Soviets and their allies are defeated, Israel will require a period of seven years of peace to fulfill the prophecy in Ezekiel 39:9. During those seven years Israel will erect its Temple in Jerusalem (if not already completed), have its dedication, and build a new community over the Soviet grave sites. It will take seven months to bury the dead after this confrontation (Ezekiel 39:12). The full seven years will be required to dispose of the military debris (verse 9). The second half of this seven-year period will be under the new peace agreement with Antichrist.[2] However, Antichrist will break his peace agreement with Israel after it has been in effect only three and one-half years. At this point in time, the False Prophet, in cooperation with Antichrist, will direct the Islamic nations against Israel, and Antichrist will sit in the Temple built in Jerusalem declaring that he is God incarnate. This will be done with the canonical blessing of the False Prophet as the leader of Islam. Israel fears these two madmen (Jeremiah 48:2), so it will flee the Holy Land and find places of haven in countries referred to as "righteous nations." Most of these righteous nations will organize as the "kings of the east" to topple the two kingdoms which have united as Babylon.

THE APOSTATE CHURCH

The confederacy of Islamic nations will include other religions. On September 1, 1979, ten major religions of the world met for one week at the Princeton Theological Seminary in Princeton, New Jersey, for their second "World Conference of Religion and Peace." The president was Angelo Fernandes, a Roman Catholic Archbishop of India. The religions involved were Christian, Buddhist, Confucianist, Hindu, Jewish, Sikh, Shintoist, and Zoroastrian. Eighty percent of these are Eastern religions which theologically, at least, are anti-Christian and accept idolatry, in accordance with the requirements under the Babylonian regime (Revelation 17:15).

The False Prophet is the scarlet-colored Beast of Revelation, chapter 17, and the seven heads are Rome (verse 9), which shows that both the Antichrist and False Prophet finally are located in the same city, thus making it the center of the united Babylonian Empire of the last days. It also reveals that this kingdom is divided into seven territories, as is the Antichrist's. However, in this confederacy of the False Prophet the nations retain their borders, and are ruled by a ten-member council of clergymen (Revelation 17:12) which initially makes frontal assaults against Christians. Later, with the political and military backing of Antichrist, upon whom they will have conferred the honor of *God* (ALLAH) *incarnate,* these assaults also will include Israel.

FIRE OVER LEBANON

Events cast their shadows before their occurrence, so one of the first signs of Syria's emergence as the predicted "little horn" will be its role in Lebanon. In Zechariah 11:1-6 we read in part: "Open thy doors, O *Lebanon,* that the fire (bombings, destruction) may devour thy cedars . . . the pride of *Jordan*[3] (next to equally suffer destruction) is spoiled . . . I will deliver the men, every one into his neighbor's hand (Syria), *and* into the hand of his king (False Prophet) and (Antichrist) shall smite the land, and out of their hand I will deliver them." Syria will become the dominant nation of the Islamic Arab world in forming a successful confederacy in the Middle East and Western Asia.

The PLO, according to Scripture, will move into Lebanon and later continue into Jordan, bringing bombings and destruction upon both nations. At their headquarters in Beirut, Lebanon, the PLO had established launching sites to attack Israel militarily. Between 1975 and 1981, they killed over 60 thousand people and wounded hundreds of thousands. At least 42 different militia, with an estimated 15 thousand armed men, have been in Lebanon. Compounded with all of this, there are similar hostile militia in other countries neighboring Israel. Since the Moslems are anti-Christian, the Christians who are in control of Lebanon's government are fearful that should the Moslems gain control, they will all be massacred. It is, therefore, easy to understand why God must intervene in Israel's behalf since it also is a genocidal target of Islam. Isaiah 10:34 suggests that Lebanon will not be a nation of peace again, but it will be a constant threat to Israel; nor will Israel eliminate the hostile forces there. Only Christ, the "mighty one," will succeed in doing so when He returns to earth.

In time, the PLO will have sanctuaries in *Jordan* comparable to those of Lebanon. Over one-half of all displaced Palestinians live in Jordan,[4] a viable reason why the PLOs will relocate there for their final assaults upon Israel. This will not occur until after the unsuccessful invasion by the Soviets, and the successful invasion under the Antichrist and the False Prophet. There will be seven years of peace intervening these two assaults. The latter will occur just three and one-half years before Christ's descent to the earth. It is known in Scripture as the time of "Jacob's trouble."

Israel will begin to feel that God is forsaking her when the PLO's stationed in Jordan begin to make sporadic raids upon her. The Islamic nations will start believing they have the upper hand. However, in Jeremiah 12:5-9 we read: ". . . they have called a multitude after thee (Israel): believe them not (that) I have forsaken mine house, I have left mine heritage; I have given the dearly beloved of my soul into the hand of her enemies . . . Mine heritage is unto me as a speckled bird, the birds round about against her. . . ." In the first invasion attempt God will *defeat* the invading forces; and, in the latter, God will *provide* sanctuaries for Israel.

We can expect the Soviet Union to eventually take over Iran and Turkey, and the PLO to have bases in both Lebanon and Jordan, simultaneously or in succession, but God will deliver Israel (Ezekiel 38:21-23; 39:1-5); first, from the Soviets, and later, from Babylon. After the defeat of the Soviets, the Syrian or Islamic confederacy will remain intact because of a new threat which is emerging as a result of the swift conquests of the European conqueror. Furthermore, another confederacy will form for the purpose of self-defense in Northern Africa under the leadership of Egypt (Daniel 11:5).[5]

The False Prophet, who rises in the Islamic confederacy, will collaborate with Antichrist and launch a dual genocide. It will be against the Christians first; then after three and one-half years, it will be against Israel. The latter assault will end in the Battle of Armageddon and Christ's return to earth to set up His Millennial (Manifested) Kingdom.

ISRAEL'S GREATEST MILITARY VICTORY

Before the rise of the European conqueror and the False Prophet, the Soviet Union and the Middle East confederacy headed by Syria will launch a concerted drive to destroy Israel. Israel's sentiment at that time is best expressed in Ezekiel: ". . . our hope is lost: we are cut off *for* our parts" (Ezekiel 37:11). Like butchers with meat cleavers, Israel's neighbors, together with the advancing Soviets, will make Israel's situation untenable! "Whose possessors slay them, and hold themselves not guilty . . . " (Zechariah 11:5). It is then that God in defense of His land and people will move swiftly to defeat all of the attacking and invading forces. The entire operation will be over within *one day* (Ezekiel 38:14,19; 39:8,11,13,22). The casualties among Israel's enemies will be enormous, especially within the Soviet forces. It will take seven months to bury the carcasses and seven years to purge the land of the piles of military debris and wreckage. Here will be the largest junk yard and the most human carnage ever concentrated in one place.

In the Millennium the descendants of the Christian Arabs, who were formerly Moslems, will pay great homage to the Israelis: "*The sons also of them that afflicted thee* (Israelis) shall come bending unto thee, and all they that despised thee shall call thee, The city of the LORD, The Zion of the Holy One of Israel. Whereas thou hast been forsaken and hated, so that no man went through thee, I will make thee an *eternal excellence,* a joy of many generations" (Isaiah 60:14,15). The Christian Arabs, who were previously Moslems and survived the Tribulation judgments, will have a sanctuary and a nation of their own in the land of Assyria (Isaiah 19::24,25).

A PROPHETIC SHADOW OVER EUROPE

When the European conqueror, as Antichrist, is ready to unite his reconstituted Roman Empire with the kingdom of the False Prophet to form the prophetic last day Babylonian coalition, the entire terri tory once under the ancient Empire of Rome will be in favor of this powerful arrangement.

A sign of significance is the emergence of Moslems in Western Europe. During the last decade they have established thousands of mosques, prayer rooms, and Koran schools. It is estimated that some forty percent of Berlin's young people practice Islam. The Moslem magazine, *Al Islam,* published in Munich, Germany, stated: "If the human civilization is to be preserved from destruction, then Islam is the only religion for the world of tomorrow."

Another meaningful incident in the light of prophecy is the recent assassination attempt upon the life of Pope John Paul II in St. Peter's Square. It is believed that those supporting it were Turkish Islamic fundamentalist residents in West Germany.

Certain prophetic events which involve Germany cast their shadows in advance, such as: 1) World War I and World War II which were instrumental in opening the land of Palestine to Israel and in creating the State of Israel, both unintentionally; 2) the reign of Kaiser Wilhelm II of Germany being the last Caesar of the ancient Roman Empire; 3) its being the first country as a signatory to the "Treaty of Rome," which will fall to the European conqueror, the predicted Antichrist.

[1] See page 344, footnote 3.

[2] See chart B.

[3] Four-fifths of the original Palestine was Jordan.

[4] While Jordan will be the next Middle East trouble area, it is Syria that will successfully unite the Islamic Arab world.

61

Man of Mystery

The merger of the kingdoms of Antichrist and the False Prophet will become the biblical Babylon.

The final Tribulation of the saints is divided into two parts according to Scripture (compare Revelation 6:9-11 with 7:9-17). The first portion takes place shortly after the pre-Tribulation Rapture and involves the persecution of saints living primarily in Islamic lands. Moslems will be the main followers of the false prophets and eventually will become disciples of the False Prophet, the coadjutant of the Antichrist. The second part of the Tribulation commences when their kingdoms combine into the predicted Babylon of the last days, the forecasted "Beast" of the Apocalypse.

By the time of the Peace Covenant the Christians will have increased greatly in numbers. Islam will feel threatened with the strong and bold proclamation of the redeemed that Jesus Christ and none other is the Son of God and the only Redeemer of mankind.

Antichrist at first will tolerate his mother church, which as Roman Catholic (Revelation 17:3,9), is a strong representative of Christianity. After uniting with the False Prophet, however, he turns against his church (Daniel 11:37,38), rapes the Vatican, and begins his campaign of murder, arson, and plunder (Revelation 17:16). This crusade will expand to include the Tribulation Saints, regardless of their religious affiliations (Daniel 7:21; Revelation 13:7,10).

Three and one-half years after the defeat of the Soviet forces (Ezekiel 39:9) Israel will sign a seven-year peace covenant with the European conqueror; however, the agreement will be honored only for three and one-half years (Daniel 9:27; 12:7; Revelation 13:5). Then after Antichrist breaks this peace covenant he will seek the genocide of all Israelis.[1]

Why does Antichrist suddenly turn against Christianity, both Catholic and Protestant, since the attack upon his life comes from the sword of Islam? Why is Islam taken off the black list of religious adversaries?

MAN OF MAGIC

The False Prophet at this time performs amazing feats and deceives Islam into accepting him as the latest in their line of prophets since Mohammed. Other non-Christian faiths also will be attracted to this new prophet, as predicted in God's Word: "And he doeth great wonders, so that he maketh fire come down from heaven on the earth in the sight of men, And deceiveth them that dwell on the earth by the means of those miracles which he had power to do *in the sight of the beast* (European conqueror now incarnate as the Beast) . . . " (Revelation 13:13,14). The reason that God is allowing Satan to make these strong delusions is plainly recorded in 2 Thessalonians 2:9-12.

Obviously the European conqueror, who survives the assault upon his life, emerges a different person. He is now called the "Beast" because he is incarnated by Satanic power. Because of his fascination with this new Islamic prophet, he calls for a conference. The result of this meeting is the merger of the kingdoms of Antichrist and the False Prophet as the biblical Babylon of the last days. This conference, lasting only "one hour" (Revelation 17:12), heals the breach between the two kingdoms.

CHANGING SIGNALS

The impact of this agreement that heals the rift becomes apparent as the False Prophet accepts the European conqueror as "God" (Allah) incarnate and as the messiah who is to unite the entire world into an Islamic millennium. Christianity's belief that Jesus Christ is the Son of God is anathema to them. The Roman Catholic Church, which is the mother church of the European conqueror (Daniel 11:37; Revelation 17:5), is temporarily tolerated as an accommodation while she is riding on the back of the Beast. However, this peace alliance with the voice of Christianity is soon broken and all adherents to Jesus Christ, in dogma or experience, become targets for genocide (Daniel 7:21; Revelation 13:7; 17:15,16,17).

Every nation in the world will be infiltrated by mercenaries, spies, and terrorists who direct their polemics toward ridding the earth of all those claiming an association with Jesus Christ. It will be wholesale massacre on an unbelievable scale (Revelation 6:9-11; 7:9-17; 13:10; 14:12; Daniel 11:32-35).

RAPING OF THE VATICAN

The raping of the Vatican and all of the far-flung churches is described as follows: "And the ten horns (provinces in the kingdom of the European conqueror) which thou sawest upon the beast, these shall hate the woman (Catholicism, Revelation 17:3,9,16,18), and shall make her *desolate* and *naked,* and shall eat her *flesh,* and *burn her with fire*" (Revelation 17:16).

The European conqueror, who was ruler over his kingdom, will become the "Beast" in authority over the Babylonian alliance, together with the False Prophet. He will be declared God incarnate and will not tolerate Christianity which resists the honor of deity conferred upon him by the False Prophet. Christians will declare from God's Word that he is an impostor and a paranoiac, and for this they will pay dearly with their lives.

JACOB'S TROUBLES[2]

Israel, during its seven years of peace after the defeat of the Soviet forces, will remove the carcasses and military debris (Ezekiel 39:9,12), and construct a grave site city (verse 16). At the end of these seven years of peace the forces under Antichrist and the False Prophet will invade

Israel, and as "madmen" (Jeremiah 48:2), they will seek its genocide.

The Tribulation Temple will be constructed in Jerusalem (2 Thessalonians 2:4; Malachi 3:1; Daniel 11:31; 12:11; Matthew 24:15; Mark 13:14) and the Ark of the Covenant (Testimony) will be situated in this new temple (Daniel 9:27; 11:30).

Before Antichrist sits in the Tribulation Temple in Jerusalem, we know that all of its furnishing will be removed, as indicated by the word "desolate" (Daniel 9:17,27, 11:31; 12:11; Matthew 24:15; Mark 13:14). Many of the Israelis will flee into Egypt with these possessions (Isaiah 19:20; 66:19).

When Christ returns to earth and brings an end to Armageddon, He will call all of Israel throughout the world to return to the Promised Land (Zechariah 10:8; Isaiah 11:10-16). This will be Israel's second experience of crossing the Red Sea on dry ground as well as its time of crossing the seven streams of water in Egypt.

Because of the ferociousness of Armageddon and the unbelievable destruction befalling the earth, the Lord Jesus will need to return and bring Armageddon to an end: "And except those days should be shortened, there should no flesh be saved: but for the elect's sake those days shall be shortened" (Matthew 24:22). "Alas! for that day is great, so that none is like it: it is even the time of Jacob's trouble; but he shall be saved out of it" (Jeremiah 30:7).

JUHAD! MOSLEM'S HOLY WAR

We have already shown how the "sword" of Islam vanquishes the European king and how he recovers as the "Beast" of Revelation, chapter 13. We notice in Revelation 6:4 that the rider on the red horse who is making war has "a great sword," and in Revelation 13:10 we read that the False Prophet from Islam will lead the major assault upon the Tribulation saints. Perhaps Antichrist feels that this is a religious problem to be solved by the False Prophet: "He that leadeth into captivity shall go into captivity: he that killeth with the *sword* must be killed with the *sword*. Here is the patience and the faith of the saints."

ANTICHRIST IN THE WESTERN HEMISPHERE

"Thus saith the Lord God; it shall *also* come to pass, that at the same time (that the Soviets invade the Middle East) shall things come into thy mind, and thou shalt think an evil thought: And thou shalt say, I will go up to the land of unwalled villages (Western Hemisphere); I will go to them that are at rest, that dwell safely, all of them dwelling without walls, and having neither bars nor gates (no nuclear shelters as preparations for an attack);

"To take a sp*oil,* and to take a prey; to turn thine hands upon the desolate places that are now inhabited, and upon the people *that are gathered out of the nations,* which have gotten cattle and goods, that dwell in the midst of the land.

"Sheba (Africa), and Dedan (Arabia), and the merchants of Tarshish (Anglo-Saxons), with all the young lions thereof (descendants), shall say unto thee, Art thou come to take a sp*oil?* hast thou gathered thy company to take a prey? to carry away silver and gold, to take away cattle and goods, to take a great sp*oil?*" (Ezekiel 38:10-14).

Today we are seeing communism fervidly being infiltrated into Central and South America by the Soviets. The insurgents and revolutionaries, trained in Libya and Ethiopia, are being sent into that part of South America to operate under Cuban advisors.

When the Soviet forces in the Middle East are defeated by God, the presence of the terrorists and insurgents in Central and South America will become a question of grave concern. It appears probable that the communists in the Western Hemisphere may rally to the cause of the European conqueror who will be the great power in the reconstitution of the ancient Roman Empire, thus precipitating the first persecution and massacre of Tribulation saints in the Americas.

The prophet Ezekiel makes no mention of these insurrectionists in his prophecy. Nevertheless, since they are communists and anti-Christian, we have every reason to believe they will spearhead a major anti-Christian movement in the Western Hemisphere. Perhaps they will be most effective by terrorist infiltration. According to prophecy they will not conquer the Americas but will be a source of great irritation and trouble.

The increasing presence of the Russians in Iran is cause for grave concern since Scriptures predict Iran's fall to the Soviets (Ezekiel 38:5). Military supplies, advisors, increased trade with the Soviets, and training camps in Khorasan Province are ominous signs of what is happening in Iran.

[1] See chart J.
[2] See chart B.

62

The Lord's Royal Court[1]

The 144,000 will become the attendants in the Lord's Royal Court.

The 144,000 sealed Israeli Christians are subsequent to the martyrdom of the *first* segment of the Tribulation saints: "*After* these things . . ." (Revelation 7:1). Also, they are contemporary with the *second* segment of Great Tribulation saints when the Antichrist, or "Beast," unites with the False Prophet under Babylon, and launches his final assault upon all Christians throughout the world: "*After* this (the sealing of the 144,000) I beheld, and, lo, a great multitude . . . which came out of great tribulation . . ." (Revelation 7:9).

At the last trumpet, the 144,000 will be translated along with the remnants of the Great Tribulation saints. We read: ". . . which were redeemed (the 144,000) from among men, being the firstfruits unto God and to the lamb" (Revelation 14:3,4). They are seen standing with Christ on "Mount Sion" (in heaven; compare Romans 11:26) where they are "harping with their harps" (verse 2). The 144,000 will not be present during the final wrath of God (I Thessalonians 5:9; Luke 21:35,36) when Babylon will make its final assault upon the Zionists (Jews), because they also must appear at the Judgment Seat of Christ (2 Corinthians 5:10).

The 144,000 will be the initial ones identified with the tribes of Israel, and when Christ returns to earth to set up the judgment of the nations, all remaining members of Israel will learn their tribal identity (Matthew 25:31-46). It is a historic fact that the lost tribes migrated westward and became integrated into the commonwealth of nations by intermixing with the Gentiles. They have *not* lost their tribal heritage. Thus the prophecies from Genesis 22:17 and Hosea 1:10 that Israel will multiply as the sand of the sea are being fulfilled: "Yet the number of the children of Israel shall be as the sand of the sea . . . in the place where it was said unto them, Ye are *not* my people, *there* shall it be said unto *them,* Ye are the sons of the living God" (compare Romans 9:25,26; 11:25,26).[2]

There were more Israelites who made their exodus from Egypt than there are Jews (the tribe of Judah) in the world today; therefore, the dispersed tribes must be included in the Biblical prophecies which refer to Abraham and Sarah as the progenitors of *"many (multitude, company) nations"* (Genesis 17:4,5,16; 35:10,11; 48:10; Romans 4:17,18). What a surprise will come to many when it is revealed that they are *blood brothers and sisters of the Jews* (Judah), because Abraham and Sarah are their ancestors. Any anti-semitism will certainly boomerang at that revelation! Christ was of the tribe of Judah.

Paul compares the dispersing of the lost tribes among the nations to "grafting" because the tribal genealogical lines are kept intact by their implantation to the parent stock. They retain their original identification, even though some traits may be different. Since the name "Jew" today is used in a wider sense, it is probable that many so-called "Jews" belong to one of the other tribes of Israel. Isaiah 49:22 shows that all of the tribes will be identified by the time Christ returns to earth.

UNIVERSAL WITNESSING

"And this gospel must *first* be published among *all nations"* (Mark 13:10). "But he that shall *endure unto the end* (of the Tribulation), the same shall be saved. And this gospel of the kingdom shall be preached in all the world for a *witness unto all nations;* and then shall the *end* come" (Matthew 24:13,14). It is the Lord's commission that the *Church* is to reach *every nation* with the Gospel (Matthew 28:19,20), and this will be accomplished during the *Great Tribulation:* ". . . even to the *end* of the (age)."

The Jews (Judah), together with all Israeli Christians, will reach *every creature* during the *Millennium:* ". . . In those days (Millennium) it shall come to pass, that ten men shall take hold of all languages of nations (even ethnic groups), even shall take hold of him that is a *Jew,* saying, We will go with you: for we have heard that God is with you" (Zechariah 8:23). ". . . Go ye therefore into all the world, and preach the gospel to *every* creature . . . " (Mark 16:15).

VIRTUOSOS

". . . These are they which follow the Lamb wheresoever he goeth . . . " (Revelation 14:4).

The 144,000 sealed ones will become the attendants in the Lord's Royal Court. They will be the entourage in His travel over the earth during the Millennium. They have a theme song and they supply their own accompaniment. Evidently they are virtuosos and accomplished musicians. Harps are purely representative of their many musical talents. God, as a God of variety, hardly will limit them to one musical instrument.

[1] See chart L.

[2] The omissions of Dan and Ephraim suggests these probabilities: The amalgamation of Dan and Ephraim into the Gentiles made possible God's promise to Abraham being fulfilled for a universal body of redeemed — the Church — from the descendants of Abraham (Romans 9:24,25; Matthew 16:18; I Corinthians 12:13). Ephraim means *increase*; Dan means *judge*. The Royal Redeemed some day will judge the nations of the world (Matthew 19:28; Luke 22:30). The Man*dan* (North Dakota) American Indian Tribe is believed to descend from Dan. They are called "white" Indians because of their blue eyes and blond hair. Amos 9:9 shows Israel will be scattered and their tribes lost among all nations of the world.

63
Child of Destiny

If there is a last *trumpet sound heralding the rapture of the redeemed, it stands to reason there must also be a* first *trumpet sound.*

The contrast of "man" and "child" is neither accidental nor incidental to the truth "Manchild" (Revelation, Chapter 12) conveys. Christ is God's "Man" and one who is redeemed is God's "child."

The "manchild" is to rule with a "rod of iron" (Revelation 12:5), symbolizing that which Christ and the Church will accomplish together (Revelation 2:26,27; 19:15). The Church will reign with Christ (Revelation 5:10), but before it can reign with Him there must be a common bond—immortality! In order for the Church, as the Body, to equate with Christ, the Head, both must possess immortality. One cannot possess a body of mortality while the other has a body of immortality. Consistency and unity are essential if they are to rule as a compound unit. There will be a marriage of the Lamb of God to the Church (Revelation 19:7-9) to effect this union. It will not be marriage of Christ to a child bride, thus indicating that the Church must arrive at the stature of the fulness of Christ (Ephesians 4:11-14; I Corinthians 13:10-12).

AN AMALGAMATED BEING

The birth of the manchild is into the state of immortality. Christ is referred to as "the firstborn from the dead," an obvious reference to His receiving immortality (Revelation 1:5), and "the firstborn among many brethren" (Romans 8:29; Colossians 1:18). Christ incarnate (in flesh) is known as "the first begotten *into* the world" (Hebrews 1:6).

The redeemed also are begotten sons of God and firstfruits: "Of his own will begat he us with the word of truth, that we *should be* (future) a kind of firstfruits of his creatures" (James 1:18). This common bond of immortality is essential for this marriage union. This is the message conveyed in the allegory of the "manchild."

Both Christ and the Church come through the same faith line—from Abraham through Isaac as depicted in Romans, chapter 4. This is the "woman" giving birth to the manchild (Galatians 3:7,16,29; Matthew 1:20; 12:48,49).

ASSAULT ON THE MANCHILD

When Christ was born, Satan hoped the Christ Child would be destroyed by Herod's massacre of infant males. The tribulation periods of the Church in history have shown similar vicious attacks upon the redeemed. The final Great Tribulation will be Satan's last desperate attempt to prevent the completed Body (Church) from being united to Christ in immortality! "For he (God) will finish the work, and cut it short in righteousness: because a short work will the Lord make upon the earth . . . blindness in part is happened to Israel, until the fulness of the Gentiles be come in. And so all Israel shall be saved . . . " (Romans 9:28; 11:25,26).

The Body of Christ will be completed as God planned in the eternal past. It will not be an incomplete Bride (Church) to which Christ will be wed throughout eternity.

LAST TRUMPET SOUND

According to I Thessalonians 5:4,9, the moment that Antichrist breaks his seven-year peace covenant with Israel, the final last day

Tribulation saints will be raptured: "But ye, brethren, are not in darkness, *that that day* (when the covenant is broken with Israel and their lives are imperiled) should overtake you as a thief . . . For God hath not appointed us (the redeemed) to wrath (synchronized with Israel's Tribulation), but to obtain salvation (rapture) by our Lord Jesus Christ."

We see that two harvest Rapture days are scheduled in God's calendar—the pre-Tribulation and the post-Tribulation ones—both of which are included in the "first resurrection."

If there is a *last* trumpet sound heralding the Rapture of the Royal Redeemed, it stands to reason there also must be a *first* (preceding) trumpet sound. This distinction becomes obvious by comparing I Corinthians 15:52 with I Thessalonians 4:16, which reveals two distinct events with seven years of separation. The seven trumpet judgments between these two Rapture trumpets are "judgment" trumpets, not "rescue" trumpets.

We must recognize that the phrase "first resurrection" embodies both events. It includes all of the redeemed both preceding and including the Great Tribulation. All must be resurrected and raptured into immortal bodies if they are to appear at the Judgment Seat of Christ *before* the Marriage of the Lamb of God to become the Royal Redeemed (Revelation 20:6; 2 Corinthians 5:10).

Those who are not in the "first resurrection" must wait for the final resurrection, which will be into *mortality* (not immortality). This means they will appear at the White Throne Judgment at a much later time (John 5:22-30). At this point the "second death" will take effect upon those who are *lost;* and those whose names are still retained in the book of life will be discharged in their *mortal* bodies to the New Heaven to make their decision for Christ.

After all of the resurrections and raptures of the redeemed under the *first resurrection* have taken place, those who remain and accept Christ as their personal Saviour will go to the judgment at the White Throne since they also missed the Judgment Seat of Christ: "Blessed and holy is he that hath part in the first resurrection (which includes only those redeemed up to that time), *on such the second death hath no power...*" (Revelation 20:6). In the meantime, if they have not died during the last three and one-half years before the Lord returns to the earth, they will enter the Millennium in their natural bodies. They will not receive their immortal bodies until all of the redeemed who come in after the first resurrection receive theirs. This will occur when death is destroyed (John 5:22-29; I Corinthians 15:26).

STORMCLOUDS OVER AMERICA

The approach of the "day of the Lord" (three and one-half years before the Millennium) is the scenario covered in Revelation, chapter 12. When Satan and his fallen angels are cast out of heaven onto the earth, they combine with Antichrist and the False Prophet to bring their full hatred against Israel because it is the only remaining inheritor on the earth. To punish the countries in the Western Hemisphere for providing sanctuary to the fleeing Israelis ("woman"), Satan unleashes torrential rains upon the "wilderness" (a probable reference to America), because in John's day it was just that. God will react by opening up crevices in the ground to quickly drain the flooded waters.

64

Last Harvest Day Rapture

The time of the final Rapture is the regathering of all of the redeemed both in heaven and on earth (Mark 14:27).

"For the day of vengeance is in mine heart, and the year of my redeemed is come . . . And I will tread down the people in mine anger, and make them drink in my fury, and I will bring down the strength in the earth" (Isaiah 63:4,6).

This final harvest resurrection and rapture of the last day *Tribulation* saints will immediately precede the pouring out of God's wrath upon the earth. While there will be mass martyrdom under Antichrist (Beast), there still will be many survivors. God at this time brings the Church Age to a close by calling unto Himself the final gleaning of saints from the earth. This will culminate the "first resurrection."

The final rescue of the remnants of the Church will bring all of the redeemed into heaven to appear at the Judgment Seat of Christ. Those who survive that judgment will become the corporate wife of Christ: ". . . his wife hath made herself ready" (Revelation 19:7).

The final harvest day Rapture is called the time of the "last" trumpet: "Behold, I shew you a mystery; We shall not *all* sleep, but we shall *all* be changed, In a moment, in the twinkling of an eye, *at the last trump:* for the trumpet shall sound, and the dead shall be raised incorruptible, and we shall be changed (into immortality)" (I Corinthians 15:51,52).

The *last* trumpet is thus called, not because this event falls under the seventh trumpet in Revelation 11:15-19, but because the last of the redeemed are to be resurrected and raptured into bodies of immortality to complete the "first resurrection." In verse 18 we read: "And the nations were angry, and thy wrath is come, and the time of the dead (redeemed) that they should be judged (at the Judgment Seat of Christ), and that thou shouldest give reward unto thy servants the prophets, and to the saints, and them that fear thy name, small and great; and shouldest destroy them which destroy the earth."

When Paul spoke of "the last trump" in I Corinthians 15:52, the book of Revelation had not yet been written, therefore I Corinthians cannot be referring to the seventh trumpet in the Apocalypse because it was written much later.

COSMIC ALL-STARS

Also at this time the long subdued voice of Michael (Jude 9) will be raised to summon the hosts of God's holy angels. They will stream out of their principalities in the planetary heavens with their Chief Prince Michael (Daniel 10:15) to do battle in the lower cosmos against Satan and his fallen angels (Revelation 12:9): ". . . At that time thy people shall be delivered, *every one* that shall be found written in the book of life, And many of them[6] that sleep in the dust of the earth shall awake . . ." (Daniel 12:1). Satan is cast out at this particular time to prepare his allies, both those in spiritual and physical bodies, for the final battle of Armageddon to seek possession of the earth.

The final persons gleaned at the last harvest Rapture will join the martyred and deceased believers who passed on before them in the pre-Tribulation Rapture of the Saints. All of the final Tribulation saints will obtain their bodies of immortality and will join with the victorious holy angels in this triumphant response: "Therefore rejoice ye heavens, and ye that dwell in them (Tribulation saints as well as the holy angels). Woe to the inhabiters of the earth and of *the sea.* For the devil is come down unto you, having great wrath, because he knoweth that he hath but a short time" (Revelation 12:12).

It is at this moment that Satan will influence not only Abaddon (Apollyon), who already has incarnated the man, Antichrist, and turned him into a veritable beast, but he also will persuade the "prince of the kingdom of Persia," who has incarnated the False Prophet, to now direct his full measure of assault upon Israel. Why Israel? It is because all of the potential rulers of the earth (Royal Redeemed) already have been transported into heaven. The only remaining inheritors on the earth under God's plan are the people of Israel. This explains why Antichrist tries so desperately to obtain possession of the golden Ark of the Covenant from the Israelites in Egypt. It is a divine symbol of God's protection. The remnants of Israel are scheduled in God's plan to be the sure possessors of the earth: ". . . blindness in part is happened to Israel, *until* the fulness of the Gentiles be come in. And so all Israel shall be saved: as it is written, There shall come out of Sion the Deliverer, and *shall* turn away ungodliness from Jacob" (Romans 11:25,26).

The wrath of God, which is to be poured out particularly upon the last day Babylon, as well as the combined kingdoms of Antichrist and the False Prophet, is designed to begin the end of these kingdoms. The flight of Israel from its land will, therefore, prove a blessing in disguise as it finds sanctuaries outside of the Babylonian territories. Although many Israelis will become martyrs at that time, there will not be genocide because a "remnant" will survive.

END OF THE CHURCH AGE

". . . and then shall the *end* (of the Church Age) come" (Matthew 24:14). ". . . but he that shall endure unto the *end* (of the Church Age), the same shall be saved (to enter the Millennium if he successfully emerges from the Judgment of the Nations)" (Mark 13:13).

TWO-ACT DRAMA

"Well, you will recall now *what* it is that restrains him (Antichrist) from being revealed before his appointed time. For the secret force of lawlessness is at work already; only it cannot be revealed till *he* who presently restrains is removed. Then shall the Lawless One be revealed, whom the Lord shall destroy with the breath of his lips" (2 Thessalonians 2:6-8, *Moffatt*).

Two restraints to the appearance and revelation of Antichrist are mentioned. The word "what" refers to an *event*—the Rapture! The word "he" to the *subject*—the redeemed Church! (Compare Revelation, chapter 12; and see Chapter 63).

The Holy Spirit never leaves the earth for He is absolutely essential to the redemption of souls (I John 5:6; Galatians 4:6).

There are three massive spiritual awakenings that will occur *after* the first harvest day Rapture: first, that which will create the Tribulation Church; second, the last harvest day Rapture of the Tribulation Church that will create the "righteous nations;" and finally, the visible return of Christ as He pleads from His podium in the skies. This will bring in the remnants of those whose names are still in the book of life (See chapter 67).

The Church Age, however, will *end* with the last harvest day Rapture and the redeemed Church becoming the wife of Christ (Compare Matthew 24:13,14; Acts 2:16-21). These two Raptures will complete the "first resurrection" (Revelation 19:9; 20:6) and usher in *all* the redeemed Church to the Judgment Seat of Christ and to the Marriage of the Lamb (2 Corinthians 5:10; Revelation 19:7-9).[1]

REVELATION OF ANTICHRIST

Antichrist, according to 2 Thessalonians 2:4, will not be revealed to the world until he sits in the Temple of God in Jerusalem being declared "God" personified. Of course, he will have been recognized by the Tribulation Christians, for they are not deceived (Matthew 24:24; 2 Thessalonians 2:9-13; Romans 1:23-32; Revelation 13:12-18; 14:9-13). This revelation of Antichrist to the world will not come until all of the Tribulation saints have been rescued by being transported into glory. It is at this juncture that Antichrist turns his wrath upon Israel in a genocidal endeavor but a "remnant" of Israel will be preserved (Isaiah 10:24-27; 11:10-16).

The last three and one-half years of the wrath of God, without any mixture of mercy (Revelation 14:10), occurs because the completed Church is gone. Those saved after this post-Tribulation Rapture are called "the meek: for they shall inherit the earth" (Matthew 5:5). These will be the natural people in mortal bodies who have survived the Tribulation period and the Judgment of the Nations (Matthew 25:31-46).

We read in Zephaniah 2:3: "Seek ye the Lord, all ye *meek* of the earth . . . it may be that ye shall be *hid* in the day of the Lord's anger."

A SUCCESSFUL FIRE ESCAPE

The "first resurrection," which will include all of the Royal Redeemed coming into immortality, will finalize only when the last of them have been garnered during the final harvest days. Their Rapture (including the 144,000) will occur when Antichrist, as the Beast, moves into the Jewish Tribulation Temple in Jerusalem and establishes his image for worship in lieu of the brazen altar (Daniel 12:11). The two witnesses will be revealed *before* this final Rapture as a positive sign that the Church Age is at an end.[2] They also must appear at the Judgment Seat of Christ, for there are no exceptions.

It is necessary that we make a contrast between this final harvest day and the first one:[3]

FIRST HARVEST DAY RAPTURE	*FINAL HARVEST DAY RAPTURE*
1. Appearance of Christ in clouds	1. Appearance of Christ in a cloud
2. Sound of archangel	2. No sound of archangel
3. Deceased redeemed will come from Paradise	3. Deceased redeemed will come from under the altar
4. Redeemed will escape "all" of the last day tribulations	4. Redeemed will escape just the final wrath of God

5. Believer's "hope"	5. Believer's "comfort"
6. Redeemed will receive crowns	6. Redeemed will receive palms
7. Christ is seen uncrowned	7. Christ is seen crowned

For references compare: I Thessalonians 4:17; Luke 21:36; I Thessalonians 5:9; Revelation 4:4; 5:6; 7:9; 14:14-16.

UNDER FIRE

All of the Tribulation saints are *not* martyred as evidenced by the following texts: "patient and endure" (James 5:8,11); "wake" and "not appointed to wrath" (I Thessalonians 5:9,10); "salvation," "deliverance" (I Thessalonians 5:9); "not all sleep" and "shall be changed" (I Corinthians 15:51,52); "deliverance" (Daniel 12:1); "snare" and "shall deliver" (Luke 21:35,36).

I Thessalonians 5:9, refers to the Tribulation Rapture: ". . . not appointed unto wrath." This wrath of God comes *after* the seven seals are opened as seen in Revelation 11:18: "thy wrath is come."[3]

RIGHTEOUS NATIONS

Will anyone be converted to Christ after the last of the Tribulation redeeemed have been ushered into immortality and the Church Age *per se* has come to its "end"? Certainly, many will be included among the "righteous" individuals and nations at the Judgment of the Nations (Matthew 25:31-46) and this will enable them to enter the Millennium under the reign of Christ. This also will be the time of Israel's spiritual and national revival.

"Blessed and holy is he that hath part in the first resurrection: on such the second death hath no power . . . " (Revelation 20:6). The misfortune is that those "righteous" individuals will have been deprived of so much by not having been in the "first resurrection." They will have missed awards at the Judgment Seat of Christ, participation as the corporate wife of Christ, and life in the Millennium should they die between the last harvest day Rapture and the Millennium. Perhaps they even will have been involved and slain in the battle of Armageddon. However, they will be resurrected and reunited with their physical bodies to appear at the White Throne Judgment. Then they will go on into the New Heaven where they will remain in their mortal bodies until death is destroyed. All of the redeemed at that time, including those in the New Heaven, will receive bodies of immortality and go on into the Father's Manifested Kingdom.

Those who die during the Millennium must also wait to appear at the White Throne Judgment in bodies of mortality, after which their destiny will be determined. The Christians who die during the Millennium as a result of unfortunate circumstances will go into Paradise to await their appearance at the White Throne Judgment. In the New Heaven the trees of life will be available to them to insure their survival until they receive their immortal bodies.

HEAVENLY VACATIONISTS

When the final harvest Rapture day arrives, there will be a regathering of all of the redeemed both in heaven and on earth: ". . . shall gather together the elect from the four winds, from the uttermost part of the *earth* to the uttermost part of *heaven*" (Mark 13:27). This is synchronized to occur just *before* the Judgment Seat of Christ where all of the redeemed in the book of life must appear (2 Corinthians 5:10).

This final Rapture and resurrection complete the "first resurrection," thereby bringing all of the Royal Redeemed into their bodies of immortality (Revelation 20:6).

MARRIAGE OF THE LAMB OF GOD

"Let us be glad and rejoice, and give honour to him (God): for the marriage of the Lamb is come, and his wife (the redeemed) hath made herself ready. And to her was granted that she should be arrayed in fine linen, clean and white: for the fine linen is the righteousness of the saints. And he saith unto me, Write, Blessed are they which are *called* unto the marriage supper (the guests) of the

Lamb . . ." (Revelation 19:7-9).

The drama of the Marriage of the Lamb of God (Jesus Christ) is better described in Matthew 22:1-14. The redeemed in immortal bodies, who are in the "first resurrection," will be in the corporate wife of Christ. All of those in Paradise who are not yet redeemed, but whose names are still in the book of life (such as little children), will be "guests." These "guests" will appear at the wedding in their "garments of salvation" (distinguished from the "robe of righteousness") taken from one of the palatial ivory wardrobes.[4]

After the "guests" appear at the wedding they will return to Paradise to await their appearance at the White Throne Judgment, because their names are still in the book of life. From there they will go into the New Heaven where they must make their decision for Christ (Revelation 20:15; 22:14).

The redeemed, who have disqualified themselves by their unsavory works, are denied the privilege of being the wife of Christ or even being "guests" during the wedding ceremonies. In addition, they are retired to "outer darkness" as a necessary disciplinary measure. They will emerge rehabilitated at the White Throne Judgment, but at what a great eternal loss, because at the Judgment Seat of Christ they will have been found to be "unprofitable servants" (Greek: worthless or useless) (Matthew 25:14-30). "For it had been better for them not to have known the way of righteousness, than, after they have known it, to turn from the holy commandment delivered unto them" (2 Peter 2:21). "But many that are first shall be last; and the last shall be first" (Matthew 19:30).

Even though such persons may be defrocked or divested of their wedding garments, their souls will not be lost (I Corinthians 3:15). Their wedding garments are "spotted by the flesh" (Jude 23), implying serious misdemeanor and misconduct in putting their wills above the will of God. Thereby their testimonies become shoddy before the world.

The importance of the Judgment Seat of Christ is indicated by the following Scriptures:

"And now, little children abide in him; that, when he shall

appear, we may have confidence, and not be ashamed before him at his coming" (I John 2:28).

"I counsel thee to buy of me gold tried in the fire, that thou mayest be rich; and white raiment, that thou mayest be clothed, and that the shame of thy nakedness do not appear; and anoint thine eyes with eyesalve, that thou mayest see" (Revelation 3:18).

"Behold, I come as a thief. Blessed is he that watcheth, and keepeth his garments, lest he walk naked, and they see his shame" (Revelation 16:15).

"That he (Christ) might sanctify and cleanse it (the Church) with the washing of water by the word,

"That he might present it to himself a glorious church, not having spot, or wrinkle, or any such thing, but that it should be holy and without blemish" (Ephesians 5:26,27).

After the testing by fire at the Judgment Seat of Christ, Christ will "elect" from those remaining to be His wife. Although those with unacceptable conduct are excluded, they will be disciplined and rehabilitated for the sake of the salvation of their souls.

"Keep yourselves in the love of God, looking for the mercy of our Lord Jesus Christ unto eternal life.

"And some have compassion, making a difference:

"And others save with fear, pulling them out of the fire; hating even the garments spotted in the flesh.

"Now unto him that is able to keep you from falling, and to present you faultless before the presence of his glory with exceeding joy . . ." (Jude 21-24).

SEVEN SEALS - SEVEN TRUMPETS - SEVEN VIALS[5]

Seals:

The book of seven seals is in Christ's hand. The opening of the seals begins immediately after the pre-Tribulation Rapture and ends with the post-Tribulation Rapture of the saints.

Trumpets:

The trumpets are associated with angels who are responding to the prayers of the first group of Tribulation saints (Revelation

6:10; 8:3-6). They begin to sound their trumpets during the period of the Great Tribulation, which is contemporary with the 144,000 sealed Israelites upon the earth (Revelation 9:4). They end their trumpet judgments when the remnants of the Great Tribulation saints have been raptured (Revelation 14:14-16; 15:1). An example of the nature of these judgments is found in Revelation 8:10,11: "And the third angel sounded, and there fell a great star from heaven, burning as it were a lamp, and it fell upon the third part of the rivers, and upon the fountains of waters; And the name of the star is Wormwood: and many men died of the waters, because they were made bitter."

Astrophysicists inform us that comets are full of poisonous cyanides. If a large comet were to plunge onto the earth its cyanogenous fallout would pollute the earth's waters. Scripture refers to this as "Wormwood. . . and many men died of the waters, because they were made bitter."

Five hundred and sixty-seven comets have been catalogued by scientists to date. These comets have appeared to earthlings at intervals of approximately every twenty-five years to every sixteen thousand years.

Vials:

Like poison used for the purpose of exterminating pests, these seven last judgments fall on the remaining unredeemed on the earth. The major portion of them descend on the combined kingdoms of Antichrist and the False Prophet, known as Babylon. The last Gentile kingdom is vanquished during the Battle of Armageddon (Revelation 16:16-21; Daniel 2:44).

[1] No Scriptures show the Church on earth during "the day of the Lord." See I Thessalonians 5:1-11 and chart Z.

[2] See chapter 68.

[3] The phrases "sudden destruction" and "that day" both refer to "the day of the Lord" (I Thessalonians 5:2-4), which will immediately follow the final harvest Rapture.

[4] See page 202.

[5] See chart I.

[6] The word "many" shows that not all the Tribulation saints will have been martyred at the time of the post Tribulation Rapture.

65

Post-Tribulation Christians

If the post-Tribulation Rapture were eliminated, not all of the Church would make an appearance at the Judgment Seat of Christ (2 Corinthians 5:10).

If the post-Tribulation Rapture were eliminated, not *all* of the Church would make an appearance at the Judgment Seat of Christ (2 Corinthians 5:10). In that case, Christ would have only a partially completed bride to become His wife at the Marriage of the Lamb of God; therefore, the "first resurrection" must include *all* of the Royal Redeemed in their bodies of immortality.

Not only will the pre-Tribulation and Tribulation saints, in *immortal* bodies, enter the Millennium, but also the post-Tribulation saints in *mortal* bodies. They will later constitute the "righteous nations" over whom Christ and His wife will reign. These are the nations, called "the kings of the East" (Revelation 16:12), that fight against the Babylonian regime at the Battle of Armageddon.

Those who enter the Millennium as natural people obviously are not converts to Christ previous to the final Rapture or they would have made the "first resurrection" events. It is equally true that they have not taken the "mark" of the Beast (Revelation 14:9,10; 19:20). They are converted to Christ during "the day of the Lord," the time of Jacob's trouble when the final terrifying wrath of God is poured upon the earth. This confirms the fact that the Holy Spirit will still be present upon the earth (I John 5:6; Galatians 4:6; see page 390).

How are so many people influenced to accept Christ as their Redeemer during this crisis period? Whereas the *first* harvest day Rapture ignited the spiritual flame that created the Tribulation Church, the *last* harvest day Rapture of the Tribulation Christians created the "righteous nations" that organized a military resistance to bring down Babylon. Upon the visible return of Christ to earth, He sets up His podium in the skies to make a final plea to the Gentiles and the tribes of Israel. Thus, He gathers in the remnants of those whose names still remain in the book of life. Jeremiah 25:31 states: "A noise shall come even to the ends of the earth; for the LORD hath a controversy with the *nations, he will plead with all flesh;* he will give them that *are wicked to the sword,* saith the LORD." In Malachi 4:2: "*But* unto you that fear my name shall the Sun of righteousness arise with healing in his wings; and ye shall go forth and grow up as calves of the stall."

FLASHING SWORD IN THE SKY

Antichrist and the False Prophet, as Beasts, will be cast into the lake of fire (Revelation 19:20) and Satan will be bound in the bottomless pit (Revelation 20:1). But what about those who have taken the mark of the Beast? We read that they will be destroyed by Christ Jesus as He returns to the earth: "And the *remnant* were slain with the sword of him that sat upon the horse, which sword proceeded out of his mouth: and *all* the fowls (worldwide) were filled with their flesh" (Revelation 19:21).

Bodies will lie by the millions, not only at the site of Armageddon, but throughout the earth: "And the slain of the Lord shall be at that day from one end of the earth even unto the other end of the earth: they shall not be lamented, neither garnered, nor buried; they shall be dung upon the ground." (Jeremiah 25:33).

The Scriptures dramatically portray vast numbers fleeing across the open countryside searching for effective hiding places, only to be slain in their flight; and those who successfully reach places of refuge find their shelters crumbling down upon them as a result of the great earthquake (Revelation 6:12-17). There are the searing death rays from the flashing sword in the sky,[1] penetrating to reach them (Luke 17:33). The huge, dazzling sword strikes like lightning as the Lord moves from east to west over the earth (Matthew 24:27). It is the same brightness that will destroy the Antichrist "without hand" (Daniel 8:25; 2 Thessalonians 2:8).

TERROR IN THE NIGHT

"... *wailing* shall be in the streets... Woe unto you that desire the day of the Lord! To what end is it to you? The day of the Lord is *darkness* and not light... Shall not the day of the LORD be *darkness*, and not light? even *very dark*, and *no brightness* in it?" (Amos 5:16-20). The frightful fact is that the persons seeking places of refuge from the flashing sword must stumble and grope in panic through darkness.

Those who successfully survive the judgments of God upon the earth will represent a greatly reduced population. Among those who do remain, one half will have taken the mark of the beast, which spells their absolute doom (Revelation 14:9-11). Thus we see that the influence of the Babylonian last day regime will be widespread. Consequently, we read in Luke 17:30-37: "... one shall be taken (in death), and the other left (to survive)." Concerning those taken we read: "Wheresoever the body (carcass) is, thither will the eagles (vultures) be gathered together." It is called: "... the supper of the great God" (Revelation 19:17).

[1] See chapter 77.

66

The Judgment Seat of Christ[1]

Christ's expertise in divine jurisprudence gives Him the wisdom to execute the righteous justice of God.

"For we must all appear before the judgment seat of Christ; that every one may receive the things done in his body ('the works'), according to that he hath done, whether it be good or bad.

"Knowing therefore the terror of the Lord, we persuade men; but we are made manifest unto God; and I trust also are made manifest in your conscience" (2 Corinthians 5:10,11).

"And if the righteous scarcely be saved, where shall the ungodly and the sinner appear?" (I Peter 4:18).

". . . The Lord shall judge his people" (Hebrews 10:30).

The Judgment Seat of Christ will occur immediately after the Rapture of the final Tribulation saints. The White Throne Judgment will commence at the conclusion of the Millennium.

A distinction must be made between the Judgment Seat of Christ and the White Throne Judgment. The first is exclusively for all of the redeemed within the mystery Kingdom of Heaven (or God) whose names are in the book of life. Those in Paradise who were prematurely cut from life will appear only as "guests" at this judgment, but they will make their final appearance at the White Throne

Judgment. This latter judgment will include the "wicked" who will have infiltrated into the mystery Kingdom as intruders and enemies of Christ.

REMITTING AND RETAINING SINS

"Whose soever sins ye remit, they are remitted unto them; and whose soever sins ye retain, they are retained" (John 20:23).

"Who shall lay anything to the charge of God's elect? It is God that justifieth" (Romans 8:33).

". . . Who can forgive sins, but God only?" (Luke 5:21).

Friend and foe in the mystery Kingdom, the Church of Jesus Christ, will meet some day in judgment. All too frequently within this spiritual (mystery) Kingdom there arises acrimony among believers because their carnality has been given precedence over the will of God, whether by overt or covert operation. Malice, grudges, and bitter feelings are permitted to grieve the Holy Spirit. Frequently these frictions bring blight upon the testimony of the Church and dull its effectiveness. But if there is true reconciliation, the sins committed between the parties will have been remitted (forgiven by God). If not, the sins are retained and all parties involved in a state of unforgiveness and irreconciliation must face Christ for His final determination at the tribunal for Christians. It could result in serious losses, such as being disrobed of the wedding garment, as well as further disciplinary measures. There are those who must be escorted out from the wedding because they do not have on the wedding garment (Matthew 22:11:44).

Christ's expertise in divine jurisprudence gives Him the wisdom to execute the righteous justice of God. While it is incumbent upon any church body to exercise disciplinary action in preserving the local spiritual body, its decisions, including excommunication, can only be interlocutory since the destiny of souls is not in its power any more than are the decisions regarding inscription or deletion of names in the book of life. All judgments have been committed into the hands of Jesus Christ by the Heavenly Father (John 5:22).

UNCIVIL SERVANTS

The children of the mystery kingdom who are escorted out of the Wedding of the Lamb are addressed in a different manner than the wicked. Jesus called them His "friends" and "servants" (Matthew 22:12; 25:14-30). Although they have responded to the Gospel message and are redeemed, they cannot be chosen for the greater positions of reigning as "kings" and ministering as "priests" under Christ in His manifested kingdom (Revelation 1:6). In the light of eternity, this should sober every Christian and make an indelible imprint as to his motives regarding the necessity for faithfulness. It is important for one to evaluate his conduct and his dedication.

The question arises as to whether the Great Tribulation saints have an advantage when they go into the Marriage of the Lamb: "But and if that servant says in his heart, My lord delayeth his coming (for the Tribulation Saints); and shall begin to beat the menservants and maidens, and to eat and drink and to be drunken; the lord of that servant will come . . . and will appoint him his portion with the unbelievers" (Luke 12:45,46). Faithfulness in Christian service has great compensation, but may be difficult for the newly redeemed to attain because of the pressures of those days.

It is true that many last day Tribulation saints will fare better than those who make the pre-Tribulation Rapture: "And many that are first shall be last; and the last shall be first" (Matthew 19:30). In fact, many newly redeemed persons quickly terminated from life may possibly retain their wedding garments, whereas those serving Christ for some time may conceivably lose their wedding garments.

At the Judgment Seat of Christ the full impact of damages incurred against the Lord's cause are on record and will be evaluated. After the true motives and causes behind each situation have been candidly revealed, a final verdict will be rendered. This can be serious and costly to the guilty parties: "Whosoever shall fall upon that stone (Christ) shall be broken; on whomsoever it shall fall (in judgment), it will grind him to powder" (Luke 20:18). "Some men's sins are open beforehand, going before to judgment (already settled and forgiven);

and some men they follow after (that is, to the tribunal). Likewise also the good works of some are manifest beforehand; and they that are otherwise cannot be hid" (I Timothy 5:24,25).

This tribunal of the redeemed before Christ is a day of revelation. It is not only the time of reward, but also the time of serious consequential loss. The guilty will suffer as a result of rewards being transferred to the offended or innocent parties, while the latter will be enriched with greater honors. Important positions within the Lord's Manifested Kingdom will be awarded or lost on the basis of what takes place at this tribunal for the redeemed.

SPIRITUAL BANKRUPTCY

"Every man's work shall be made manifest: for the day shall declare it, because it shall be revealed by fire: and the fire shall try every man's work of what sort it is.

"If any man's work abide which he hath built thereupon, he shall receive a reward. If any man's work shall be burned, he shall suffer loss: but he himself shall be saved; yet so as by fire" (I Corinthians 3:13-15).

"And now, little children, abide in him; that when he shall appear, we may have confidence, and not be ashamed before him at his coming" (I John 2:28. See also: Matthew 24:42-51; 25:14-30; I Peter 4:17,18; Revelation 3:18).

The disciplined followers of Christ will emerge from "outer darkness" truly rehabilitated, but at what a permanent loss! If God did not graciously wipe away tears of remorse from their eyes, their irretrievable losses would be unbearable and their memories of past demeanors would continue to haunt them forever: ". . . and the former shall not be remembered, nor come into mind" (Isaiah 65:17).

TEARS IN HEAVEN

"And, behold, I come quickly, and *my reward is with me, to give* every man according as his work shall be" (Revelation 22:12).

"For it had been better for them not to have known the way of righteousness, than, after they have known it, to turn from the holy commandment delivered unto them" (2 Peter 2:21).

If these undisciplined followers of Christ had been obedient, they would be among the Royal Redeemed; or at least, like the "guests," they could remain in the heavenly Paradise. Since they grieve the Holy Spirit (Ephesians 4:30), they will be unable to receive a glorified body at the time. The Holy Spirit is the earnest of the inheritance in the future (Ephesians 1:13,14). It is the Holy Spirit that gave Jesus Christ His new body (Romans 8:11).

There are numerous rewards and honors to be distributed among the redeemed at the time of the Rapture, but whether they will be retained or increased will be determined at the Judgment Seat of Christ. Those who are stripped of honors, which are transferred to others more worthy (for these rewards will not be discarded), will be humiliated. Tears will flow copiously down unhappy, grieved, and disappointed faces. There will be cries and whimpers. The losses will remain permanent and irrecoverable. One wonders how contentions between Christians would be resolved if the accusers would realize that they simply are enriching the rewards of the accused.

Those who flirt with the powers of darkness ("wicked") will be denied not only their position in the corporate wife of Christ, but they will be denied an appearance at the Wedding of the Lamb of God. They will be held in the regions of the lost for future appearance at the White Throne Judgment where their final disposition will be determined (Matthew 13:42,50). Here there will be "wailing and gnashing" of teeth in uncontrollable grief, symbolizing the full measure of their sorrows. Scientific studies show that among the earthly creatures, human beings are the only ones capable of shedding emotional tears because of anger, elation, or grief.

INTERLOCUTORY JUDGMENT

The Judgment Seat of Christ is not the final judgment for those incarcerated in "outer darkness." After their disciplinary period, they must make an additional appearance at the White Throne Judgment where the results of their rehabilitation will determine their future status: "For the time is come that judgment must begin at the house of God: and if it first begin at us, what shall the end be of them that *obey not* the gospel of God? And if the righteous scarcely be saved, where shall the ungodly *and* the sinner appear?" (I Peter 4:17,18).

The Apostle Paul in I Corinthians, Chapter 5, refers to a person in the church at Corinth who was guilty of fornication. Paul admonished the church to take disciplinary action and "... deliver such an one *unto Satan* for the destruction of the *flesh* (not soul), that *the spirit may be saved* in the day of the Lord Jesus Christ" (verse 5).[2] This Christian's untimely death did not erase that fornicator from the book of life. Here is a classic example in I Corinthians 3:15: "... he shall suffer loss; *but he himself shall be saved; yet so as by fire."* The Apostle was not about to pass premature judgment; he would leave it to God. We are not yet the Royal Redeemed sitting in the royal jury box (I Corinthians 6:2,3); therefore, "Judge not, that ye be not judged. For with what judgment ye judge, ye shall be judged..." (Matthew 7:1,2). Paul's recommendation concerning the person's soul was disciplinary; not judgmental.

TURNING TEARS INTO LAUGHTER[3]

In both the paradisiacal and immortal bodies there reside emotional beings who are capable of expressing love, excitement, praise, and worship. This confirms the fact that both bodies are tangible entities (Revelation 7:17; 21:4; Isaiah 25:8). When God wipes away their tears He will eradicate any deep-rooted emotions which are embedded in the soil of sin, as well as bad memories of a past negative life.

BEYOND THE GATES OF HELL

"Gather ye together, *first* the tares, and bind them in *bundles* to burn them: but gather the wheat into the barns" (Matthew 13:30).

The fate of the religious "wicked" in the mystery Kingdom of Heaven will be determined first. The word "bundles" implies that they will appear before Christ according to the respective cult or religious persuasion that runs counter to God's redemptive plan. Their disposition will be resolved by God's holy angels (verses 38-42).

Who are these religious "wicked"? "Not every one that saith unto me, Lord, Lord, shall enter into the kingdom of heaven (that is, the Manifested Kingdom): but he that doeth the will of my Father which is in heaven. Many shall say to me in that day (i.e. Judgment Seat of Christ) Lord, Lord, have we not prophesied in thy name? and in thy name have cast out devils? and in thy name done many wonderful works? And then will I profess unto them, *I never knew you:* depart from me, ye that *work iniquity"* (Matthew 7:21-23).

The test of reality is not sincerity or the exploitation of the name of Jesus. Christ warned of this kind of deception: ". . . Take heed that no man deceive you. For many shall come in my name, saying, I am Christ (or, I am a Christian); and shall deceive many" (Matthew 24:4,5). The only accurate criterion is the acceptance of Jesus Christ as one's personal Saviour. He lived, died, and rose again. The mere admission that He was a great prophet or good man will not suffice (Romans 10:9; I John 4:1,2).

TWILIGHT ZONE

". . . I know you not whence ye are" (Luke 13:25). This group differs from the previous one in that they have "a form of godliness, but denying the power thereof . . . " (2 Timothy 3:5). To be closely associated with born anew Christians is not enough. This group enjoyed fellowship with God's people, but they never gave up their life of sin. They said, ". . . We have eaten and drunk in thy presence, and thou hast taught in our streets." Christ's reply was: ". . . depart from me, all ye workers of iniquity." No one gets in by proxy because of his relationship and association with Godly reborn people (Luke 13:26,27).[4]

BLIND MEN WITH LANTERNS

Simply being a proclaimer of the Gospel, by whatever clerical title, is no insurance that all is satisfactory: "But he answered and said, Every plant which my heavenly Father hath not planted, shall be rooted up. Let them alone: they be blind leaders of the blind. And if the blind lead the blind, both shall fall into the ditch" (Matthew 15:13,14).

"Woe unto you, lawyers (interpreters of the Word of God at that time)! for ye have taken away the key of knowledge: ye entered not in yourselves, and them that were entering in ye hindered" (Luke 11:52).

And even greater condemnation will be faced by spiritual leaders who understand the truth of redemption and do not respond accordingly: "Jesus said unto them, if ye were blind, you should have no sin: but now ye say, *we see;* therefore, your sin remaineth" (John 9:41). They do not have the courage of their convictions.

[1] See chart R.

[2] This scripture also shows that the "spirit" entity no longer retains its own identity when it returns to God (the Spirit Being).

[3] Luke 6:21; I Thessalonians 2:19; Job 8:20,21.

[4] The saddest words are, "What mght have been!"

67
The Lord's Olivet Discourse

There must be a distinction between the Tribulation Temple and the Millennial Temple. Israel will build the first Temple and Christ will erect the second one.

The Lord's Olivet discourse answers three questions which Jesus' disciples asked Him.

WALLS COME TUMBLING DOWN

The first question raised by the disciples was concerning the future of the temple of Jerusalem (Matthew 24:2; Mark 13:2; Luke 21:6-23). The tragedy that befell the temple, as well as Jerusalem, in 70 A.D. was a result of Israel's rejection of their Messiah. They had been warned by John the Baptist: "And now the axe is laid unto the root of the trees: therefore, every tree which bringeth not forth good fruit is hewn down, and cast into the fire" (Matthew 3:10).

Jesus prophesied over Jerusalem saying: ". . . if thou hadst known, even thou, at least in this thy day, the things which belong unto thy peace! . . . For the days shall come upon thee, that thine enemies shall . . . lay thee even with the ground . . . and they shall not leave in thee one stone upon another; because thou knewest not the time of thy visitation" (Luke 19:41-44).

Josephus, a renowned Jewish historian, reported that the Chris tians had made a successful escape from the city of Jerusalem before it fell. Prior to the impending disaster they had been forewarned by Jesus to flee Jerusalem. If they heeded those signs He promised: "...there shall not a hair of your head perish" (Luke 21:18-24). Eusebus also disclosed that as the result of a prophetic warning the Christians escaped by fleeing beforehand to Pella.

Josephus reports that after Jesus Christ was crucified, the Jews grew impatient for the promised Messiah, so they took it upon themselves to openly rebel against the Roman government in 66 A.D. On the day of Pentecost in 70 A.D., Titus, the Roman Emperor, with his army of thirty thousand, laid siege to the city which was crowded with six hundred thousand visitors. The Jewish army numbered twenty-four thousand and there was infighting among their ranks within the walls. After five months the walls were battered down; the temple and the city were burned and destroyed. As Jesus had predicted, not one stone was left upon another because much gold and precious wealth were discovered in the walls and plundered by Titus' army. Starvation was rampant and human blood ran deep in the streets. Over one million people were killed. Josephus, one of the ninety-five thousand captives, said the city appeared as though it had never been inhabited.

THE CHURCH'S LIMITED MISSION

The second inquiry by the disciples concerned the dispensation of the Church. Since the Lord's disciples were Christians, as well as Israelis, the Lord was interested not only in the nation of Israel, but also in the Church: "What shall be the sign of thy coming?" The Church would end in the following manner: "But he that shall endure unto the end, the same shall be saved. And this *gospel* of the kingdom shall be preached in all the world for *a witness unto all nations;* and then shall the end (of the Church Age) come" (Matthew 24:13,14. See also verses 4-12; Mark 13:5-13; Luke 21:24-28).

THE MISSING DIMENSION IN KNOWLEDGE

"Who will have all men to be saved, and to come unto the *knowledge of the truth*" (I Timothy 2:4). "The *beginning* of the gospel of Jesus

Christ, the Son of God" (Mark 1:1). "Then cometh the *end,* when (Christ) shall have delivered up the kingdom to God, even the Father, when he shall have put down all rule and all authority and power . . . The last enemy that shall be destroyed is death" (I Corinthians 15:24,26).

The mystery of the Kingdom of God *began* with Jesus Christ and will *end* when Christ turns the Kingdom to the Father. This will happen when "every creature" and not just "every nation" will be reached with the Gospel of Jesus Christ, including all those deceased who never heard the gospel and those who died in innocence. So long as mortality reigns the gospel message will go forth, for the ultimate goal is immortality for all the redeemed when the last enemy, death, is destroyed.

The missing dimension in knowledge is the assumption that the Gospel of Jesus Christ is for this time only, when God is trying to reach "every creature" with the Gospel. Nothing could be further from the truth of God's Word. 2 Corinthians 6:2 by no means limits salvation to the present day. The "manifested kingdom" does not end the preaching of the Gospel of the Kingdom of God.

FINAL TRIBULATION PERIOD

The third question pertained to the final Tribulation period, particularly the time of Israel's Tribulation, after the dispensation of the Church comes to its end. It was as follows: "What shall be the sign . . . of the end of the age?" It will bring an end of all Gentile rule and Christ will return to earth to set up His Millennial Kingdom (Matthew 24:15-31; Mark 13:14-27).

REBUILDING THE TRIBULATION TEMPLE

"When ye therefore shall see the abomination of desolation, spoken of by Daniel the prophet, stand in the holy place, (whoso readeth, let him understand:) Then let them which be in Judaea flee into the mountains: . . . For then shall be great tribulation such as was not since the beginning of the world to this time, no, nor ever shall be" (Matthew 24:15-26).

There must be a distinction between the Tribulation Temple and the Millennial Temple. Israel will build the first temple and Christ will erect the second one. The first temple will be built adjacent to the Temple of Omar, the golden-domed Mosque, according to Revelation 11:2: "But the court which is without the temple leave out (for) it is given unto the Gentiles. . . . " This Scripture identified the site of the Tribulation Temple, and prophesies that a structure is already there. On the other hand, Ezekiel foretells that the Millennial Temple, with its architectural grandeur, will replace both the temple of Omar and the temple of Israel.

The temple to which Christ will return is the Tribulation Temple (2 Thessalonians 2:4; Malachi 3:1).[1] This is the temple which Antichrist pollutes when he enthrones himself there and declares himself to be God: "And from the time that the daily sacrifice shall be *taken away,* and the abomination that maketh desolate *set up,* there shall be a thousand two hundred and ninety days" (Daniel 12:12).[2] In Daniel the sequence of events relating to the Tribulation Temple is as follows: First, there will be the dedication of the Temple. It would appear in the light of I Kings 8:62, that the customary thing for Israel to do would be to give "peace offerings . . . unto the Lord" after their great victory over the Soviets and their allies. Second, there will be the removal of the altar to make room for the image of the Antichrist on that precise location: "And (Antichrist) shall take away the daily sacrifice, and they shall place the abomination that maketh desolate" (Daniel 11:31). Forty-two months later the Antichrist will sit in the Tribulation temple, void of its furnishings ("desolations"), after which he will set up his image on the site of the brazen altar, just outside the Temple, to be worshipped. It is at this precise moment that Israel will flee into sanctuaries outside of the Babylonian regime which are beyond the direct control of the Antichrist.

The martyrdom of the two prophets in Revelation 11 will encourage Antichrist to invade the land of Israel and take over the Temple (Daniel 11:21). Several facts are vital in establishing the time of the last two prophets on the earth: 1) They are associated with the

Tribulation Temple (Revelation 11:1,2; Zechariah 4:2,3); 2) They are referred to as "two olive trees" standing beside the Temple candlestick; 3) They are given special power from God against their enemies who are chiding over the construction of the Temple on the Temple Mount; 4) They are associated with the outpouring of the Holy Spirit in the last days (Zechariah 4:6; compare Acts 2:1-21), which lends courage to the Great Tribulation Christians (Daniel 11:32-35); 5) Their message is one of God's wondrous "Grace" (Zechariah 4:7); 6) They prophesy forty-three months, not forty-two, and are martyred about one month before the final resurrection and rapture of the Tribulation saints, after which Antichrist enters the Tribulation Temple and severs the covenant of peace with the Zionists; 7) They also must appear, together with the 144,000, and all of the Royal Redeemed, at the Judgment Seat of Christ (2 Corinthians 5:10).

WISE ADMONITIONS

The Olivet Discourse ends with wise admonitions from the Lord Jesus relative to all of the subject matter covered. It contains counsel applicable to both the Israelis and Christians—and it covers that wide range from Pentecost, or the birth of the Church, to the time of the Millennium, or the maturation of the Church (Matthew, chapters 24 and 25).

As a gesture of good will, Antichrist will make a seven-year peace covenant with the Zionists at the time Israel completes and dedicates the Tribulation Temple. Jesus Christ warned Israel in John 5:43 that this good will gesture would be consummated with someone other than their Messiah whom they are rejecting: "... another shall come in his own name, him ye will receive." According to I Thessalonians 5:3, it will be an uncertain agreement: "For when they shall say, Peace and safety; then sudden destruction cometh upon them, as travail upon a woman with child; and they shall not escape." Jeremiah 6:14 states: "They have healed also the hurt ... of my people slightly, saying, Peace, peace, when there is no peace."

RESULTS OF FIVE MAJOR WARS IN THE LAST DAYS

WORLD WAR I

Palestine was opened for Israel's return to the promised land.

WORLD WAR II

Israel became a nation.

THE SOVIET INVASION

After the defeat of the Soviets, Israel will have seven years of peace to bury the war casualties and debris, to erect the graveside city, and to rebuild its temple.

ARMAGEDDON

The Messiah, the Lord Jesus Christ, will return to earth.

FINAL ASSAULT ON THE CITY OF JERUSALEM

It will end all organized resistance to Christ's righteous government.

[1] See chart Q.

[2] It is conceivable that the False Prophet will sit simultaneously either in the Temple of Omar or in the Vatican.

68

Last Two Prophets on Earth

There must be nations during the Millennium over which Jesus Christ and the Royal Redeemed can reign.

". . . These are the two anointed ones that stand by the Lord of the whole earth" (Zechariah 4:14).

Some theologians believe that these two witnesses, Moses and Elijah, were the ones that were predicted by Zechariah, and fulfilled in Revelation 11:1-13, as well as the same two that appeared at the transfiguration, resurrection, and ascension of Christ (Mark 9:4; Luke 24:4; Acts 1:10). Whoever they are, they have a vital and important message for this final critical time, especially to Israel. Jesus described this crucial period as follows: "For then shall be great tribulation, such as was not since the beginning of the world to this time, nor shall ever be. And except those days should be shortened, there should no flesh be saved: but for the elect's sake those days shall be shortened" (Matthew 24:21,22; compare Daniel 12:1).

The two witnesses proclaim the gospel of God's grace through the Lord Jesus Christ: ". . . Not by might, nor by power, but by my spirit, saith the Lord of hosts . . . with shoutings, crying, Grace, grace unto *it*" (Zechariah 4:6,7). The word "it," in verse 7, refers to the "headstone" of the Church, who is Christ.

FROM THE ALTAR TO THE IMAGE

Immediately after the final harvest day Rapture of the Tribulation saints, Antichrist makes his appearance at the Temple in Jerusalem. All of the Tribulation saints will escape his assaults because they will have been transported into heaven. This so infuriates Antichrist and the False Prophet, that they break the peace covenant with the Zionists, then relentlessly and ferociously pursue the genocide of all Israel. It is at this point, with a show of military force, that they take over the Temple site: "And arms shall stand on his part, and they shall pollute the sanctuary (Temple) of strength, and shall take away the daily sacrifice (remove the brazen altar), and they shall place (substitute) the abomination that maketh desolate (that is, the image as described in Revelation, chapter 13)" (Daniel 11:31). This prophecy clearly identifies the precise spot where the image of Antichrist will be set up for worship.

During the persecution of the Tribulation saints, the two witnesses will make an appeal to all nations to receive Jesus Christ as their Redeemer at this critical juncture in man's history. Antichrist, perhaps irritated by this brazen ministry of the two prophets, will plan to break the covenant with Israel and relocate his center of power to the Tribulation Temple in Jerusalem.

God allows the subversive plans of Antichrist to assassinate the two prophets. This will occur just prior to the final harvest day Rapture. The effectiveness of the two prophets and the shocking disappearance of the Tribulation saints, will cause many nations to forge a military force under the "kings of the East" to destroy the combined kingdom of Babylon. These are also the "righteous nations" that will make their appearance at the judgment of the nations.

Antichrist, after the martyrdom of the two prophets, will break the peace covenant with the Zionists, move into the Tribulation Temple, and set up his image just outside on the Temple grounds. The Levites, in their flight from Antichrist, will take with them the Temple furnishings and the golden Ark of the Covenant as they flee into Egypt. After Christ descends to the earth, these furnishings and the Ark of the Covenant will return to Jerusalem as they leave Egypt the second time under the mighty power of God (Isaiah 11:12-16).

The two prophets make their appearances at the Judgment Seat of Christ and the Marriage of the Lamb. This supports the position that their prophesying will occur prior to the final harvest Rapture.

According to Daniel 8:14 and 12:12, it will take forty-five days for Armageddon to end, for the Judgment of the Nations to be completed, for the Temple furnishings to be in place and for the Levites to be in position for the reconsecration and dedication of the Temple in the presence of the Messiah, Jesus Christ. This will officially launch the beginning of the Millennium.

69

United States in Bible Prophecy

While the Bible does not claim to be a book of science, yet whenever it touches science, it is indisputably accurate.

Mankind spread out slowly from the cradle of civilization, totally unaware of the broad expanse of land area existing upon the earth. However, this movement was impeded when the single land mass, called the "Pangaea," purportedly broke apart into continents and islands as the result of a violent earthquake in the days of Peleg (Genesis 10:25).

"ISLES" OF THE GENTILES

Some geographers claim that South America and Africa were once joined. Historians inform us that many of the ancient ruins and pyramids in Central and South America can be traced to the same time period during which the pyramids in Egypt were constructed. This occurred before the Western and Eastern hemispheres were separated.

The land masses existing outside of the Old World were called "isles" or "wilderness." In the Scriptures, when God spoke and revealed the future of the Western hemisphere, and the United States, in particular, those terms were employed for identification purposes.

In Scripture, the first mention of the New World is found in Genesis 10:5: "By these were *the isles of the Gentiles* divided in their lands; every one after his tongue (that is, after Babel), after their families, in their nations." When the prophets of God spoke, they were not cognizant of what their prophetic utterances involved, but they did speak as the oracles of God. But God, who brought the New World into existence, was fully aware of its future course in world events.

BEYOND "THE SEA"

The Old World population was distributed around the Mediterranean Sea while other bodies of water remained very much a mystery. The ancient name for the Mediterranean was simply "The Sea" or "Great Sea." Thus the prophets spoke in their time in the light of the knowledge that was comprehended by their contemporaries. Here is a graphic example: "Listen, *O isles,* unto me; and hearken; ye people *from far . . .*" (Isaiah 49:1). ". . . The kings (rulers) of the *isles* which are *beyond the sea*" (Jeremiah 25:22).

EUROPEAN HOLOCAUST

Isaiah 17:10,11 depicts the desolation of Palestine as the result of Israel's rebellion against God: ". . . thou hast forgotten the God of thy salvation." Further recorded is the trickling return of the Israelis to their land and the response of the land to their labors: "In that day shalt thou make the plant to grow, and in the morning shalt thou make thy seed to flourish . . . the harvest shall be a heap *in the day of grief and desperate sorrow.*"

In verses 12-14 the nations of the world are depicted in the end time as being in a state of turmoil. Verse 12 foretells the war in all its fury: ". . . and to the rushing of nations, that make a rushing like the rushing of mighty waters!" Foretold are the tragic events that devastated the Jews during World War II when millions perished in concentration camps.

Then follows a turning point in the war: "God shall rebuke them, and they shall flee far off, and shall be chased as the chaff of the mountains before the wind..." (verse 13). God used the United States and its allies in producing a great victory favorable to Israel. Moreover, President Harry S. Truman played a vital role in assisting Israel to become a sovereign nation.

MYSTERY NATION

The entire eighteenth chapter of Isaiah is devoted to the Western Hemisphere, and especially the United States. Not until the discovery of America in recent centuries was this chapter understood. It remained an enigma to Bible scholars in ancient times. The Bible expositor, Bishop Louth, admitted his inability to comprehend its ambiguous language. He said, "This is one of the most obscure prophecies; the people to whom it is addressed, the person He sends, the ambassadors, and the nations to whom they are sent are extremely doubtful."

It is comprehensible that this prophecy was clouded in mystery until the discovery of the United States. This passage becomes obvious only when it is understood as a prophecy of this nation. It is also necessary to examine the Hebrew root words to ferret out their rich contents.

WORLDWIDE SANCTUARIES

The nineteenth chapter of Isaiah reveals events that will transpire just prior to the Lord's return to earth and which are known as the time of "Jacob's trouble" under Antichrist.

Isaiah 19:16-20 predicts that great numbers of Israelis will flee to Egypt, and will dwell in five cities where they will be protected from Antichrist, not by Egypt, but by the golden Ark of the Covenant which they have in their possession: "In that day shall Egypt be like unto women: and shall be afraid and fear because of the shaking of the hand of the LORD of hosts, which he shaketh over it... In that day shall there be an altar to the LORD in the midst of the land of Egypt, and a *pillar* (compare Exodus 40:36) at the border thereof to the LORD. And it shall be for a *sign* (see Joshua 4:6,7) and for a *witness* (see Numbers 4:5; 17:7,8) unto the LORD of hosts in the land of Egypt. . . . " This will occur when Antichrist breaks his seven-year peace covenant with Israel and seeks its extermination (Daniel 12:1,7; Matthew 24:15-26; Jeremiah 30:7).

In Isaiah 19:4, Antichrist is described as "a cruel lord; and a fierce king." He will conquer the northern portion of Africa because, according to the prophecy of God's Word, he shall possess all of the territory within the confines of the ancient Roman Empire. These five cities, as sanctuaries for the Israelis in Egypt, will lie just south of the kingdom of Antichrist barely outside of his reach, as designed by God.

It was under the Camp David Agreement that Israel and Egypt came to amicable peace terms, largely through the efforts of an American President. This may help to pave the way for Egypt to give sanctuary to Israel in its flight from Antichrist. If not, the presence of God symbolized in the Ark of the Covenant will be primarily responsible for the Israelis' decision to locate in Egypt (Isaiah 19:16). This also will keep Israel out of the area where the battle of Armageddon will be fought.

BEYOND THE OLD WORLD

Ethiopia was the southernmost boundary of the empires of the ancient world. Practically nothing was known of Africa below this point. In fact, all of Africa's unexplored and unknown territory in antiquity was known as Ethiopia.

Sailing vessels avoided going down the western coast of Africa beyond Morocco. To do so would be to enter "the green sea of darkness," an unnavigable swamp full of monsters, as reputed by Arab geographers. If one dared venture out across the Atlantic, he might drift near the equator, where men turned black permanently or would lose their lives. Once a ship dared enter the open uncharted wastes of the ocean, it would be subjected to the mercy of the weather.

THE VAST TERRA INCOGNITA

To Europeans of medieval times, the world consisted of just four continents—Europe, Africa, Asia, and the vast Terra Incognita which was vaguely understood, and believed to be situated somewhere in the Southern Hemisphere. On contemporary maps, these land masses pressed the earth's surface and crowded out the oceans. The Atlantic was a narrow waterway which suddenly dropped into an abyss. There was no Pacific, and Terra Incognita made the Indian Ocean but an island sea.

THE UNKNOWN AFRICAN CONTINENT

To the Europeans, Africa was a land of fantasy. They believed that somewhere a river of gold emptied into a seething, tropical sea that no man could reach. Over it hovered legless birds which spent all of their lives in endless flight. There were sheep as large as oxen; giants that could wade the ocean and seize sailing vessels with one hand; women with eyes of precious gems who could slay intruders with a single glance; and men with such enormous feet that, when held aloft, they served as shelters. These and many other legends abounded since none of them could be challenged by observers. Europeans were almost entirely ignorant of both Asia and Africa.

Discovery of Africa was obstructed by prevailing geographical hazards. Life could not be supported close to the equator. Cape Nun, on the northwestern coast of Africa, at latitude 29 degrees north, received its name because legend insisted that if any seaman dare venture past it, he would never return.

Beyond Cape Nun, they were told, was a boiling sea. The searing sun would destroy all who were not already black. Beyond that lay the Antipodes where only monsters lived. Since all men descended from Adam, no man could pass the tropics where these hazards existed. It was known as the Terra Incognita, and was joined to Africa; therefore, it could not be circumnavigated anyway.

BIBLE AND ANTIQUITY

While the Bible does not claim to be a book of science, yet wherever it touches science, it is indisputably accurate. Of all of the literature up to and including the medieval times, only the Bible contains the truth of God's creation and excludes the fantasies and tales of ignorant men. It reports nothing of Atlas holding up the earth. It declares that the stars were as the sands of the seashore, instead of supporting the Greek astronomer, Hipparchus, who calculated the number of stars in the heavens as limited to 1,080. The Bible states that the earth was spherical in shape as opposed to contemporaries of the day who said it was flat. In Proverbs 8:27 the word "compass" means circle, a clear indication of spherical shape. In Isaiah 40:22, the Scripture refers to God sitting above the "circle" of the earth, a word meaning sphere.

Moses, by divine inspiration, said in Deuteronomy 4:19 that God divided the stars upon all of the regions under the heavens, implying a round earth because all of the nations would see different stars. And our Lord Jesus, in speaking of His second advent in Luke 17:34-36, pictured people in various parts of the earth simultaneously sleeping, grinding, and harvesting—occupations for night, early morning, and midday. Could there be any other explanation than the rotundity of the planet? The Bible did not make errors as did the contemporaries of the times.

HOLY SCRIPTURES: INDISPUTABLY ACCURATE

While in ancient times men believed it was impossible to cross the Atlantic Ocean, or that any additional continents existed, the Word of God in unmistakably clear language dared to venture, not only the fact of the existence of other continents but that of the United States, in particular. It also predicted its origin as a nation, its history, and its destiny.

Isaiah, chapter 18, is not only prophetic, but it contains the most challenging of all eschatological truths. Without a knowledge of eschatological Scriptures, one is at a loss to understand world changes. Fulfilled prophecy is an indisputable argument for the inspiration of God's Word, which we know as the Holy Bible.

70
Land of Outstretched Wings

The prophet Isaiah identified the geographical location of the unnamed nation (United States) as "beyond the rivers of Ethiopia."

As previously indicated, Isaiah, chapter 18, has challenged and confused translators and commentators because they did not associate it with the United States. Consequently, the Authorized King James Version is obscure at several vital points, not because of inaccuracy of the original text, but as a result of the ambiguity of the translators. The Old Testament was written in Hebrew and the translators were required to find English words which they believed best expressed the original text. Because they did not recognize the description of our country and because of lack of Biblical precedent, they did not examine this chapter in the light of the United States fulfilling this enigmatic passage. We must, therefore, refer to the Hebrew roots in the light of our present knowledge in order to understand this chapter.

SALUTE TO THE NEW NATION

The prophet saluted the new nation of his prophecy thuswise: "Woe to the land shadowing with wings, which is beyond the rivers of Ethiopia." "Woe" is derived from the Hebrew word "howy," mean ing a salute. The proper translation is "Ho" and not "woe." "Woe" implies judgment, as if God were about to visit the new nation with His wrath. The Hebrew word "howy" in our vernacular is "howdy" or "hello."

The prophet Isaiah identified the geographical location of the unnamed nation as "*beyond* the rivers of Ethiopia" (Isaiah 18:1). The word "beyond" is taken from the Hebrew "eber" meaning *on the opposite side, across, or straight.* Standing in Jerusalem, as the prophet was looking "across or straight," his prophetic vision must have logically swept the full length of the Mediterranean Sea and across the Atlantic Ocean, before it fell upon the United States.

NATIONAL EMBLEM

The prophet described the United States as "the land shadowing with wings." Our national emblem is the king of birds, the bald eagle with its outstretched wings, which is on some of our currency such as the quarter, the half dollar, and our one dollar bill. In contrast, the Mexican eagle has its wings closed and the German eagle is double-headed. England's emblem is the lion, the king of beasts. The phrase, "shadowing with wings," as interpreted in Jamieson, Fausset, and Brown reads, "barks with wings like sails" and "land of the clanging sound of wings." The prophet is saying, "Ho, to the land of the noise of wings, the roar of winged craft."

America is a land of wings in more ways than one. This nation invented the outstretched wings of the flying machine, and it possesses the most powerful air force on earth, which has been a major influence in changing the history of the world.

AIR POWER

World War I delivered Jerusalem and the Promised Land from centuries of Turkish rule and gave Israel a home in Palestine. This was the result of the introduction of aircraft into major warfare by the Allied Forces and the event was considered significant enough to deserve mention in the Scriptures: "As birds *flying,* so will the Lord of hosts defend Jerusalem; defending also he will deliver it; and passing over he will preserve it" (Isaiah 31:5). This was declared by the same prophet who prophesied concerning the United States

The description of American air power continues. The prophet says: "That sendeth ambassadors by the sea even in vessels of bulrushes upon the waters, saying, Go ye swift messengers" (Isaiah 18:2). "Vessels of bulrushes" is prophetic language describing lightweight, compact planes that fly with a roaring sound. The prophet says that they came by sea, beyond the rivers of Ethiopia; that is, beyond the Old World. How forcefully this prophecy was fulfilled, especially in World War II, as great fleets of planes were ferried across the Atlantic and Pacific to bring the might of America against its enemies. America has not been invaded by a foreign power; she has always fought her major wars on foreign soil. "Go ye swift messengers" depicts rapid transportation to the far-flung continents of the world from the land of outstretched wings.

MERCHANT FLEETS

The prophet Isaiah anticipated an invention by more than twenty-five centuries. He pictured huge steamships drawing sea water into condensers, thereby converting the water into steam to propel the vessels. These liners were traveling to the Old World from the nation of outstretched wings and were "water-drinking vessels" from the Hebrew word "gama" which means to "swallow or drink."

BEYOND THE HORIZON

The prophet used the word "hitherto" to show that this new nation is far removed from the Old World. The Hebrew word is "halah" and means "beyond, forward, thenceforth, yonder, or to the distance."

The history of this new nation, the United States, has been predominantly Christian, and it undoubtedly shall emerge from the final Tribulation period as one of the "sheep" nations to enter the Millennium (Matthew 25:31-46).

71

Early American History

The United States represents the "times of the Gentiles" in their finest hour... Antichrist in its worst period in history.

In Isaiah 18:2, the prophet directs his prophecy to the terrain of the new nation. He says it is a "land the rivers have spoiled." The Hebrew word is "baza," meaning "to cleave or divide." The word is taken from the ritual of the Old Testament priests who spoiled, or cut the animal from head to tail and across the middle, dividing it into quarters for sacrifice. Likewise, we see the Mississippi River bisecting the nation from north to south; the Ohio River and its tributaries dividing the eastern half; and the Missouri River and its tributaries partitioning the western half. Thus, the United States is also quartered, but by rivers. Many of these rivers and their tributaries are the borders of states.

WATERSHEDS

The prophet tells us that the land of outstretched wings is a land of mountains (Isaiah 18:3). Two great ranges of mountains stretch from north to south at both the eastern and western ends of the United States—the Rocky Mountains and the Appalachian Mountains. They provide the nation with tremendous water reserves and drain age systems, thus making it ideal for bountiful crops. Therefore, the prophet describes the land as "like a clear heat upon herbs, and like a cloud of dew in the heat of harvest" (Isaiah 18:4). God speaks of seeing this land from His "dwelling place."

REDSKINS

In describing the inhabitants of the new land, the prophet tells us it is "a nation scattered and peeled, to a people terrible from their beginning hitherto" (Isaiah 18:2). The word "scattered" originates from the Hebrew root, "mashak," which means "to be tall, stately, stretched out, extended, scattered." "Peeled," means "smooth countenance" or beardless. Isaiah said, "a nation tall and smooth-skinned," a literal description of our Native American Indians.

THE MELTING POT

"Scattered" in the Hebrew root means "drawn out." Not only has God "drawn out" this new nation from among the countries of the world, but also He has "drawn out" from the peoples of the earth all nationalities under its outstretched wings. This nation represents the "times of the Gentiles" in their finest hour. Antichrist will unify the world to form a system that will cause the "times of the Gentiles" to be the worst period of history.

The prophet described the inhabitants as a "people terrible from the beginning." We see prophesied here the difficulties that the early settlers and the Indians encountered with one another. There was extreme cruelty and bloodshed on both sides in establishing this new land. The prophet described this "budding" of the nation as "sour grapes. . . ripening in the flower" (verse 5). Those were difficult and hard days.

72

Independence Day

Any American citizen may become a landowner. This generally was not true in the Old World.

"The sour grape ripening in the flower" includes not only the Indian battles, but also the Revolutionary War and the Civil War: one war to gain its independence from the Old World, and one to fulfill its God-given destiny.

The prophet says the new nation "shall cut off the sprigs with pruning hooks, and take away and cut down the branches." Pruning is necessary for better fruit, so the nations of the Old World had to be cut off. Thereby, the new nation by this pruning process not only gained its independence, but by this action became strong and fruitful

METES AND BOUNDS ACT

Isaiah in his prophecy of the United States said that it is "a nation meted out and trodden down." The word "meted" is from the Hebrew "gav" meaning "a line, cord, or rule for measuring." It is a nation measured. "Trodden" comes from "mbuwcah," which literally rendered is "underfoot." The prophet is saying, "A nation measured by lines underfoot."

The Metes and Bounds Act, enacted by Congress at the turn of the nineteenth century, provided for surveying the land. It was the first time in world history that a government decreed its land be surveyed according to the North Star. The public land was carved into mile square sections on a base line running east and west. The United States is divided into sections, half sections, and quarter sections in line with the North Star. Thus it is measured by lines underfoot.

In the Old World land was measured according to local landmarks, such as from a rock to a tree; from a bridge to a gully; or from a brook to a hilltop.

Any American citizen may become a landowner. This generally was not true in the Old World. Major landowners were "lords" or important government personnel, and serfdom was prevalent. No one in America is denied the right to own property. It is the privilege of every citizen.

The words "trodden down" express how the land of outstretched wings will be inhabited and trafficked. It is a land unexcelled in means of travel. It leads the world in miles of highways and railroads, as well as in ports for its ships and airlines. It is a land of immense traffic. It is a nation meted out and *trampling* (according to one translation).

The American citizen not only travels extensively in the land of outstretched wings, but he is one of the leading travelers throughout the world, The United States is a nation on wheels, and its people are *trampling* the world.

BLESSED OF GOD

In directing attention to America's favored position among the world of nations Isaiah said, "For so the Lord said unto me, I will take my rest, and I will consider in my dwelling place like a clear heat upon herbs, and like a cloud of dew in the heat of harvest."

"I will take my rest" literally means, "I will be still and look on." When Israel rejected its Messiah, it could not, as a nation, fulfill its important role to the rest of the world, so God deferred it until a later time. In the meantime, God turned to a Gentile nation to carry forth His purposes in the last days. The United States is now fulfilling that interim period.

73
The Macedonian Call

Israel will become the greatest missionary nation of all times. It will bring the gospel to the remotest ethnic areas of the world.

God must have had America in mind when Paul in a vision heard the call: ". . . Come over into Macedonia, and help us" (Acts 16:9,10). He had tried twice to enter Asia with the gospel, but was forbidden of the Holy Spirit (verses 6,7). Therefore, Paul went westward and from Europe the gospel came to America with our Pilgrim fathers.

When Israel, as a nation, rejected Jesus Christ as its Messiah, Paul said, ". . . Lo, we turn to the Gentiles" (Acts 13:46). Paul did not realize the significance of his statement regarding the great dispensational change that was to occur.

With Israel about to be scattered over the globe, and temporarily shelved as a nation that God might use, the Lord reached among the Gentiles for a nation which would carry forth the gospel in the end-time (Matthew 28:19; Romans 9:24-28). Like Esther in her day (Esther 4:14), we believe that the United States has been raised for such a time as this until Israel again can assume its responsibilities under God:

". . . I will call them my people, which were not my people; and her beloved, which was not beloved.

"And it shall come to pass, that *in the place* where it was said unto them, Ye *are not my people:* there shall they be called the children of the living God.

". . . For he will *finish the work and cut it short* in righteousness, because *a short work* will the Lord make upon the earth.

"And as Esaias said before, Except the Lord of Sabaoth (host) *had left us a seed,* we had been as Sodom, and been made like unto Gomorrah.

"What shall we say then? That the Gentiles, which followed *not* after righteousness, have *attained* to righteousness, even the righteousness which is of faith" (Romans 9:25-30).

The Lord said, ". . . *he* will finish the work, and cut it short in righteousness." If Israel, as a nation, will not perform her obligation, then God must turn elsewhere for a nation of people who will fulfill the divine commission. God said it would be "a short work . . . upon the earth."

A MISSIONARY NATION

Seventy-five percent of all foreign missionary work is being performed or sponsored by citizens of the United States who are sending the gospel throughout the world by radio, television, Christian literature, and trained native workers. From no other country in the world has so much gospel been dispersed during the last several decades as from this nation. Any American citizen who fails to support this worldwide ministry is betraying his God-given heritage as an American citizen. We also must recognize how much other Western nations have contributed toward sending out the gospel. They certainly will be among the "righteous" nations to exist during the Millennium.

America has been the main bulwark against communistic attempts to take over the world. It has spread the gospel, and has prevented an untenable situation from engulfing the world. America is God's gift to mankind, especially in the last days. It is an oasis in a world of sin. It seems incredible, therefore, that the United States should not have been mentioned in God's Word. We believe that it is unmistakably included in the eschatological Scriptures of the Bible.

When God commissioned Paul to bring the gospel westward, Asia succumbed to many strange and powerful pagan religions, but Europe embraced Christianity. From Christian lands have emanated the greatest discoveries and inventions to bless mankind. Progress has been phenomenal in these countries.

The United States has emerged as the world leader among nations within a relatively short time.[1] When the mighty reformation was born in Europe, the gospel brought its redemptive power to bear upon those who were long steeped in superstition and fear. Great spiritual revivals rapidly followed, bringing millions of souls to rebirth in Christ Jesus. America is yet a young nation, and thus far in her brief existence has made significant accomplishments in disseminating the gospel to the world.

ISRAEL'S EMERGENCE

The United States in many ways is truly representative of the "times of the Gentiles" (Luke 21:24; Romans 11:25). The culture of America is unlike any in the world as it is the melting pot of all nationalities. Here Christianity has found its strongest base of operation; it is the mainstream of religious life and is sanctioned by the government. In fact, the American government was founded upon Christian principles.

It is evident from Scriptures that the United States, by divine decree, will continue its important role until Israel resumes its rightful place as a nation in the economy of God. Jesus indicated the time of this transition: ". . . and Jerusalem shall be trodden down of the Gentiles *until* the times of the Gentiles be *fulfilled*" (Luke 21:24).

The reoccupation of all of Jerusalem by Israeli troops in the "six-day miracle war" of June 1967, signaled the beginning of this transition. Subsequent to that time, according to Jesus, many disturbing factors in the Gentile world will accelerate. The escalation of wars, riots, and terrorism is only one facet of many to perplex the world. There also will be an intensification of natural disturbances and heavenly phenomena such as UFOs (Acts 2:19; note word "wonders").

As God has blessed America in the past to fulfill its designated role according to God's Word, so God is enabling Israel to end its dispersion and to emerge permanently as the center of world power and influence. The problem in the Middle East now centers around Israel and the city of Jerusalem, which is at the moment of transition indicated in Zechariah 12:2: "Behold, I will make Jerusalem a cup of trembling unto all the people round about. . . . " God is calling us to pray for the peace of Jerusalem. So long as this city is not at peace, the world will not have peace (Psalm 122:6).

UNITED STATES AND ISRAEL

The United States must undergo a difficult period as it makes the transfer from leadership to a second-rate position—second to Israel, where the world's future capital will be located. It is from Jerusalem that Jesus Christ will reign in His Temple over all mankind.

Paul, in Romans 11:25,26 said, "For I would not brethen, that ye should be ignorant of this mystery, lest ye should be wise in your own conceits; that blindness in part is happened to Israel, *until* the fulness of the Gentiles come in. And so *all Israel shall be saved,* as it is written, there *shall come out of Sion the Deliverer,* and shall turn away ungodliness from Jacob."

WHEN GOD WHISTLES[2]

After the fulness of the Gentiles, something will happen to Israel as a nation which has never happened to America. Never has the United States enjoyed having everyone of its citizens converted to Christ. However, when Israel resumes its divine place under the Messiah, Jesus Christ, the entire nation will be converted to Christ without exception.

During Christ's millennial reign on earth,[3] Israel will become the greatest missionary nation of all times. It will bring the gospel to the remotest ethnic areas of the world. Then, and only then, will the commission of Christ be fully completed: ". . . Go ye into *all* the world and preach the gospel to *every creature*" (Mark 16:15). However, it will take one thousand years under world peace to accomplish this enormous task.

Not everyone, of course, will be converted to Christ (Revelation 20:7-9), but the work of evangelization to reach every creature in the world will have been achieved. The Church will succeed only in accomplishing its outreach to *all of the nations:* "And this gospel of the kingdom shall be preached in all the world for a *witness unto all nations;* and then shall the end (of the Church Age) come" (Matthew 24:14).

At the end of the Millennium all of the ungodly who refused the gospel will be destroyed (Revelation 20:9), including those born during that time. Only the saved will enter into the New Heaven (Revelation 21:24).

[1] It represents only 6% of the world's population.
[2] God will whistle unto scattered Israelis: "I will hiss unto them" (Zechariah 10:8). Hebrew for "hiss" is *sharak,* meaning "whistle."
[3] See Chapter 62, "The Lord's Royal Court."

74

Future of the United States

It will take only "one day" for God to destroy the Soviet Union.

The United States has been like "salt," thus preventing the world from becoming as Sodom and Gomorrah which were destined for God's judgments. However, when the salt's "savor" is removed via the Rapture of the Church, the world will become, as Jesus said, "... thenceforth good for nothing, but to cast out, and to be trodden under foot of men" (Matthew 5:13).

The United States has been a "light" resisting the creeping darkness of communism and the influence of evil. When the "lights" of Christians are removed, the midnight darkness will blanket the earth as on a moonless night. At that time the impact of the United States and other Christian nations will be diminished measurably and the lights will flicker as if being extinguished. God no longer will be obligated to shield America from judgments.

CONTINENTALISM

The prediction of a confrontation between the Soviet and American forces will be fulfilled. God said, "And I will call for a *sword* against him (Soviets) throughout all my mountains (Israel)..." (Ezekiel

38:21). This confrontation will unify the Western Hemisphere. Its united strength, which will be felt by the Soviets, will terminate the Soviet Union as a great power and end communism forever.

NUCLEAR ARMS PROLIFERATION

Therefore, when we read in Ezekiel 39:6, "And I will *send a fire on* . . . them that dwell *carelessly* in the isles and they shall know that I am the Lord," we have a clear prediction that the Western Hemisphere will be bombed. It is possible that these bombings may come upon the United States and its allies from Soviet puppet governments in the *Americas.* This offensive action against the American nations undoubtedly will be timed to blunt any military retaliation against the Soviets when they move into the Middle East. The word "isles" in the plural suggests that many of the Western nations are possible targets of bombings by the Soviets and their Western Hemispheric puppet nations. The end result will be the strengthening of continentalism in the Western anti-Communist nations. This ultimately will contribute greatly to the defeat of Babylon, which will consist of the combined kingdoms of Antichrist and the False Prophet, and will cover one-third of the land surface of the earth (Revelation 8:7-12; 9:15).

The word "carelessly" subtly suggests that these bombings will not occur until *after* the pre-Tribulation Rapture of the redeemed, but we may see increasing signs of Soviet influence in the nations of the Western Hemisphere with communistic leanings. It should serve as an obvious warning of the imminent translation of the redeemed.

GOD'S INTERVENTION

The Soviet advance into the Middle East will be met with stiff resistance, not only from its natural enemies, but primarily from Almighty God: "And I will rain upon him (bogging down his military machines), and upon his bands, and upon the many people that are with him (Soviet allies), *an overflowing rain, and great hailstones,* fire, and brimstone" (Ezekiel 38:22).

". . . There shall be a great shaking in the land of Israel" (Ezekiel 39:19). This earthquake will be localized first in the Middle East, and will bring huge boulders cascading down the mountainsides. These will crash into the Soviet forces and their equipment, opening crevices that will swallow up men and machines. In fact, such crevices may be the interment ground for Israel's enemies.

This earthquake also will create great tidal waves sweeping across the Mediterranean Sea (verse 20), known as "The Sea" in ancient times. In addition, this earthquake will trigger a chain of other earthquakes throughout the world as Jesus Christ predicted: ". . . and earthquakes in divers (many) places" (Matthew 24:7).

BERLIN WALL COMES DOWN

No reliable reference will support the premise that Gog and Magog, in Ezekiel, chapters 38 and 39, refer to the Soviet Union *per se*. The U.S.S.R. will not exist as a nation during the Millennium (Revelation 20:8), so consequently Gog cannot apply to this nation. Since Gog simply means an *enemy of God*, in that sense it fits the prophetic picture.

Magog is the father of the Scythians and Tartars, and the ancient lands they occupied are located where modern Russia now exists. Ezekiel's prophecy indicates that the land of Magog is in the "north parts" (39:2).

The Soviet Union also will be bombed, as will the Western nations: "And I will send a fire on Magog . . . " (Ezekiel 39:6). The Soviet military move into the Middle East will cost them dearly in enormous casualties on their home front as well as on the battlefield. First, there will be bombings; second, there will be devastations wrought by God directly through nature's power; and third, there will be terrible confusion among their troops as they destroy each other in the melee; ". . . every man's sword shall be against his brother" (Ezekiel 38:21).

These combined efforts to destroy the Soviet Union's military power will be overwhelmingly successful. God says, "And I will turn thee back, and *leave but the sixth part of thee* . . . " (Ezekiel 39:2). The bedraggled and frightened remnants will flee northward. The Soviet satellite nations will vent their hatred upon the Soviets who will flee for their lives. The Berlin Wall will come down and the satellite nations will once again declare their independence.

God will call in carnivorous fowl to pick the flesh from the carcasses (Ezekiel 39:17-22). It will take the nation of Israel *seven months* to bury the dead on its soil (Ezekiel 39:11-16) and *seven years* to dispose of the piles of military debris (Ezekiel 39:9,10). These verses show that this war will not be fought with primitive weapons of wood; only metal can be smelted and reprocessed into hardware to militarily strengthen the arms of the forces of Israel.

SEVEN YEARS OF PEACE FOR ISRAEL[1]

For Israel to accomplish the task of cleansing the land from the effects of the war with the Soviets and its allies, it will be necessary for it to have at least seven years of peace after the defeat of the Soviet Forces shortly after the Rapture. It will take only *"one day"* for God to destroy the Soviet Union in this conflict (Ezekiel 39:14,19; 39:8,11, 13,22), which means there must be at least three and one-half years after the Rapture before the signing of the seven-year peace covenant between Antichrist and the Zionist leaders. The people of Israel will not necessarily be in agreement with this unholy alliance, as the word "many" suggests (Daniel 9:27), for "many" does not mean "all." This also will give Israel time to build the Tribulation Temple in which both Antichrist and Christ will be seated, in that order.

The Tribulation cannot commence until the redeemed are *raptured and present* in heaven witnessing the opening of the seven-sealed book that contains the Tribulation events; however, Christ informs us we shall "... be accounted worthy *to escape all these things* that shall come to pass, and to *stand before the Son of man"* (Luke 21:36; compare Revalation 5:1-10).

The Tribulation of the saints (Revelation 6:9-11; 7:9-17; Daniel 7:21; 9:27; 11:32-37), therefore, must continue a full seven years. The first segment will extend from the pre-Tribulation Rapture to the signing of the seven-year peace covenant. This covenant will be broken by Antichrist after it has been in effect only three and one-half years (Daniel 9:27). Following this, Israel will become the target for annihilation by Antichrist and the False Prophet (Daniel 12:1,7).

The assault upon the Christians during the Tribulation days will reach even into the United States. Presently the Moslems are increasing their migration into both Europe and the Americas where they are erecting mosques (temples). This will place them into strategic positions, after the pre-Tribulation Rapture, to launch their persecution and martyrdom of the Christians at the command of the False Prophet.

[1] See Chart B.

75
World War III

It is a common misconception that the Moslem Mosque (in Jerusalem) will have to be razed before the Tribulation Temple is built.

Ezekiel, chapters 38 and 39, does not refer to the Battle of Armageddon, but to the Soviet military advance into the Middle East. This first war will be fought in "the valley of Hammon-gog" (Ezekiel 39:11). The second war will be fought against Antichrist in "the valley of Megiddon" (Zechariah 12:11), and is called "the Battle of Armageddon." These two battles are widely separated both as to location and time. The first will be fought right after the pre-Tribulation Rapture of the redeemed. The second will be fought *just before* the Millennial reign of Christ on earth. The first battle will be a turning point in Israel's spiritual life: "So the house of Israel shall know that I am the LORD their God from that day and forward" (Ezekiel 39:22). Out of that spiritual awakening will come the 144,000 sealed Israelites (Revelation 7:1-8; 14:1-4).

A GRAVEYARD CITY

An interesting prediction is that a city will be built upon the massive grave sites of Israel's enemies. It will be called "Hamonah": "And also the name of the city shall be Hamonah. Thus shall ye cleanse the land" (Ezekiel 39:16). There is no such city today nor has there been before. Since this city cannot be built and/or named until after the seven-month task of burying the dead has been completed at this grave site, the exact location of this city cannot be geographically identified.

It will take God only *one day* to gain victory over Israel's enemies (Ezekiel 38:19; 39:8,11,13,22). Their defeat will occur shortly after the pre-Tribulation Rapture: "Watch ye therefore and pray always, that ye may be accounted worthy *to escape all these things...*" (Luke 21:36).[1]

Immediately *after* the defeat of the Soviets and its allies, Israel will engage in a mass burial effort. Extensive trenches will be dug to bury numerous carcasses as the defeat will be overwhelming in casualties: "And I will turn thee (Gog) back, and leave but a sixth part of thee, and will cause thee to come up from the north parts, and will bring thee upon the mountains of Israel... thou, and all thy bands, and the people that is with thee... Thou shalt fall upon the open field: for I have spoken it, saith the Lord GOD" (Ezekiel 39:2-5).

THE TEMPLE MOUNT[2]

The Mosque of Omar on the Temple Mount in Jerusalem is located adjacent to the place where Solomon's Temple previously was situated; however, the latter was destroyed by the Babylonians in 587 B.C. It is the same site where the restored Temple was razed by Titus in 70 A.D.

According to Revelation 11:2, the Tribulation Temple will be erected on this spot in time for the Antichrist to make his appearance here (2 Thessalonians 2:4), and it is the location to which Christ will return after He comes to earth the second time (Malachi 3:1).

There is a common misconception that the Moslem Mosque will have to be razed before the Tribulation Temple is built. In Revelation 11:2 we read: *"But the court which is without the Temple* leave out, and measure it not; for it is *given unto the Gentiles....*"

During the seven years of peace, Israel will bury the dead of its enemies, cleanse the land from the debris of military hardware, construct the grave site city, and rebuild its Temple (Ezekiel 39:9,12,16).

These seven years of peace for Israel are a result of God's great victory against the Soviet military machine: "Thus will I magnify myself and sanctify myself; and I will be known in the eyes of many nations, and they shall know that I am the Lord . . . So will I make my holy name known in the midst of *my people Israel;* and I will not let them pollute my name any more; and the (Gentiles) shall know that I am the LORD, the Holy One in Israel" (Ezekiel 38:23; 39:7).

Because of the seven years of peace promised Israel after the defeat of the Soviets, we know that the seven-year peace covenant between the Zionists and Antichrist cannot possibly take place immediately after the pre-Tribulation Rapture. That covenant will be broken after three and one-half years (Daniel 9:27; 12:1,7) and Israel will become the target for genocide.

The persecution and martyrdom of the first segment of Tribulation (not "Great Tribulation") saints will occur in response to their fervent and successful worldwide evangelistic efforts in influencing masses of people to accept Jesus Christ as their personal Saviour. Their persecution will not come from Antichrist, who is just rising as the European conqueror, but from religious leaders bitterly opposed to Jesus Christ as the only Redeemer. The prophetic Word shows their persecution will come from religions embracing idolatry as well. These will all unite to form a religious confederacy or kingdom under a ten member body of ruling clergy, which later under the False Prophet will accept Antichrist as God incarnate and erect an image of him to be venerated and worshipped.

Opposition to Antichrist will emanate primarily from Western countries and their allies who cannot endorse this anti-Christian theology and embrace idolatry as a form of worship. At first they will passively resist, but will become actively resistant when the combined kingdoms of Antichrist and the False Prophet threaten the stability of the world, not only religiously, but politically, economically, and militarily. This growing balance of power shifting in favor of Antichrist (now prophetically called Babylon) must be stopped at all costs in order to avoid the subjection of people to the iron rule of these despots.

Many in the world, however, will believe that Antichrist is the anticipated "messiah" who is to bring world peace and prosperity. Unfortunately, many (not "all") in Israel, presumably their leaders, will make a peace covenant with him (Daniel 9:27).

TWO "GREAT TRIBULATIONS"[3]

"...Thus shall Babylon sink, and shall not rise..." (Jeremiah 51:64).

The Babylon of the last days will not be the ancient city that is to be restored. It will be the city of Rome (Revelation 13:1; 17:9,10).

When the major religions under the Islamic False Prophet combine into a confederacy with the Antichrist (prophetically called also a "kingdom" - Revelation 17:12,17), this coalition is called "Babylon." It is for this reason that Antichrist, with the aid of the False Prophet, successfully extends his influence to head the world-wide religious confederation.

This alliance will produce two massive persecutions extending over a seven-year period. The first persecution, under the Babylonian coalition, will be against Christians, whom Scriptures call the "Great Tribulation saints" (Revelation 7:9-17). These must be distinguished from the first segment of Tribulation saints before the coalition (Revelation 6:9-11). During this period of the Great Tribulation against the saints under Babylon, Israel will be under the protection of a peace covenant with Antichrist. After Antichrist breaks his covenant with the Zionists, Israel will face genocide called the time of "Jacob's trouble" (Jeremiah 30:7).

Scriptures refer to two separate "Great Tribulations": First, against Christians under the Babylonian coalition (Revelation 7:9-17); and second, against Israel under the same coalition (Matthew 24:21) three and one-half years later. These two Tribulations do not run concurrently.[4]

Before the Babylonian coalition begins its efforts of genocide against Israel, all of the Tribulation saints will have been raptured

(Revelation 14:14-16; I Thessalonians 5:1-11; I Corinthians 15:51-53). Each of these two "Great Tribulations" will last three and one-half years (Revelation 13:5-7; Daniel 12:7).

EUROPEAN CONQUEROR

While the Soviet forces and its allies are being defeated by God, the European conqueror will continue his victories without any interference from God. NATO and the EEC will have collapsed.

Immediately after the defeat of the Soviet Union, its satellite nations in the eastern portion of the ancient Roman Empire (the Comecon Nations included) fall to Antichrist. In a relatively short time Antichrist takes over much of the ancient Empire: ". . . they shall fly as the eagle, that hasteth to eat. They shall come all for violence: their faces shall sup up as the east wind (toward the Middle East), and they shall gather the captivity as the sand" (Habakkuk 1:8,9).

These rapid conquests of the European victor will cast fear in the Middle East, Western Asia, and in Northern Africa. The Middle East and Western Asia will quickly form their confederacy, as will Northern Africa, for the purpose of defending themselves.

There will be a charm about this European conqueror who has been so successful and who has not encountered divine opposition. The ungodly will accept this an an "omen" that he is the "messiah" bringing world peace: ". . . who is able to make war with him?" (Revelation 13:4). He has great charisma and is not a religious person; in fact, he shows hostility to all religions. Although he apparently is a renegade from the Church at Rome (Daniel 11:37), he carries her on his back and tolerates her for a brief time; then seeks to destroy her (Revelation 17:3,16-18).

COVENANT OF PEACE[5]

Once this conqueror has consolidated his kingdom, he will be crowned king. Since he wisely will discern God's protection upon Israel in the defeat of its enemies, he will make a seven-year peace covenant with Israel. This will be synchronized with the dedication of the completed temple of Israel in Jerusalem (see chapter on "Two Witnesses"). But it is an uncertain peace: "For they have healed the hurt of the daughter of my people *slightly*, saying, peace, peace: when there is no peace" (Jeremiah 8:11). The covenant will be broken after three and one-half years (Daniel 9:27; 12:1,7; I Thessalonians 5:3,4).

The message is clear. When the covenant with Israel is broken, the time of "Jacob's trouble" commences. At that precise moment the remaining Tribulation saints will be raptured (Revelation 14:14-16; I Corinthians 15:51,52; I Thessalonians 5:1-10).

Immediately after the pre-Tribulation Rapture, many false Christs will arise. The word "Christ" means "anointed one." These will be false anointed ones—deceivers. They will claim to have the divine unction and knowledge to guide people through this time of psychological crisis. They will teach people to look for the Messiah upon the earth. Since the word is in italics in the Bible, meaning that it was supplied by the translators, these may well be non-religious persons who will direct their interested audiences to accept the European conqueror, and later the False Prophet, who will have Satanic powers, to reinforce the deception (2 Thessalonians 2:10-12; Revelation 13:11:18),

SECOND FLANDER'S FIELD

America and its allies have the greatest percentage of Christians therefore the pre-Tribulation Rapture will have a devastating effect upon them. Massive numbers of redeemed persons, together with innocent children, will be missing (Luke 21:35; Matthew 18:1-10); Luke 18:16). Hundreds, possibly thousands, of vital positions in the political and military establishments will become vacant instantly. Communication may be tragically disrupted by absent personnel. The death of a single person in an important position can greatly affect the efficiency of a nation. It will indeed be a crisis situation for Christian nations. They will find it difficult to handle any immediate danger of invasion to themselves or their friends.

Non-Christian countries will suffer the least impairment of

their operational effectiveness by the pre-Tribulation Rapture; and according to Bible prophecy they will take advantage of the immobilized situation of Christian nations. Terrorists who are aligned with the European conqueror and with Islam are basically anti-Christian and anti-Israel as are the Soviet Union and Communist countries. Conversely, the pre-Tribulation Rapture will leave them virtually intact politically, militarily, and communication-wise. From four directions Israel will be threatened with invasion as symbolized by Scripture: Persia (Iran) coming from the *east;* Ethiopia (Cush) from the *south;* Libya from the *west;* and the Soviet Union from the *north* (Ezekiel 38:4-9). At this point in time Israel will prophetically cry out: ". . . our hope is lost: we are cut off for our parts" (Ezekiel 37:11). But God will miraculously save her by defeating her enemies.

PROPHETIC NAMES AND PLACES

We have mentioned the fact that in Scripture the United States and the Western Hemisphere were referred to as "isles" and "wilderness" places at the time of the prophecy. The existence of the Western Hemisphere was known by God: ". . . behold he taketh up the isles as a very little thing" (Isaiah 40:15).

Naturally, with successive generations, names and places are changed. Old names disappear and new ones enter the scene. The prophets of old used the names and places which were familiar to them when they presented their prophetic picture. They were accustomed to symbolizing future events and places. We must, therefore, search for the history of those names and places to understand why the prophets used them as symbols.

Let us consider the name "Tyrus." It was a city on the Mediterranean Sea which had been founded in B.C. 2750. It was a center of Old World trade with extensive traffic in gold, silver, iron, tin, copper, wheat, ivory, jewelry, oil, and dyes. It was used by the prophets as a symbol of the prophetic Babylonian megalopolis under Antichrist, a coming world figure. The city best represented the prophetic situation of the future; it represented the center of Antichrist's world power—Rome! (Revelation 17:9,18; 18:10,21).

The prophetic eye of Ezekiel observed this great future city and the world leader who will control this world's commerce: "Son of man, say unto the prince of Tyrus, Thus saith the Lord GOD; Because thine heart is lifted up, and thou hast said, I am a God (compare Revelation 13:4); I sit in the seat of God (see 2 Thessalonians 2:4), in the midst of the seas (worldwide power); yet thou art a man, and not God . . . " (Ezekiel 28:2). Many prophecies are given using Tyrus to describe the world trade center of Antichrist. These are found in Ezekiel 29:18; 27:32-36; Jeremiah 51:49).

TRADE WITH ANTICHRIST

Antichrist, the coming world ruler, will have worldwide commerce, reaching far beyond the Old World. His kingdom is called Babylon because it is so vast and powerful. It encompasses the areas around the Mediterranean Sea, mainland Europe, portions of the Soviet Union, as well as the Middle East Moslem countries that extend into Western Asia: " . . . Tyrus . . . a merchant of the people for *many isles* . . . " (Ezekiel 27:3,15).

[1] See chart A.
[2] See chart E.
[3] See chart G.
[4] See chart G.
[5] See chart H.

76
United States in the Final Tribulation

The "kings of the east" represent the combined might of all righteous nations to bring down the wild man of the Babylonian regime.

When Antichrist, as the "wicked one," breaks his covenant with Israel and turns his attention to destroying it, America will offer Israel sanctuary within its borders: "For, behold the Lord will come . . . I will send those that escape . . . *to the isles afar off* . . . And they shall bring all your brethren . . . out of all nations . . . to my holy mountain Jerusalem . . . " (Isaiah 66:15-21). "Hear the word of the Lord, O ye nations, and . . . *isles afar off,* and say, He that scattered Israel will gather him . . . For the Lord hath redeemed Jacob, and ransomed him from the hand of him (Antichrist) that was stronger than he. Therefore they shall come and sing in the height of Zion, and shall flow together to the goodness of the Lord . . . And there is *hope in thine end,* saith the LORD, that thy children shall come again to their own border" (Jeremiah 31:10,15-17; see Revelation 12:6,14). "Wherefore glorify . . . the Lord God of Israel *in the isles* of the sea" (Isaiah 24:15). "As I live, saith the Lord GOD, surely with a mighty hand, and with a stretched out arm, *and with fury poured out* (during the final Tribulation upon the kingdom of Antichrist), will I rule over you. And I will bring you out from the people, and I will gather you out of the countries wherein ye are scattered . . . And *there* will I plead with you face to face, like as I pleaded with your fathers in the wilderness of the land of Egypt, so will I plead with you, saith the Lord GOD" (Ezekiel 20:33-36). This sounds as though there will be a great spiritual revival among the Israelis in the lands outside of the Babylonian kingdom during the period called "Jacob's trouble."

COUNTRIES UNDER GOD'S DIRECT WRATH

Certain countries will suffer infinitely worse than others, but many will survive to enter the Millennium. The Arab lands are going to experience incredible devastation (Obadiah 6-12; Joel 3:19-21; Amos 9:11-15) because under Islam they will lead the massacre upon Christians and Israelis. The "confederacy" of the Arabs will be organized by Syria in the Middle East in collaboration with Western Asia. It will defer to the leadership of the False Prophet (Obadiah 7; Ezekiel 30:1-8) and extend to include the southern confederacy initiated by Egypt. These countries will be destroyed (Ezekiel 29:9-12), and except for the Egyptians, no survivors will be left to enter the Millennium: ". . . and there shall not be any remaining of the house of Esau (Arabs) for the Lord hath spoken it" (Obadiah 18). Their lands will be occupied by Israel during the Millennium (verse 20), with the exception of a portion called "Assyria," which will accommodate the Islamic or Moslem followers who become Christians and survive (Isaiah 11:6; 19:23). Only the Egyptians will be able to return to their homeland, which will be restored after forty years (verse 13).

ORPHANS OF BABYLON

The reaction of the United States and its allies is seen when Antichrist extends his kingdom into a worldwide Babylonian style dynasty. This will be mainly through economic control and religious tyranny: "Thus saith the Lord GOD of Tyrus, Shall not the *isles shake at the sound of thy fall,* when the wounded cry, when the slaughter is made *in the midst of thee?* . . . Now shall the *isles tremble* in the day of thy fall; yea, the *isles* that are in the sea (merchant ships in the Mediterranean Sea) shall be *troubled at thy departure*" (Ezekiel 26:15-18).

The millions who will have taken the mark of the Beast, as devoted followers of Babylon, will be scattered throughout the world to promote the policies of the coalition. Terrorism will be commonplace under the Antichrist and the False Prophet, and in the course of events, the megalopolis world capital will be destroyed: ". . . I shall make thee a desolate city . . ." (verse 19). ". . . Alas, alas, that great city (Rome) . . . for in one hour is she made desolate" (Revelation 18:19). Rome, as Babylon, will be no more and its faithful will be like orphans experiencing an agonizing ordeal.

Immediately upon the heels of the destruction of Rome, Christ will return to the earth. One-half of the world's population will have taken the mark of the Beast and will be destroyed by the brightness of the Lord's presence as He descends from east to west. Their corpses will lie on the ground throughout the world (Jeremiah 25:30-33; Luke 17:24-37), and vultures will be called by God to consume the flesh in an event called "the supper of the great God" (Revelation 19:17).

KINGS OF THE EAST

". . . I will establish the throne of (David's) kingdom *for ever* . . . And thine house and thy kingdom shall be established *for ever* before thee . . ." (2 Samuel 7:13,16; compare I Kings 2:4; 8:25; 2 Chronicles 6:16; Jeremiah 33:17).

The so-called ten tribes of Israel *are not lost, but merely hidden,* as revealed in various names, such as: Scythian (meaning Scotland) where some migrated; *Hebri*des (meaning Hebrew), islands off the west coast of Scotland; *Ebro* River (meaning Hebrew), located in Spain where the Celts migrated enroute to the British Isles; and others. These tribes of Israel originally were settled in the area of the Caspian Sea (now Iran). When the tribe of Dan migrated from there it left a trail like a *serpent:* "Dan shall be an adder by the way . . ." (Genesis 49:17). The route can be detected through place names containing syllables such as "Dan," "Den," "Din," "Don," and "Dun." Historians have traced the following: Jor*dan,* Dar*dan*elles (Turkey), *Don* River (Russia), *Dan*ube River (through Bulgaria,

Rumania, Yugoslavia, Hungary, Austria, Germany), *Dan*zig (Poland), *Dan* (Germany), *Den*mark, Swe*den*, *Dun*dee and E*din*burgh (Scotland), Lon*don*, and others. Some descendants have been traced through Su*dan* (Africa) and then through *Dun*kerque, France into England. The other nine tribes disappeared in the nations of Italy, Spain, France, Belgium, Switzerland, Ireland, as well as Britain, where they reappeared as the Celts. Eventually Dan and the Celts migrated to the New World and merged with the Gentiles. These tribes are *not* Jews (Judah).

Historians have traced British royalty back to King David. This line occasionally has included a "daughter of Zion." Thus the promise of Jehovah to David of an *unbroken lineage of rulers* continues to this very day as evidenced by the British Empire, the Scandinavian monarchies, and several other European nations. Many of the people of the United States originate from these descendants of the ten tribes who have merged with the Gentiles, thereby fulfilling the prophecies in Isaiah 63:16; Hosea 1:10; Romans 9:25-33. Great Britain and the United States have become the main repositories of the lost tribes. This is not an endorsement of the "British Israel" movement with its extreme and unbiblical doctrines which they have injected into some true historical data.

Denmark, England, and Ireland cannot be included as part of the predicted revived Roman Empire ("ten horns"); therefore, three others must yet come into the EEC to fulfill the prophecies. The Scandinavian countries, Great Britain, and its commonwealth of nations, as well as the United States and its allies, will never be included in the last day Babylon. They will be part of the resistance forces in an alliance called the "Kings of the East" which will overthrow the Babylonian regime and culminate in the battle of Armageddon before the Millennium. These nations are assured a secure place with national boundaries as "righteous nations" in the Millennial reign of Christ, according to Matthew, chapter 25.

"I have waited for they salvation, O LORD" (Genesis 49:18).

"And they shall wander . . . *from the north* even to the east . . . to seek the word of the LORD . . . " (Amos 8:12).

These scriptures appear to predict a spiritual awakening among the descendants of *Dan* in the northern nations. Perhaps many given to ecclesiolatry, who have not had a personal relationship with Christ, like an awakening spiritual giant will comprise a large portion of the Tribulation Saints.

WAR AGAINST THE BABYLONIAN COALITION

"But tidings out of the *east* and out of the *north* shall trouble him (Antichrist): therefore he shall go forth with great fury to destroy and utterly to make away many . . . yet he shall come to an end, *and none shall help him*" (Daniel 11:44,45). These last words predict a *revolt* within the kingdom of Antichrist and the False Prophet. Those who have *not* taken the mark of the Beast will join the Scandinavian invasion from the *north* against Antichrist to topple the Babylonian regime in Europe and to drive it eastward toward Megiddo. Simultaneously, the forces from the *east* will come against the False Prophet and drive his military might westward toward Megiddo for the battle of Armageddon (Zechariah 12:11; Revelation 16:14-16; 14:20). The only recruits of Antichrist and the False Prophet will be those who have taken the mark of the Beast since for them the die already will have been cast.

Because the Soviets no longer will occupy their former territory, the "kings of the east" can come overland and down from the north into the Middle East with their military expeditionary force. With naval power in the Arabian Gulf (Daniel 11:30), the Babylonian regime is going to be confronted with an invasion force from several fronts (See map). Thus this international military alliance with the aid of Jesus Christ will overthrow the hegemony of Antichrist (2 Thessalonians 2:8; Joel, chapter 2; Zechariah, chapter 14).

The descendants of the tribe of Dan will share in this great victory over the Antichrist and the False Prophet: "Dan *shall be . . . an adder* in the path, that biteth the horse heels (prophetic of military power), so that his *rider* (compare Revelation 6:2) *shall fall backward.*" Antichrist not only will fail to conquer the descendants of Dan, but they also will be part of the "righteous natons" to over throw the Babylonian regime. The name "Dan" means "judge" (Genesis 49:16) so this tribe will aid in the administration of justice during the Millennium.

Revelation 9:13-21 is not a reference to a two hundred million army under the "kings of the east": 1) They are "plagues" (verse 20); 2) Revelation 16:16 occurs, not under the sixth trumpet, but later under the sixth vial (verse 12), which means the "kings of the east" are involved at a much later time; 3) Under the seventh trumpet there is an announcement of coming judgments, but under the seventh vial there is the actual judgment, which is much later when the "kings of the east" organize to move militarily; 4) It certainly is not a reference to China exclusively, although it may be included.

ROAD TO ARMAGEDDON

"Blow ye the trumpet in Zion, and sound an alarm in my holy mountain: let all the inhabitants of the land tremble: for the day of the LORD cometh, for it is nigh at hand" (Joel 2:1).

"For they are the spirits of devils . . . which go forth unto the kings of the earth and of the whole world, to gather them to the battle of that great day of God Almighty" (Revelation 16:14).

It is Satan who obviously stirs up war fever (Revelation 20:7-10).

The scenario of events that leads up to the battle of Armageddon is as follows: 1) Satan knows that Christ is about to descend to the earth; 2) He knows that Christ will land on the Mount of Olives near Jerusalem; 3) He wants to destroy the city from which Christ plans to reign; 4) Satan, through the Antichrist and the False Prophet, will have established his world capital at Rome, prophetically called Babylon; 5) He wants to destroy the "righteous nations" which are to be Christ's allies when He returns to the earth; 6) He is thwarted in his attempt to seize the Ark of the Covenant which will have given the Zionists such great victory over the Soviet forces. This Ark will prevent the genocide of all Zionists and Jews (Judah) and it will assure them further victories (Daniel 11:30,33; Isaiah 11:10-16); 7) Antichrist wants all of his forces near Jerusalem so that they may confront not only the "righteous nations" invading his territory, but also the might of Jesus Christ when He descends to the earth; 8) the loss of life at Armageddon will be enormous (Revelation 14:20; Zechariah 12:3,4,11; Luke 17:29,30,37), with his forces reduced to a "remnant" as revealed in Revelation 19:21; 9) Christ will completely exterminate the forces of Antichrist and the false Prophet: ". . . I will seek to destroy all the nations that come against Jerusalem" (Zechariah 12:9).

THE BROKEN ARROW

"Keep silence before me, *0 islands,* and let the people renew their strength; let them come near; then let them speak: let us come together to judgment (Matthew 25:31-46), Who raised up the righteous man from the east (Jesus Christ; compare Matthew 24:27; Luke 17:24), called him to his feet, gave the nations before him, and made him ruler over kings? . . . The *isles saw it,* and feared; the ends of the earth were afraid, drew near, and came. They helped every one his neighbour; and every one said to his brother, be of good courage. So the carpenters encouraged the goldsmith . . ." (Isaiah 41:1-7). "Behold my servant (Jesus Christ), whom I uphold; mine elect, in whom my soul delighteth: I have put my spirit upon him: he shall bring forth judgment to the Gentiles . . . *the isles shall wait for his law* . . . I the Lord have called thee in righteousness . . . and give thee for a covenant of the people, for a light of the Gentiles . . ." (Isaiah 42:1-6). ". . . Men shall worship him (Christ) every one from his place, even all the *isles* of the (Gentiles)" (Zephaniah 2:11; see Isaiah 42:10-13; 35:1). "They that dwell in the *wilderness* (same as "isles") shall bow before him: and his enemies shall lick the dust. The kings of Tarshish and the *isles* shall bring presents: the kings of Sheba and Seba shall offer gifts. Yea, all kings shall fall down before him: all nations shall serve him" (Psalm 72:9-11).

WEALTH FLOWS FROM AMERICA TO CHRIST

God has blessed the Western Hemisphere, especially North America, and it will demonstrate its generosity toward Christ when He returns to the earth. Fleets of planes will be loaded with wealth, materials, and construction crews to help rebuild the areas devastated during the judgments of God and Armageddon: "Who are these that *fly as a cloud,* and as doves (symbols of peace) to their windows (airports)? Surely the *isles* shall wait for me (Christ) . . . to bring thy sons (Israelis) from *far,* their silver and their gold with them, unto the name of the LORD thy God, and to the Holy One of Israel, because he hath glorified (blest) thee: for in my wrath I smote thee, but in my favor have I mercy on thee" (Isaiah 60:9,10).

In the Western Hemisphere there will be a spiritual awakening among all Israelis and they will return to the Holy Land with gifts for Christ: "From beyond the rivers of Ethiopia my supplicants (worshippers) . . . " (Zephaniah 3:10). "In that time shall the present (offering) be brought into the LORD of hosts of a people scattered and peeled . . . to the place of the name of the LORD of hosts, the mount Zion" (Isaiah 18:7).

ALL SEA LANES LEAD TO JERUSALEM

The great sea lanes across the waters, with their boundless flow of commerce, are depicted by another prophetic name, "Tarshish." Tarshish was the grandson of Noah and he and his descendants migrated to western Italy where a city was named after him. The port city was known for its mighty ships that made long voyages. Scriptures predicted that huge merchant and cargo ships would cross the waters from distant ports in the New World, but the sea lanes would be altered because of the effects of the Pangaea.[1] Tarshish will come into the picture only after Christ returns to earth, at which time He will receive an abundance of goods and wealth.

The distinction between Tyrus and Tarshish is that prophetically the former refers to the inhabitants of the Old World; whereas the latter ("daughters of Tarshish") refers to descendants from the Old World who migrated to the Western Hemisphere. They will be the ones who will traffic to Christ after He descends to earth (Isaiah 60:9; Jeremiah 10:9; Ezekiel 27:25; Isaiah 2:12,16,19-21).

[1] Psalm 136:6, "To him that *stretcheth* out earth . . . ."

77

Sword in the Sky[1]

The bodies of the Antichrist and False Prophet will have no caissons to convey them, no honor guard, and no crowds of weeping admirers.

"And out of his (Christ's) mouth goeth a sharp sword . . . " (Revelation 19:15).

" . . . He will give them that are wicked to the (Christ's) sword . . . " (Jeremiah 25:30).

For as the lightning cometh out of the east, and *shineth* even unto the west; so shall also the coming of the Son of man be" (Matthew 24:27).

"Behold he cometh with clouds, and *every eye shall see him* . . . and all kindreds (families) of the earth shall *wail* because of him . . . " (Revelation 1:7).

"For *my sword* shall be bathed in heaven: behold, it shall come down . . . upon the people of my curse, to judgment: the *sword* of the LORD is filled with blood . . . and a great slaughter . . . their land shall be soaked with blood . . . For it is the day of the Lord's vengeance . . . and the streams thereof shall be turned into pitch (oil), and the dust thereof into brimstone, and the land thereof shall become burning pitch" (Isaiah 34:5-9).

BATTLE FOR POSSESSION OF THE EARTH

The Planet Earth is a "pearl" in the cosmic ocean. It will be sole survivor in the universe; the battleground and confrontation between God and Satan, between righteousness and unrighteousness.

Satan knows the earth will survive the conflagration of the universe but that his habitations in cosmos are scheduled for annihilation (Ephesians 6:12). He knows that unless he can somehow get possession of the earth he cannot save himself and his cohorts from their eternal fate in the lake of fire. Since by divine decree the Royal Redeemed and Israel are the potential rulers and inheritors of the earth, Satan must destroy them. That is the reason for the final Tribulation.

Satan knows Biblical prophecies. From his principalities in space he knows his eviction notice has been served. He knows that the day is coming when his powerful allies, who are now in the bottomless pit, will be released upon the earth. In spite of the predicted outcome in favor of God's newly appointed rulers, Satan has no alternative but to make his final stand for possession of the earth. He will be like a vicious wild beast with his back to the wall (Revelation 12:12).

The battle for possession of the earth will be a live-or-die situation. It will engage not only spiritual forces consisting of fallen angels against holy angels, but it will also involve mortal beings under Satan's command against mortal beings of righteous nations. The latter are those who refuse to accept Satan's ultimatum to worship him. Those under Satan's command already will have taken the mark of the beast. They are equally aware of their fate, which is the lake of fire (Revelation 14:9-11). Thus the Battle of Armageddon will be a fiercely fought war.

ARMAGEDDON[2 and 3]

The Battle of Armageddon (Revelation 16:16) will be fought at the southern rim of the plain of Esdraelon, called the valley of Jezreel, near the Sea of Galilee. The blood will flow like a river for about 170 miles as described in Isaiah 63:1-6; Joel 3:9-16; Zephaniah, chapter 1; Revelation 14:20.

Zechariah 14:1-4,9 gives a description of the Lord's return from heaven to earth at which time He turns the tide of battle. The casualties will be enormous and mounting. Jesus Christ will descend and circle above the earth from east to west (Matthew 24:27). As He arrives over the horizons from the east, the brightness of His coming will pale out the sun and destroy not only Antichrist and the False Prophet (2 Thessalonians 2:8), but also millions throughout the world who have taken the mark of the Beast. Vast numbers already will have been destroyed by massive earthquakes, tidal waves, plagues, and the scorching heat.

During His first circuit over the earth, Christ will make His plea to those who have not taken the mark of the Beast. This is the point at which many devoted followers of Islam will accept Jesus Christ as their Redeemer. Consequently, they will be permitted to reside during the Millennium in a place called Assyria. The remaining wicked who have taken the mark of the Beast will be destroyed. Those who survive will appear at the Judgment of the Nations (Matthew 25:31-46). First, "... he will *plead* with all flesh; (next) he will give them that are wicked *to the sword* (of Christ)" (Jeremiah 25:31; compare Revelation 19:19; Luke 17:33-37).

Panic will grip the wicked: "Whosoever shall seek to save his life shall lose it..." (Luke 17:33). "But I will remove far off from you the northern army (the last day Babylon), and will drive him ('... the foe from the north...' *Moffat*) into a land barren and desolate... because he hath done great things" (Joel 2:20).[4] The newly redeemed will stand their ground, remain calm and confident with peace in their hearts, ready to receive Christ with open arms: "... and whosoever shall lose his life (give it up to Christ) shall preserve it."

THE EAGLES HAVE LANDED

One-half of all living human beings will be destroyed: "And the *slain of the LORD* shall be at that day from one end of the earth even unto the other end of the earth: they shall be dung upon the ground"

(Jeremiah 25:33; see also verses 30-38). Obviously, these are the ones who have taken the mark of the Beast (Revelation 14:9-11).

"Then shall two be in the field; the one shall be taken (slain), and the other left. Two women shall be grinding at the mill; the one shall be taken (slain), and the other left . . . For as lightning cometh out of the east, and shineth even unto the west; so shall also the coming of the Son of man be. For wheresoever the carcass is, there will the eagles (vultures) be gathered together" (Matthew 24:27, 28,40,41).

"And I saw an angel standing in the sun; and he cried with a loud voice, saying to all the fowls that fly in the midst of heaven, Come and gather yourselves together unto the supper of the great God; That ye may eat of . . . the flesh of all men, both free and bond, both small and great" (Revelation 19:17,18). No animals except horses are referred to in connection with destruction. In light of the extra-celestial white horses coming to earth with Christ, will these replace all mortal horses upon the earth? Horses have been symbolic of military power throughout history in times of war and in prophecy.

Peace will come to the earth when Christ returns in His brightness destroying planes, machines, and military equipment, which will form piles of debris. This equipment will then be converted for use in farming and the production of foods adaptable to the environment of the Millennium.

"And I will punish the world for their evil, and the wicked for their iniquity . . . I will make a man more precious than fine gold; even a man than the golden wedge of Ophir" (Isaiah 13:11,12; compare Isaiah 24:6).

The gigantic sword that will extend from Christ (Revelation 19:15; Jeremiah 25:31; Habakkuk 3:11,12) will presumably reach over a vast area of the sky covering miles across the heavens. *This visible manifestation of the Holy Spirit* will strike with devastating blows, as streaks of lightning zeroing in upon those with the mark of the Beast. The sword shows *intelligence* in its selection of targets. Those fleeing into the mountains and into places of hiding will find they are not shielded from the omniscience and penetrating power of this flashing sword which will reap a bloody harvest (Revelation 14:17-20).[5]

SATAN TAKEN HOSTAGE

Satan will be apprehended, wrapped in chains, and incarcerated in the bottomless pit by an ordinary angel who conceivably could be one of the Royal Redeemed demonstrating to Satan that he is now one of the real rulers of the earth (compare Revelation 17:1 to 19:10; 21:9 to 22:8,9).

Satan will be defeated so thoroughly that all spirit of resistance will vanish from him. As he enters the bottomless pit in his humiliation, there is heard the resounding wail of all his chieftains and angelic followers: "The underworld is astir to greet you, astir to meet you, rousing all the ghosts for you, that on earth were captains, moving monarchs of the world from their thrones, to hail you, one and all of them, to cry to you. So you are *weak now* as we are? So you fare as we fare? Your pomp has passed down to the world below . . . " (Isaiah 14:9-20, *Moffatt*).

How will all this blood that has soaked the ground be removed? The glory of God will "burn" up the blood and so "purge" the ground (Isaiah 4:4).

The Babylon of Antichrist and the False Prophet will fall: "As Babylon hath caused the slain of Israel to fall, so at Babylon shall fall the slain of all the earth" (Jeremiah 51:49). This prophecy has not been previously fulfilled; it is prophetic of the last days.

THE VANISHING DUO

"And the beast (Abaddon - Apollyon) was taken, and with him the False Prophet (prince of Persia) that wrought miracles before him, with which he deceived them that had received the mark of the Beast, and them that worshipped his image. These both were cast *alive* into a lake of fire burning with brimstone" (Revelation 19:20). ". . . And shouldest destroy them which destroy the earth" (Revelation 11:18). The bodies which they will have inhabited (Antichrist and the False Prophet) will lie on the ground (Isaiah 14:19,20).

After Abaddon (Apollyon) and the Prince of Persia have been cast into the lake of fire, the bodies of Antichrist and the False Prophet will suddenly disappear from the face of the earth as if vaporized. The Spirit of God will transport them *alive* into the lake of fire. All of those who have been deceived into taking the mark of the Beast will be lying by the millions as carcasses all over the earth as prophesied: ". . . they shall not be lamented, neither gathered, nor buried; they shall be dung upon the ground" (Jeremiah 25:33).

This marks the end of the Babylonian regime with all of its paganistic citizens. Those who will accept Christ at His descent will be called "righteous" nations (Matthew 25:31-46). Since they accept Christ Jesus *after* the "first resurrection" they will enter the Millennial Age in their physical bodies.

HEAVEN DECLARES WAR

As the closing period of the Tribulation under Antichrist approaches, there is from heaven a declaration of war that "there should be time no longer," meaning that no longer will God's final judgments be delayed (Revelation 10:6).

It is at this precise moment that the holy angels in heaven will be sent to regather the Christian tourists who are scattered in heaven (Matthew 24:31). According to Mark 13:27, God simultaneously gathers the remainder of the Tribulation saints: "And then he shall send his angels, and shall gather together his *elect* from the four winds, from the uttermost part of the *earth* to the uttermost part of *heaven."* (Note in Revelation 14:15 that an "angel" called for Christ to bring in the remnants of the Tribulation saints to meet Him in the "cloud," not "clouds").

THE SEVEN THUNDERS[6]

The seven-sealed book is now completely open. There are no more seals to be broken (Revelation 10:2). At this point in time, Christ, the Lion of the tribe of Judah, will raise His voice as "a lion roareth" (verse 3). Then the "seven thunders uttered their voices."

It is evident that storm clouds are now banking on the prophetic horizon. There occasionally are weak sounds of distant thunders with faint streaks of lightning, but as the time narrows and the dark clouds of the "day of the Lord" come, there is the ominous, frightening realization of the terrifying storm that is approaching.

"And when the seven thunders had uttered their voices... I heard a voice from heaven saying unto me, Seal up those things which the seven thunders uttered, and write them not" (Revelation 10:4).

A little book from Christ is transferred into the hands of Michael, the "mighty and great" angel. When this big angel receives the book, the seven thunders are heard as a prelude to the seven last plagues about to be poured forth upon the earth. Even though the small book is unsealed, the thunders remain sealed because the Royal Redeemed are not yet ready to be gathered in before the storm strikes in all its fury upon the earth.

It is under the seventh trumpet, after the last of the seven thunders has been spent, that we hear "the mystery of God should be finished." The Church, which is the mystery of God (Ephesians 3:1-10), has come to its end and the last of the final Tribulation saints are to be gathered in heaven safely away from the final judgments of God and Armageddon.

SCENARIO OF THE LITTLE BOOK

"And I (John) took the little book out of the angel's hand, and ate it up; and it was in my mouth sweet as honey: and as soon as I had eaten it, my belly was bitter" (Revelation 10:10). As John receives the little book, it is difficult to stomach, but the words of prophecy are sweet to utter because they light the way with warning beacons concerning the tragedies that lie ahead. It seems logical that God would give advance warnings for travelers on the road of life so they can take adequate precautions and seek protection. The sealing of the thunders means the prophetic messages of warning must go forth until the end of the Church Age when all of the Royal Redeemed will be brought into God's ark of safety to be forever rewarded and united with Christ Jesus in the royal family of God.

The small book which John received represents possession by the Church. Its mysteries are no longer hidden behind the seven seals, but as predicted in Daniel: ". . . the wise shall understand" (Daniel 12:9,10). This little book of seven seals in John's hands now passes to the Father, then to Jesus Christ, and finally to Michael, the archangel. Next is the actual time of the final judgments to be poured upon the earth. The Royal Redeemed take full possession of the earth as the eviction process begins, symbolized by the mighty archangel with his one foot on the land and the other on the sea. In the meantime John, who represents the Church is told: ". . . Thou must prophesy again before many peoples, and nations, and tongues, and kings" (Revelation 10:11). The archangel's feet on "the sea" (Mediterranean Sea) and the land (compare Revelation 13:11) are planted on the areas where Antichrist and the False Prophet arise and their kingdoms form the Babylonian coalition.

The final outcome is clear. There will be new rulers upon the earth, known as the Royal Redeemed, to serve as kings, priests, judges, and lords. There no longer will be prophets because no new prophecies are in God's agenda, and all unfulfilled prophecies are now scheduled for final and total fulfillment.

HOW CHRIST SOLVES THE PALESTINIAN PROBLEMS

God's Word states that Assyria, which is in Islamic Arab territory, will be set aside for newly redeemed Arabs. Its location will be north of the land of Israel but its precise boundaries are not specified: "In that day shall Israel be the third with Egypt and with Assyria, even a blessing in the midst of the land: Whom the LORD of hosts shall bless, saying, Blessed be *Egypt my people* (redeemed), and Assyria the work of my hands (redeemed), and Israel mine inheritance (a nation redeemed in one day)" (Isaiah 19:24,25).

Many in Islam expect a Messiah or Mahdi ("divinely guided one") from the family of Mohammed, who will restore pure religion and usher in a golden age.[7] There are other Moslems who believe that Jesus will descend *with* their Mahdi and kill the Antichrist.

How disappointed those Moslems will be who are deceived into believing Antichrist is the Mahdi because of their insidious leader whom the Scriptures call the "False Prophet."

But there are yet even others among Islam who believe that Jesus will return as Mahdi, citing a *hadith* (traditional saying of Mohammed) asserting that "there is no Mahdi save Isa iha Maryam (Jesus the Son of Mary)," a declaration with which Christians would readily agree.

Therefore, there will be many in Islam who will convert to Jesus Christ when He makes His appeal from the skies as He descends to earth. These converts will survive the genocide of all Islamic Arabs, according to Obadiah 18, and will be accorded a place in the prophetic land of Assyria. How merciful God is to those whose conscience awakens and who realize that their theological posture is diametrically opposed to the real truth.

People of many religions will convert to Christ because in their darkened hearts is the expectation of a great world Messiah: The "...desire of *all nations* shall come..." (Haggai 2:7). Under the right circumstances they will turn away from their theological errors and enthusiastically embrace the Lord Jesus Christ.

SUNRISE STRIP

Jesus Christ referred to the Millennium as the first day of summer on the earth: "And he spake to them a parable; Behold the fig tree and all the trees; When they now shoot forth, ye see and know of your own selves that summer is now nigh at hand. So likewise ye, when ye see these things come to pass, know ye that the kingdom of God is nigh at hand" (Luke 21:29-31).

There will be few individual survivors and only three nations remaining in the Middle East to enter the Millennium. All of the descendants of Esau (Arabs) will be decimated except for those who abandon the heretical and deceptive teachings of Islam in favor of accepting Jesus Christ as their personal Saviour: "For thy violence against they brother Jacob (Israel), shame shall cover thee, and thou shalt be *cut off for ever*... there shall not be any remaining of the house of Esau (Arabs); for the LORD hath spoken it" (Obadiah 10,18). Even their ayatollahs and spiritual leaders will be destroyed: "Shall I not in that day, saith the LORD, even destroy the *wise men* out of Edom (Arab lands), and understanding out of the mount of Esau" (Verse 8).[8]

Psalm 83 prophesies that the Arabs will be under a Confederacy formulated primarily to destroy Israel in the land of promise, as well as the Christians toward whom they also have great hostility. Israel will occupy all Arab (Islamic) lands during the Millennium (Obadiah 17).

SURVIVAL OF EGYPT

Egypt will be the only surviving nation in the Middle East besides Israel and the predicted Assyria. However, it will be punished during the first forty years in the Millennium (Ezekiel 29:9-11). Its people will be scattered throughout the world (Ezekiel 29:12), and the land will become uninhabitable (Joel 3:19). Then after the forty years of dispersion, Christ will regather them into their own land of Egypt, which again will have become productive and suitable for habitation (Ezekiel 29:13-15).

For four hundred years up to the time of Moses, Egypt enslaved Israel. God will punish the Egyptians ten years for each of the one hundred years they inflicted Israel. Although we see the kindness of God leaning toward mercy, that period of history cannot go unpunished (Joel 3:19).

Why is Egypt permitted to survive when the Arab nations are not? The reason is that it will be a sanctuary for the people that the Antichrist persecutes in Israel. Such multitudes are shielded from the wrath of the Antichrist and his coadjutant to the extent that they populate five Egyptian cities (Isaiah 19:18). After Egypt's dispersion we read of God saying: ". . . Blessed be Egypt my people, and Assyria the work of my hands, and Israel mine inheritance" (verse 25). "In that day shall Israel be the *third* with Egypt and with Assyria . . . " (verse 24).[9]

The Lord will have a great highway constructed from Egypt to

Assyria directly through the land of Israel: "In that day shall there to be a highway out of Egypt to Assyria . . . " (verse 23). Following the obliteration of Israel's hostile neighbors, Millennial peace will envelop the world under the reign of Jesus Christ from Jerusalem in the Millennium. The land of Israel will extend from Egypt to the Euphrates River (southern border of Turkey - future Assyria) and will include all of the Arabian Peninsula (Genesis 15:14-18; Exodus 32:13; Ezekiel 47:13-23). The land will be divided for each of the twelve tribes of Israel, because the world population will be relatively small at the beginning of the Millennium. As the population increases, the ten so-called "lost tribes" will expand to repopulate the earth. The present walls of the city of Jerusalem will be razed (Zechariah 2:4) and will occupy twelve square miles (an indication of the greatly reduced population), a miniature of the New Jerusalem (Ezekiel 48:1-35).

The return of the Pangaea will cause a realignment of all national boundaries. The righteous nations will absorb the territory of former nations and will possess the lands recovered from the ice and snow. All island nations will become non-existent (Revelation 16:20).

WORLD WAR V—THE LAST BATTLE

". . . The devil . . . a murder from the beginning . . . " (John 8:44). Satan will be loosed upon the earth at the end of the Millennium and quickly will seize the opportunity to rally support for his army. This will come from rebellious people among the Millennial nations who have been chafing under the strict enforcement of the righteous laws (Revelation 20:7-10). Satan will find a leader among them called "Gog," whom he can incarnate, and who will assemble an insurgent army called "Magog." They will march against the city of Jerusalem to destroy the multitude of natural righteous people assembled there. This will be Satan's final stand against God for all time. The fire of God will fall upon "Gog" and his wicked army and reduce it entirely to ashes, thus will be fulfilled the prophecy: " . . . I will bring thee to *ashes* upon the earth in the sight of all them that behold thee. All they that know thee among the people shall be astonished at thee: thou shalt be a terror, *and never shalt thou be any more*" (Ezekiel 18:18,19). "For behold, the day cometh, that shall burn as an oven; and all the proud, yea, and that do wickedly, shall be a *stubble;* and the day that cometh *shall burn them up,* saith the LORD of hosts, that it shall leave them neither root nor branch (end of all their offspring)" (Malachi 4:1). "... And the fire came down from God out of heaven, and devoured them" (Revelation 20:9).

SURVIVALISTS

"To open the blind eyes, *to bring out the prisoners* from the prison, and them that sit in darkness *out of the prison house* (Isaiah 42:7). "Seek ye the LORD, all ye meek of the earth ... it may be ye shall be *hid* in the day of the LORD's anger" (Zephaniah 2:3). "... And the opening of the *prison* to them that are *bound* (Isaiah 61:1). "... I have sent forth *thy prisoners* out of the *pit* wherein *is no* water" (Zechariah 9:11). "... *Ye prisoners of hope*" (verse 12). The first mission of the Royal Redeemed when they return to earth with Christ will be to seek out the post-Tribulation Christian survivors and gather them for the Judgment of the Nations where the guilty will face them to be charged and punished (Matthew 25:31-46).

Very few humans will remain as a result of the devastation of the earth from various sources such as the judgments of God, the battle of Armageddon, the massacre of millions of Christians, the disappearance of other millions by the two Raptures, and the destruction of half of the people remaining on the earth by the Lord's brightness as He returns to earth: "... the inhabitants of the earth are *burned* (by His brightness) and *few men left* (Isaiah 24:6). Some who will have taken the mark of the Beast will be badly burned but will recuperate sufficiently to stand on trial at the Judgment of Nations. It will not be difficult for Christ to gather the survivors from the nations for this judgment (Matthew 25:31-46). No wonder we read: "... Woe, woe, woe, to the inhabiters of the earth ..." (Revelation 8:13).

"The LORD . . . cometh to judge the earth" (Psalm 96:13). "And he shall judge among the nations, and shall rebuke many people . . . " (Isaiah 2:4).

The Judgment of the Nations will take place in the valley of Jehosaphat (Joel 3:1,2,9-14), commonly known as the valley of Kidron, which lies just east of Jerusalem. The decimation of manpower is very intense according to this prophecy: "And in that day *seven women* shall take hold of *one man,* saying, We will eat our own bread, and wear our own apparel; Only let us be called by thy name, to take away our reproach" (Isaiah 4:1).

Aspersions will be directed against the final Tribulation saints from devoted followers of Antichrist and the False Prophet: "Where is the promise of his (your Messiah's) coming (to earth)?" (I Peter 3:4). The beliefs of the Tribulation Christians will be challenged as just so much wishful thinking regarding a messiah, other than Antichrist, coming to set up a Millennial reign of peace and prosperity upon the earth.

The word, "coming," in connection with the return of Jesus Christ is "parousia" from the Greek, meaning "presence of the Lord Jesus Christ." It is *not limited* to meeting Christ *above* the earth but includes joining Him also *upon* the earth in His power and majesty: ". . . at the coming (parousia) of our Lord Jesus Christ *with all his saints"* (I Thessalonians 3:13); ". . . with the *brightness* of his coming (parousia) of our Lord Jesus Christ (and) of his *majesty"* (2 Peter 1:16).

Christ will return to earth from the east and land on the Mount of Olives (Olivet). He then will proceed through the Golden Gate which is now closed, and enter the Tribulation Temple. Following this, the few remaining nations[10] will gather just east of Jerusalem in the Valley of Kidron (Jehosaphat) where Christ will judge them as they are surrounded by the white royal horses and the Royal Redeemed who rounded them up. The Tribulation Temple, which is yet to be erected just north of the Temple of Omar, will be replaced by Christ's glorious Millennial Temple that will occupy the entire Temple Mount.

DESTROY THE ROOTS

"Thus saith the Lord . . . the Amorite . . . I destroyed his *fruit from above* (spiritual loss), and his *roots from beneath* (national existence)" (Amos 2:1,9).

And now the *axe* (Christ Jesus) is laid unto the *root of the trees* (to judge the nations - Matthew 25:31-46) . . . " (Matthew 3:10).

God by foreknowledge knew that Islam would emerge from among the Arab nations to be hostile to Israel, Christians, and the Redeemer, and that it would adopt for God a strange name—Allah! Therefore, God wanted the very roots destroyed before the Islamic nations become a major threat to peace in the Middle East and to the entire world.

When Israel, as a *nation,* made peace overtures to its pagan neighbors (Judges, chapter 2), and Saul their *first king* refused to destroy the Amalekites (I Samuel, chapter 15), God decreed that Israel's descendants would perpetually be tested by wars (Judges, chapter 3). Israel suffered and went into exile, spared only because of King David's righteousness. Christ, the prophetic axe, will cut off all unrighteous nations, including the Arabs (Obadiah, Habakkuk, Haggai, chapter 2, Amos 9:10-15).

[1] See illustrations on page 613 and 622.

[2] See chart S.

[3] Joel (3:10) predicts farm equipment will be commandeered and smeltered for use as implements of war. When Christ returns to earth the process will be reversed (Isaiah 2:4; Micah 4:3).

[4] This Scripture, because of its context, must be distinguished from Ezekiel 38 and 39, which refers to the Soviet forces.

[5] This is the ultimate weapon which sends out a death ray close to the speed of light, thereby exceeding man's latest projected efforts for a "Particle Beam Weapon" in his war arsenal.

[6] See chart P.

[7] The Shiites (the revolutionary arm of Islam) believe that the 12th *Imam* (successor to Mohammed) is coming to earth; this *Mahdi*—their messiah—is about to come. Their infiltration and fanatical conquests of other lands is their divine commission to prepare the way for his return. In the year 2000 A.D. there will be 120 million Moslems in the Soviet Union. Soviet support of the radical Moslems in Syria, Iran, and elsewhere is symptomatic of what lies in the future.

[8] Obadiah 6 predicts the Middle East Arab oil wells; and Isaiah 34:9,10 predicts their destruction. They have been used for greed, extortion, and the promotion of Islamic anti-Christian and anti-Israel goals. Their sheikdoms and palaces will be destroyed (verses 12,13). The mosques will all be burned to ashes (verse 19).
[9] See illustration on page 615.
[10] Many new nations will come into existence during the Millennium (Isaiah 60:22).

78
A World of Kings

During the Millennium people will make a major adjustment from being carnivorous to once again being herbivorous.

"I appoint unto you a kingdom as my Father hath appointed unto me" (Luke 22:29).

"Fear not, little flock; for it is your Father's good pleasure to give you the kingdom" (Luke 12:32).

The word "millennium" comes from the Latin word "mill" (thousand) and "annum" (year), which gives it the meaning of one thousand years. It is also known by other descriptions such as "the days of heaven on earth" (Psalm 89:29); "the dispensation of the fulness of times" (Ephesians 1:10); "the world to come" (Matthew 12:32); "the thousand years" (Revelation 20:2-4); "in the regeneration" (Matthew 19:28); "the times of the restitution of all things (Acts 3:21); "thy kingdom come" (Matthew 6:10); "the stone kingdom" (Daniel 2:34,35); "a generation" (Psalm 22:30); "the consolation of Israel" (Luke 2:25); "the redemption of Jerusalem" (Luke 2:28); and "the sure mercies of David" (Isaiah 55:3).

The Millennial government will be a world theocratic absolute monarchy ruled by the Lord Jesus Christ, the King of kings. His Royal Redeemed, called "kings," will have ruling positions subject to Him.

The megalopolis of Jerusalem will be the world's capital. Christ has the key of David (Revelation 3:7), the symbol of authority. He alone has the right to sit upon the throne of David at that time. In addition to their Temple ministries, the twelve tribes of Israel will have specific missions to accomplish in the scheme of redemption in order to reach every person with the gospel of Jesus Christ (Mark 16:15-18; Zechariah 8:23).

DAVID THE PRINCE

Since Jesus Christ will be seated on David's throne during the Millennium (Luke 1:32), what role will David fill at that time? According to Ezekiel 45 and 46, he not only will supervise, but also he will explain the meaning of the ancient feasts and offerings during the visual presence of Christ, the Redeemer, who fulfilled all of the symbolisms.

We know David is the "prince" mentioned in Ezekiel: "And upon that day shall *the prince prepare* for *himself* and for all the people of the land a bullock for *a sin offering*" (45:22). Because Christ was sinless and was called a "Prince" by divine decree (Isaiah 9:6), this does not refer to Him.

David, as part of the Royal Redeemed, is a "king" as well as a "prince." When Christ abdicates His throne on the New Earth located in the New Heaven, and reassumes His place beside His Heavenly Father in the *New Jerusalem* within the Holy City, He will serve as Jurist at the Great White Throne during its sessions of the final tribunal. After this, David then will reascend the throne in *Jerusalem on the New Earth* in fulfillment of the prophecies in God's Word.

DRUG CONTROL

Since obnoxious and destructive weeds come from the cursed ground and are the source for most forms of drugs, these dangerous and damaging derivatives may prove difficult, if not impossible, to produce and procure during the Millennium when the curse from the ground will be lifted. When this curse is removed, there will be an abundance of food with improved taste, texture, and quality. The ground will produce profusely without the necessity for fertilization. Weeds and other spurious growths infesting the ground and damaging good plant life will find existence impossible on the blessed earth.

The word "pharmakeia" is the Greek word for "sorcerers" in Revelation 21:8 and means "drugs." It is the same word from which "pharmacy" is derived. The word implies that no drugs will be permitted in the Lord's Manifested Kingdom. There will be no need for pain pills (Revelation 21:4). Intoxication will be in violation of the established laws during the Millennium (Galatians 5:21; Romans 13:13). There will be no pharmaceutical outlets (Isaiah 33:24), optical centers, or firms making and dispensing wheelchairs, crutches, eyeglasses, and hearing aids; also no hospitals or physicians' offices (Isaiah 11:6-9; 34:16; 35:3-6; Revelation 21:4).

All manufactured goods made available to the general public will require this imprint: "HOLINESS UNTO THE LORD" (Zechariah 14:20). This will replace the trademark code under Antichrist which is forced upon the people during his reign (Revelation 13:16,17).

When Adam and Eve stole from the forbidden tree in the Garden of Eden, they established in the human race a lifestyle of stealing from God, thereby incurring His punishment. The good things of life were withdrawn and became scarce. The cursed ground not only has produced weeds, making it more difficult for man to sustain himself, but it also has produced by-products, in the form of drugs, which have added to his sorrows, as well as contributed to his well-being. These can be abused easily and often create adverse reactions in an individual. God's judgment upon the modern drug culture abuse is clear in Revelation 22:3.

SILENT RAGE

The ungodly, with an appetite for these by-products under the curse, will have only two options. One choice will be to be converted to Christ and have their selfish desires changed, and the alternative will be to live out their miserable lives and die a premature death. Unfortunately, many will prefer death and its eternal consequences in spite of the prevailing wonderful surroundings at the time (Isaiah 11:5; 26:8; 62:2,8; Daniel 9:24).

When God created mankind with a physical body He also provided the correct balance of diet. The tree of life and the other trees in the Garden of Eden were designed specifically to meet the body's dietary needs, but when mankind sinned, these provisions were no longer available. Thus, man's dietary needs were altered to require food from the cursed ground. Food would grow differently and weeds would produce by-products, such as tobacco, which could damage the physical body. Under the Mosaic Law God established the best diet which man can have while living upon the cursed ground.

There is evidence that much of the unconverted posterity will resent the type of righteous government that exists during the Millennium (Revelation 21:27; 22:11,14,15). At the end of this period of time when Satan is loosed for a season, he will find many followers and dissidents fuming inwardly, ready to rally behind Satan's leader, "Gog."

GUN CONTROL

"They shall not hurt nor destroy in all my holy mountain: for the *earth* shall be full of the knowledge of the LORD, as the waters cover the sea" (Isaiah 11:9).

"And he (Christ) shall judge among the nations, and shall rebuke many people: and they shall beat their swords into plowshares, and their spears into pruninghooks . . . " (Isaiah 2:4).

"The wild beasts of the desert shall also meet with the wild beasts of the islands . . . and find . . . rest" (Isaiah 34:14). Since the islands will be destroyed (Revelation 16:20) many of them fled to the mainland, the Pangaea.

When mankind sinned and his previous source of nourishment was removed, he neither could live indefinitely nor in harmony with nature and the animal kingdom. Because God allowed mankind to eat meat, man became a threat to the animal kingdom and terror reigned among the animals to the point that they were considered "wild" life. They no longer even lived at peace among themselves for we see the prey fleeing from the predator.

During the Millennium people will make a major adjustment from being carnivorous to once again being herbivorous, which was God's original plan when He created our first parents (Genesis 1:29; Isaiah 11:6-9). Herbs will provide the needed protein, and because of their medicinal benefits they will replace the need for pharmaceuticals: ". . . I have given every *green herb for meat . . .*" (Genesis 1:30). Unless the unredeemed are converted to Christ many will find these adjustments in their diet unacceptable and therefore prefer death to living out the one thousand years on earth and beyond.

WINE

The word "wine" in Scriptures comprises two shades of meaning—sweet and strong. Grape juice in its unfermented state, for example, is classified as "sweet" wine and frequently is used as a beverage because of bad water. Paul's admonition to young Timothy was to avoid drinking the water which gave him stomach problems. Instead, he was to take "a little wine for thy stomach's sake *and* thine often infirmities" (I Timothy 5:23). Obviously, the Apostle Paul, Timothy's mentor, would not recommend "strong" wine in contradiction to the Word of God (Leviticus 10:9; I Samuel 1:15; Proverbs 20:1; Isaiah 28:7).

Concerning the Lord's Supper or Communion (also "Sacrament" or "Eucharist"), the word "wine" is symbolic of the blood of Christ. If the wafer or bread, which must be unleavened, is a symbol of the holy body of Christ (I Corinthians 5:7,8; Exodus 12:15; 34:25; Leviticus 2:4,5), then why would that which symbolizes *the holy and incorruptible blood of Christ* have corruption (alcohol) in it? Corruption speaks of death, not life. The blood of Christ is eternally life-giving, and the body of Christ represents the redeemed Church which will attain perfection in an immortal body to be like Christ some day (Ephesians 5:27; Revelation 19:7-9). Christians are to be connoisseurs of life, not epicures of corruption or death.[1]

VINEYARDS ON THE MOUNTAINS

". . . And the mountain shall drop sweet wine, and all the hills shall melt . . ." (Amos 9:13).

"Every valley shall be filled, and every mountain and hill shall be brought low; and the crooked shall be made straight, and the rough ways shall be made smooth" (Luke 3:5).

". . . The everlasting mountains were scattered, the perpetual hills did bow . . ." (Habakkuk 3:6).

As God again consolidates the Pangaea[2] into one continent, other important changes in the topography of the earth will occur. Since only five percent of all land surface is presently suitable for cultivation and crops, and is steadily diminishing, God will increase the productivity of the land. Watersheds that are covered with snow will be lowered, and as they descend the snow will melt. Earthquakes will cause great avalanches to fill up the valleys and lower places with rocks, boulders, and rich soil. The inaccessible ("crooked") places will be opened up. The desert lands will return to productivity (Isaiah 35:6,7) as will the massive lands now under ice and snow. The frozen tundras will thaw and become productive land.

Certain areas will be set aside for water creatures: "In the habitation of dragons . . . shall be grass with reeds and rushes" (Isaiah 35:7). Springs will begin to flow over the deserts. The canopy over the earth will create a more uniform and temperate climate resulting in greenhouse conditions which simulate controlled farming.

SABBATICAL YEARS

"But in the seventh year shall be a sabbath of rest unto the land, a sabbath for the Lord; thou shalt neither sow thy fields, nor prune thy vineyards" (Leviticus 25:4).

The reinstitution of this sabbatical year for the land will occur during the Millennium. This will enable the land to revitalize itself, thus preventing erosion and blowing dust.

God created the earth and He has established laws which He knows are essential for the proper use of the earth: "The earth also is defiled under the inhabitants thereof; because they have trangressed the laws . . . " (Isaiah 24:5). Greed has caused land to be raped barren and deserts have spread like devouring monsters: "And I will bring the land into desolation . . . because it did not rest in your sabbaths, when ye yet dwelt upon it" (Leviticus 26:32,35). Christ, who will rule with a rod of iron, will see that the laws of righteousness are strictly enforced for the blessing of all of its inhabitants.

Earthquakes will have ceased as indicated: "Thus saith the Lord God, Because they say unto you, Thou *land* devourest up men, and hast bereaved thy nations; Therefore thou shalt devour men no more, neither bereave thy nations any more, saith the Lord God" (Ezekiel 36:13,14).

Because the Millennium is the seventh one-thousandth year since the earth's recovery from its once chaotic condition after Lucifer's sin, and is the Millennial sabbath, it will receive the kind of treatment it should have had in the past. This will help to make up for the deficit sabbaths, and the land will then be able to have its sabbath rests so it can produce profuse harvests: "And they shall say, This land that was desolate is become like the garden of Eden . . . " (Ezekiel 36:35).

"But ye, O mountains of Israel, ye shall shoot forth your branches and yield your fruit to my people of Israel; for they are at hand to come. For, behold, I am for you, and I will turn unto you, and ye shall be tilled and sown" (Ezekiel 36:8,9). "And the ransomed of the LORD shall return, and come to Zion with songs and everlasting joy upon their heads: they shall obtain joy and gladness, and sorrow and sighing shall flee away" (Isaiah 35:10).

If our first parents had not sinned, the garden of Eden would have expanded over the earth as the population increased. The tree of life would have multiplied, as it will on the New Earth, to accommodate the population explosion (Revelation 22:2). However, when the redeemed take possession of the earth under Christ, it will undergo a tremendous change but there will be no tree of life during the Millennium: "And (I) will do better unto you than at your beginnings: and ye shall know that I am the LORD" (Ezekiel 36:11).

"... For they (the Royal Redeemed) are at hand to come" (verse 8).

"And the ransomed of the LORD shall return, and come to Zion with songs and everlasting joy upon their heads..." (Isaiah 35:10).

"Then I will command my blessing upon you (the land) in the sixth year, and it shall bring forth for three years..." (Leviticus 25:4,6,21).

"Then (during the Millennium) shall ye remember your own evil ways, and your doings that were not good, and shall loathe yourselves in your own sight for your iniquities and for your abominations...."

"Then the desolate land shall be tilled... This land that was desolate is become like the garden of Eden..." (Ezekiel 36:31,34,35).

EARTH'S WATER CANOPY

"And God made the firmament (expanse) and divided the waters which were under the firmament *from the waters which were above the firmament* (the canopy of water) and it was so" (Genesis 1:7).

This is not a reference to rainclouds because the earth was watered after this from "a mist from the earth" (Genesis 2:6). If all of the water vapor in the atmosphere were to suddenly turn to rain, it would only add one inch of water on the earth, certainly not enough to have contributed much to the Great Flood in Noah's time.

Many scientists, who ascribe to the Biblical doctrine, believe that water, like the *rings* around Saturn, once formed a solid canopy high in the atmosphere giving the earth a comfortable greenhouse climate. The entire land mass at that time consisted of one continent and there were no huge buildups of ice and snow at the polar regions.

Much of the water that flooded the earth in Noah's time came from these canopies which were sustained by a law of physics established by the Creator. Perhaps an equal, or even greater, amount of water gushed from underground tables like millions of artesian wells. (Genesis 7:11).

When God was ready to remove the water from the inundated earth, He stopped the fountains of the deep from bubbling up and turned them into veritable vast drainage systems, sending much of the flood waters back into the underground tables.

There was still too much water remaining, so God shook the earth just enough to rush the surplus waters in huge tidal waves to the polar regions. There He quickly congealed them like jello by tilting the earth and allowing the waters to freeze. Six months of darkness and frigid cold would not have been sufficient time to freeze those massive quantities of water from the outside, so God must have taken the initiative to personally congeal them *from the center out* as He did in Exodus 15:8: "And with the blast of thy nostrils the waters were gathered together, the flood stood upright as an heap, *and the depths were congealed in the heart of the sea"* (Exodus 15:8).

Job (38:8-11) reflects upon how these polar regions of ice and snow came to exist: ". . . who shut up the sea with doors, when it brake forth, as if it had issued out of the womb? When I made the cloud the garment thereof (the canopy of water), and thick darkness a swaddling-band for it (frozen polar regions, as when the earth was once encased in ice), And brake up for it (the Flood) my decreed place, and set *bars and doors* (imprisoned the huge mass of Flood waters at the poles), And said, Hitherto shalt thou come, and no further: and here shall thy proud waves be stayed?" The word "bars" indicates that these huge mountains of ice are kept intact. The "doors" refers to the inlets and bays similar to those of the Ross Sea at the Antarctic.[3] Many animals, with flesh yet fresh, have been found imbedded deeply in the ice, thus providing evidence of this sudden freezing of the waters.

When the earth's security blanket (canopy of water) was removed at the time of the Great Flood, Noah, a righteous man, was unaware of the change that had taken place above the earth. The chemical or fierce actinic lethal rays of light which produce decay and decomposition in nature were no longer filtered out, so when Noah took of the vineyard he became intoxicated. Because of this disgraceful condition in which it left him, Noah placed a curse upon his son, Ham, who had violated his privacy. He also declared his utter abomination upon the introduction of this new evil. Noah recognized this corruption as the result of God's judgment on the earth. The fruit of the vine, which was to provide food, was converted into something detrimental for mankind's consumption.

ARSENAL OF ICE AND VOLCANIC POWER

"Hast thou entered into the treasures of the snow? or hast thou seen the treasures of the hail, which I have reserved against the time of trouble, against the day of battle and war?" (Job 38:22,23). "The first angel sounded, and there followed *hail and fire* (volcanic eruptions) mingled with blood (casualties), and they were cast upon the earth . . . and *all the green grass was burnt up*" (Revelation 8:7).

This massive arsenal of ice and snow at the polar regions can be considered God's stockpile of missiles ready to be launched from the cannons of volcanic fire power once the great earthquake breaks up the ice into approximately sixty-pound projectiles for this "day of battle and war": "Yet once more I shake not the earth only . . . " (Hebrews 12:26). ". . . And cities of the nations fell . . . And there fell upon men a great hail out of heaven, every stone about the weight of a talent (sixty pounds); and men blasphemed God because of the plague of the hail; for the plague was exceeding great" (Revelation 16:18-21).

The huge earthquakes described in Revelation 6:12-17 not only will prepare the ice missiles, but also they will loosen up chunks of ice weighing many tons. These will crash into the oceans causing massive tidal waves and leaving unbelievable devastations in their wake: ". . . the sea and the waves roaring" (Luke 21:28. See also Ezekiel 38:22; Matthew 24:7; Revelation 6:12; 8:7; 11:19; 16:21).

An unfrozen lake, which is the size of California, sits under Antarctica. Is it heated by a dormant volcano? Scientists are not sure, but if it is true, this 186,000 square mile lake, called "Polynyas," could be the potential cannon barrel sitting there armed and ready to be fired with awesome power. It could pulverize the ice cape and send massive ice projectiles over the earth in fulfillment of the prophecy in Revelation 16:21.

GOD TURNS UP THE THERMOSTAT

Once these ice capes at the poles are broken up and scattered over the seas and land, God will send heat plagues to quickly evaporate them and return them as a vapor canopy above the earth (Revelation 8:8,10). In Isaiah 24:6 we read: "... the inhabitants of the earth are *burned* and few men left."

A WATER CRISIS ENDS

Not all clouds are rainclouds. Genesis 2:5,6 unfolds the fact that although there were clouds on the original earth, they were not rainclouds: "... for the LORD God had not caused it to rain upon the earth and there was not a man to till the ground. But there went up a mist from the earth and watered *the whole face of the ground.*" In Isaiah 14:14 these clouds on the original earth were again revealed when Lucifer said, "I will ascend above the heights of the clouds...."

The first rainclouds came in the days of Noah, as judgment upon the earth, and they will remain with us until the Millennium. They are responsible for destructive floods; frightening storms with lightnings, thunders, and ferocious winds; tidal waves; and the creation of deserts. When the original underground watering system was in effect, the whole earth was watered. The advent of rain has resulted in the creation of vast deserts over the world, which comprise an area equal in size to the Western Hemisphere.

Hydrologists estimate that the United States has 59 thousand trillion gallons of underground reservoir water, forty-five times the volume of Lake Michigan. Between 1950 and 1980 the annual consumption of this underground water table more than doubled to about 25 trillion gallons. Every day the nation uses 21 billion gallons more from the underground reservoir than seeps back into it.

The thickness of the water beds varies from two hundred to one thousand feet. In some places this has resulted in the land sinking as much as twenty-nine feet. An average of 4.2 trillion gallons of precipitation falls on the contiguous forty-eight states daily. Of this, ninety-two percent evaporates quickly or runs off unused to the oceans. Agriculture accounts for eighty-seven percent of used water. The industrial age has precipitated a water crisis throughout the world.

The asteroid striking the Mediterranean Sea (called "the sea") will send up great quantities of steam (Revelation 8:8). This canopy of water will be a reflective sheen as described in Isaiah 30:26, and will replace the carbon dioxide accumulated in the upper atmosphere during the industrial revolution. It will be a time when the Lord bindeth up the breach of His people: ". . . and healeth the stroke of their wound" (Isaiah 30:26).

Carbon dioxide (CO_2) is being increased two percent annually as a result of our consumption of oil and coal. It is estimated that about ten percent of the crops are destroyed each year by carbon dioxide in the atmosphere. Science predicts that in 2000 A.D. the weather patterns undoubtedly will change and thereby alter the face of the globe.

Carbon dioxide released into the atmosphere acts as a canopy which gives the planet a greenhouse effect that may cause the temperature to rise five percent by the year 2000 A.D. This will result in continual expansion of desert areas. Also as a result of this rise in temperature, which will melt the polar ice and snow, the oceans are expected to rise as much as one hundred fifty feet or more, and will inundate all coastal cities and many island areas throughout the world. During the Millennium,[4] vortex winds blowing through the water canopy will create an acceptable greenhouse effect, much different from the canopy of carbon dioxide.

Meteorologists estimate that a rise of only two degrees would melt sufficient ice and snow at the poles to raise the ocean levels as much as forty feet or more. Antarctica has six million square miles of ice and snow; in contrast, the Arctic has nearly three million square miles. Greenland ice caps extend over 700 thousand square miles and in places are a mile and a half deep. This ice and snow in Greenland alone would raise the oceans twenty-three feet. Ice covers some ten million square miles of the earth's surface.

Birds, animals, and fish recently have been migrating closer to the polar regions, indicating that the warming trend is now in progress.

MILLENNIAL ARBOR DAYS

There will be arbor days in the Millennium (Isaiah 41:19; 55:12; Ezekiel 47:12). Carbon dioxide in the atmosphere acts like a greenhouse. It absorbs heat radiating from the earth and its atmosphere—heat that otherwise would dissipate into space. The more carbon dioxide there is in the atmosphere, the warmer the earth generally becomes. Carbon dioxide began to be absorbed into the atmosphere during the industrial revolution of the 19th century. A century ago it was 280 to 300 parts per million, but it is now 335 to 340 parts per million, and it is expected to rise shortly to 600 parts per million. The only known solutions to the problem are the extensive planting of forests to absorb the carbon dioxide or the reduction of our industrial output. Conversely, man has been destroying the forests and compounding the difficulty. During the Millennium, Christ perhaps will launch a massive global reforestation program. The rate of deforestation today is 120 acres every minute—more than 63 million acres a year. Forests affect the water table and rainfall, as well as prevent soil erosion.

Acid rain has cause for grave alarm. It is air pollution, composed mainly of sulphur dioxide and nitrogen oxides, which change chemically as they return to earth from the atmosphere as precipitation or dry particles. They affect the microorganisms in the soil. They break down organic matter, recycling it as nitrogen and carbon in the food chain, which can have "irreversible" long-time consequences.

Hundreds of thousands of tons of toxic wastes are produced annually. New chemicals are introduced and produced at a rate of approximately one thousand a year. Of these, eighty percent are dumped illegally. God's plan of waste disposal was outlined in Deuteronomy 23:13,14, but because man has violated that law and has created a great buildup of toxic wastes, God will move to rid the earth of them in one massive effort. He will accomplish this in several ways.

BENEFITS FROM THE HEAT PLAGUES

"For, behold the Lord will come with fire . . . and the *abomination,* and the mouse, shall be consumed together, saith the LORD" (Isaiah 66:15-17).

Heat plagues that punish the wicked on earth also will benefit the planet. Scientists have discovered that ultraviolet light rays cause microorganisms of toxic chemical wastes to break up and shortly thereafter disintegrate into harmless particles. The heat plagues of God in Revelation 16:8,9 not only will cleanse the planet of its poisonous chemical wastes, but also it will remove the curse of pests and insects, of which there are over one million species.[5]

GOD'S SUICIDE INFANTRY

"In that day thou shalt make the plant to grow, and in the morning shalt thou make thy seed to flourish . . . " (Isaiah 17:11).

There could not have been any destructive, biting, and disease-carrying insects until God cursed the ground: first, Adam and Eve could not have existed comfortably unclothed; second, the cursed ground resulted in sorrows, sweat, and death: " . . . cursed is the ground for thy sake; in sorrow shalt thou eat of it all the days of thy life . . . in the sweat of thy face shalt thou eat bread, till thou return to the ground . . . " (Genesis 3:17-19).

Conservatively speaking, paleontologists believe that approximately one-half billion species of creatures subsisted during earth's history, of which only two percent exist today. Over 15 thousand species of locusts alone affect about fifty countries, or one-fifth of the land's surface. In 1982 billions of caterpillars halted freight trains in Italy. Destructive insects such as mosquitoes, hoppers, and locusts breed as a result of hot, dry winds or flooded conditions, and winds carry these plagues as fast as forty miles a day. Billions of dollars are spent annually in endless global insect wars. Living insects on earth weigh more than ten times that of all living human beings.

Today man must use pesticides to help control insect infestations. For that reason scientists have been interested in controlled farming under enclosures; however, the canopy of water during the Millennium will do substantially the same thing. Deserts, swamps, tundras, and lands which are inaccessible and useless because of rocks and high mountains all will become available and fertile for cultivation and agricultural harvests (Luke 3:5).

Since Jesus Christ will change the appetites of animals (Isaiah 11:7), it is conceivable that He will do the same with pests that may survive the heat plagues. Thus, they may have their appetites altered to devour obnoxious weeds that infest the ground. When the weeds are eradicated, the pests will be unable to survive. According to Scripture, God will lift from the ground the curse that brought the infestation of weeds and pests in the first place. No chemicals, pesticides, or fertilizers will be required to assure bountiful crops during the Millennium.

The prediction that mankind would pollute the earth is seen in Isaiah 24:5,6: "The earth also is defiled under the inhabitants thereof; because they have transgressed the law . . . Therefore hath the curse devoured the earth, and they that dwell therein are desolate . . . ."

BENEFITS FROM THE EARTHQUAKES

Plutonium is the most dangerous substance remaining after uranium fuel has been spent in a nuclear reactor and it is capable of radioactivity for over 200 thousand years. By the year 2000 A.D. nuclear power plants will have produced sufficient radioactive waste to build a four-lane highway one foot deep from coast to coast. One-half million gallons of radioactive liquid will require disposal daily. However, God will bury the wastes underground deep in the crevices opened up by earthquakes (Luke 21:11; Revelation 11:13).

"Behold, the days come, saith the LORD, that the *plowman shall overtake the reaper,* and the treader of grapes him that soweth seed; and the mountains shall drop *sweet wine,* and all the hills shall melt . . . and they shall plant vineyards, and drink the wine thereof; they shall make gardens, and eat the fruit of them" (Amos 9:13,14). Several crops will be harvested each year as predicted in the above promise. The quotation, "the hills shall melt," shows that watersheds no longer will be a necessity. With the water canopy restored, as in the Antediluvian Age, a "mist" will water the ground (Genesis 2:6): "For the seed shall be prosperous; the vine shall give her fruit, and the ground shall give her increase, and the *heavens shall give their dew;* and I will cause the remnant of this people to possess all of these things" (Zechariah 8:12).

THE WINGS OF THE WINDS

". . . (The Lord) did fly upon *the wings of the wind* . . . The Lord also thundered in the heavens, and the Highest gave his voice; hail stones and coals of fire. Yea, he sent out his arrows, and scattered them, and he shot out lightnings, and discomforted them. Then the channels of waters were seen . . ." (Psalm 18:10-15).

Rainclouds, as we have said, were never God's original plan for watering the surface of the earth to maintain its vegetation and produce flourishing crops. They are erratic, undisciplined, and frequently manifest themselves through destructive storms, violent winds, torrential monsoons, floods, and lightnings and thunders sometimes resulting in devastating fires. They too often accompany Mother Nature on a rampage.

Rainclouds are influenced by mountains, forests, bodies of water, and jet streams. They often destroy lives, as well as property; leave great patches of arid land; and bring hail to destroy promising crops. There has to be a better way for watering the ground, and as described in Genesis 2:5,6 there was a superior method before the Great Flood.[6]

JUDGMENTS LONG-RANGE EFFECTS

God's judgments have a long-range effect as we see manifested in the past. There are one million species of insects in a world with five million different species of life forms. We are suffering with physical diseases and illnesses because man's life no longer comes from the tree of life but from the mortality of his blood. As a result of what happened at Babel, we have a confusion of languages and races, and a diversity of continents and islands. Israel's problem with the Arabs stems from Abraham's disobedience. Israel's dispersion for thousands of years was a consequence of the introduction of idolatry into their land. God sent Israel into exile when the Babylonians invaded the city of Jerusalem and destroyed Solomon's Temple after raping it of its gold and silver. For 2600 years Israel was without its own country and independence. Not until May 14, 1949, when its Parliament was convened, did it become a nation again. These are but examples of what happens because of man's rebellion against God. Lucifer and the fallen angels will live forever with the knowledge of God's judgments already upon them and the impact that they will suffer in their future.

THE WINGS OF A DOVE

"(Noah) sent forth a dove . . . but the dove found no rest for the sole of her foot, and she returned . . . and, lo, in her mouth was an olive leaf . . . " (Genesis 8:8-11).

" . . . Oh that I had wings like a dove! for then would I fly away, and be at rest" (Psalm 55:6).

" . . . Yet shall ye be as the wings of a dove covered with silver, and her feathers with yellow gold" (Psalm 68:13).

In Ezekiel 36:13,14 God addressed the *land* concerning a future time when there will be a major change to the earth under the reign of Christ: "Thus saith the LORD God; Because they say unto you, Thou land devourest men no more, neither bereaved thy nations any more, saith the Lord GOD." The destructive force of lava and the clouds of volcanic ash girdling the globe and affecting the climate will cease to exist during the Millennium. Psalm 107:29,30 states: "He maketh the storm a calm, so that the (tidal) waves thereof are still. Then are they glad because they be quiet; and he bringeth them unto their desired haven" (compare Romans 8:19-22).

As in the Antediluvian Age, the mist that watered all of the earth will return during the Millennial reign of Christ. Before the Great Flood the temperature upon the earth was ideal because the vortex winds once gave our planet an even and delightful climate. These winds are like huge fans sweeping over the poles and equator at two hundred miles per hour in the ionosphere, circulating over the earth and *passing through the water canopy*. The vortex winds were first discovered in 1957-58 during the International Geophysical Year (IGY), and they are recorded in Scripture: "The wind goeth toward the south, and turneth about unto the north; it whirleth about continually, and the wind returneth again according to his circuits" (Ecclesiastes 1:6).

Since the Flood, the vortex winds and water canopy have been replaced by erratic jet streams with a velocity of two hundred fifty miles an hour nearer the earth. While the vortex winds are still present, the water canopy disappeared with the Flood and is now located at the polar regions. The earth has been tilted 23-½ degrees, giving extremes in the weather patterns, thereby shortening time for harvesting in much of the world.

TAMING MOTHER NATURE

"... Satan went forth from the presence of the LORD... And, behold, there came a great wind from the wilderness, and smote the four corners of the house, and it fell upon the young men, and they are dead..." (Job 1:12,19).

Satan, as prince and power of the air (Ephesians 2:2) effectively used violent weather against mankind. When Jesus rebuked the wind and the raging of the waters (Luke 8:24; Matthew 14:26-32), He used the word "epitimo," which is the same word He used when casting out demons (Luke 8:32; 4:35), because He recognized that Satan was behind these intense storms. When Satan is bound during the Millennium, he no longer will be present to agitate Mother Nature into turbulent action. There was no recorded rampageous weather during the Antediluvian Age because the violence of nature was not released until the Flood. Prior to that time only a "mist" watered the earth (Genesis 2:6). The violence of nature came because of man's horrible transgressions (Genesis 6:11-13).

Destruction is the work of Satan. Christ tamed the raging waters and exorcised the raging spirits as the Prince of Peace. He brought nature to rest: "Because the creature itself also shall be delivered from the bondage of corruption into the glorious liberty of the children of God. For we know that the whole creation groaneth and travaileth in pain together until now" (Romans 8:21,22).

The evil Spirits, along with Satan, will be confined to the bottomless pit. There will be no more seizures by evil powers; no blindness or deafness; no more wheelchairs or crutches; no more silent tongues (Isaiah 35:5,6). The deaf and blind will be healed so they may hear and see the Word of God in action: "And in that day shall the deaf hear the words of the book (God's Word), and the eyes of the blind shall see out of obscurity, and out of darkness" (Isaiah 29:18).

CANCER: THE GENETIC TIME BOMB

Cancer was named after the "crab" during times of antiquity because the disease obstinately seizes and adheres like a crab. In the Medieval Period living parts of crabs were applied to the diseased areas of humans because they purportedly had magical curative powers. Cancer is found in both the animal and plant kingdoms. Its roots are in the very ground that was cursed. This disease will vanish when Jesus Christ returns to set up His Kingdom on the earth and removes the curse from the ground (Isaiah 32:15; 33:24; Revelation 22:3).

[1] Transubstantiation, the doctrine that the elements are changed into the actual body and blood of Jesus Christ, came into the Medieval Church from pagan practices. The word "cannibal" is derived from Greek, meaning "priests of Baal," who, when eating parts of human sacrifices, believed they were eating the flesh of "gods" that incarnated the human sacrifices.

[2] "To Him that *stretcheth* out the earth *above* the waters . . ." (Psalm 136:16).

[3] Since 1850 A.D., the glaciers have been in retreat. World climate has been slowly but steadily rising in temperature.

[4] Ecclesiastes 1:6.

[5] Scientists believe there may be up to one hundred times more species of plants, animals, and microorganisms than they have anticipated. They believe that each tree alone may harbor as many as thirty million species of insects.

[6] Weather-related catastrophies cause more annual devastation than do earthquakes, erupting volcanoes, and all other natural disasters combined.

79

Restoration of the Earth

Redemption included the recovery of the entire earth.

"Blessed are the meek: for they shall inherit the earth" (Matthew 5:5). "...The redemption of the purchased possession" (Ephesians 1:13,14). No one should have to inherit the earth in its present unredeemed condition with its pollution, infestations, climatic extremes, and undesirable and inaccessible land areas. Redemption involves the recovery of the entire earth for the Millennial reign of Christ, His Royal Redeemed, and other inhabitants in His Manifested Kingdom (Romans 8.19-25; Matthew 13:41).

With the discovery of the decay of the proton, physicists recognize the mortality of all matter in the third-dimensional world. In order for the earth to survive, its mortal particles must be altered into immortal properties in the same manner that the bodies of the Royal Redeemed will be changed. The concept of this alteration of the earth as the inheritance of all the redeemed is borne out from the ancient Hebrew transition of property by the kinsman-redeemer (Leviticus 25:24). Christ is the Redeemer-Kinsman of the redeemed who will inherit the earth (Compare Matthew 13:44, "and buyeth that *field.*" See also Nehemiah 5:5).

The Millennium will be a time of major construction. Perfection will be attained only at its end; then it will become the predicted New Earth. There will be a major cleanup operation of the casualties of Armageddon, including the removal of the massive debris of war machines. God will call in the carnivorous fowl to devour rotting flesh, which is known as the "supper of the great God" (Revelation 19:17). The bones will calcify, thus fertilizing and enriching the ground.

"And I will multiply upon you man and beast: and they shall increase and bring fruit: and I will settle you after your old estates, and will do better unto you than at your beginnings: and ye shall know that I am the LORD" (Ezekiel 36:11).

The cities will lie in ruins: "For thou shalt . . . make desolate cities to be inhabited" (Isaiah 54:3). "And they shall build the old wastes, they shall raise up the former desolation, and they shall repair the waste cities, the desolations of many generations" (Isaiah 61:4). ". . . The cities shall be inhabited, and the wastes shall be builded" (Ezekiel 36:10).

The ground and the water in the seas, lakes, rivers, and streams will be covered with the wreckage of man's achievements. The cities will be ghastly sites, looking like twisted and gnarled steel; like heaps of rubble, resulting from man's rebellion against God. It is the aftermath of the destruction resulting from Lucifer's first rebellion against God; the Flood of Noah's time; and the cities of Sodom and Gomorrah: "As it was in the days of Noah, so shall it be also in the days of the Son of man . . . likewise also as it was in the days of Lot . . . it rained fire and brimstone from heaven . . . " (Luke 17:26, 27,29). Smoking rubble and collapsing structures will remain to be razed. In fact, all of the works of men shall be removed and replaced: ". . . the works that are therein shall be burned up" (2 Peter 3:10).

REDEMPTION OF THE EARTH

According to the Mosaic law in Leviticus 25:23-25, redemption is buying back something that has been lost to the rightful owners. The redeemed are the legal heirs of the earth because Christ, the Second

Adam, replaced the first Adam who was disinherited by God because of his trangression. Satan, through lies and deception, dispossessed mankind of the earth. Satan became a usurper and all of the unredeemed are squatters without legal rights to rule. Their rule will be tolerated only until the Earth is redeemed in order that God may consummate His redemptive plan to the unredeemed. In the New Heaven, only the Royal Redeemed will be residents upon the redeemed (New) Earth (Revelation 21:24,27; 22:14,15). All others, whether redeemed or unredeemed, will reside on other habitable places in the New Heaven until the plan of redemption is completed. One reason Armageddon is fought so ferociously is that Satan and his compatriots do not want to be evicted and forever banished as illegal squatters from the earth. The new rulers will be the Royal Redeemed who will reign as "kings."

The present state of the geophysical earth represents a transformation from its primordial condition as a consequence of Lucifer's fall into sin: ". . . the same day were all the fountains of the great deep broken up . . . And the waters prevailed exceedingly upon the earth; and all the high hills, that were under the whole heaven, were covered . . . and the mountains were covered" (Genesis 2:11,19,20). The earth became perforated, ulcerated, ruptured, disoriented, and disfigured. Mountains and hills were located in places where the crust of the earth bulged like warts. When the Flood came, the earth's crust gave way under the immense weight of the water, causing new hills and mountains to come into existence. The openings in the earth's crust forced underground water to gush up throughout the earth like a sponge being squeezed.

During the Millennium these disfigurations upon the earth will be removed. The unsightly rocks and boulders, which make much of the land useless and inaccessible, will be covered with rich soil from the cascading mountainsides that will be shaken by gigantic earthquakes.

The Flood brought torrential rains which, since then, have caused further flooding, lightning, thunder, violent windstorms, and tidal waves. But in the redemption of the earth all of this will be changed: "The mountains shall bring *peace* to the people..." (Psalm 72:3). "And my people shall dwell in a peaceable habitation, and in *sure dwellings,* and in quiet resting places" (Isaiah 32:18).

ROSES WITHOUT THORNS

"Upon the land of my people shall come up thorns and briers..." (Isaiah 32:13).

"... Cursed is the ground... Thorns also and thistles shall it (the earth) bring forth unto thee..." (Genesis 3:17,18).

Jesus Christ, who stills the raging tempests, will again bring Mother Nature to blessed quietness.

Christ is not only Healer of the soul and body, but also He will be Healer of the earth. Like a garment of immortality upon the redeemed, so the earth will be clothed under redemption to become the New Earth, like a new creation in Christ Jesus: "The wilderness... and the desert shall... blossom as the rose" (Isaiah 35:1).

POKING GOD IN THE EYE

God has stated repeatedly that the earth is His land (Exodus 9:29; Deuteronomy 10:14; Psalm 24:1; 115:16; I Corinthians 10:26). God is the Owner of the earth and man is but a tenant.

God divided the earth under His foreknowledge: "And hath made of one blood all nations of men for to dwell on all the face of the earth, and hath determined the times before appointed, and the bounds of their habitation" (Acts 17:26).

God also has decreed a particular parcel of ground, the Holy Land, to be the inheritance of Israel. God said to Abraham, "... unto thy seed shall I give this land" (Genesis 12:7). Ten times God called this parcel "My land" (Isaiah 14:25; Jeremiah 2:7; 3:9; 16:18; Ezekiel 36:5.16; Joel 1:6; 2:18; 3:2).

Israel is not only a tenant, but has ownership by divine decree. If it is driven out and its land divided with illegal squatters, God says this is a denial of the rights of His beloved one and it will anger Him: "For thus saith the LORD of hosts; After the glory hath he sent me (the Redeemer, Jesus Christ) unto the nations which spoiled you: for he that toucheth you (Israel in the Holy Land) *toucheth the apple of his eye:* For, behold, I will shake mine hand upon them, and they shall be a spoil . . . " (Zechariah 2:8,9). Jerusalem will be a burdensome stone to the nations of the world if they fail to recognize Israel as God's legal tenant (Zechariah 12:3).

Every great empire that has invaded the Holy Land and denied Israel its right of occupation has fallen. God has decreed punishment upon those nations daring to divide the land (Joel 3:2). It is for this reason that many students of Scripture believe the British Empire has been greatly reduced; that the League of Nations is no more; that both the Soviets and Antichrist invading the land of Israel will face the immediate wrath of God; and that the Arabs will lose all of their lands and be decimated as a people except for those who are redeemed. God means what He says and has the power to enforce His will.

"And in that day will I make Jerusalem a burdensome stone for all people: all that burden themselves with it shall be cut in pieces, though all the people of the earth be gathered together against it. In that day, saith the LORD, I will smite every horse with astonishment, and his rider with madness . . . " (Zechariah 12:3,4; compare Revelation 14:18-20; 16:14-16).

JUDGMENT OF THE NATIONS

The Judgment of the Nations (Matthew 25:31-46) will result in a realignment of national boundaries because many nations will have been eliminated: "I will also gather all nations, and will bring them down unto the valley of Jehosaphat, and will plead with them there for my people and for my heritage Israel, whom they have scattered among the nations, and *parted my land*" (Joel 3:2; compare Zechariah 12:9).

Christ Jesus, when He returns to earth, will evict all illegal squatters and give the Holy Land back to its legal tenant (Micah chapter 4; Jeremiah 31:10,12; Zephaniah 3:19,20; Ezekiel 11:17).

SANITATION DEPARTMENT

The devastation of Armageddon before the Millennium will affect even the topography of the earth. Millions of human and animal carcasses will be strewn all over the earth. According to Revelation 6:12-17, the stars of heaven will be relocated until familiar stars and constellations will have disappeared: "Whose voice then shook the (original) earth: but now he hath promised, saying, Yet once more I shake not the earth only, but also heaven" (Hebrews 12:26).

Meteors will fall in one great crescendo like thousands of streamers from the explosion of millions of huge firework displays. As this celestial phenomena of falling meteors ends, like a scroll that is closing, all the lights of heaven will vanish: "And the heaven departed as a scroll when it is rolled together . . . the sun became black as sackcloth" (Revelation 6:12,14). "Woe unto you that desire the day of the LORD! to what end is it for you? the day of the LORD is darkness, and not light . . . Shall not the day of the LORD be darkness, and not light? even very dark, and no brightness in it?" (Amos 5:18,20).

The Millennium will be the time for cleaning up the earth (Ezekiel 39:9,12; Revelation 19:17,18; Luke 17:37; Jeremiah 25:33).

DELEGATION OF AUTHORITY

The Judgment Seat of Christ will have rewarded and honored the faithful with top positions based on their past performance as Christians on earth, as determined by their motives and works (I Corinthians 3:8-16; 2 Corinthians 5:10,11; Romans 14:10-12; I John 2:28).

These Royal Redeemed will help supervise the great reconstruction and, as "lords," will be in charge of all real property—residential, business, commercial, and industrial. Many of the Royal Redeemed in life have suffered material losses because of their loyalty to Jesus Christ, and now their reward could be as much as one hundredfold—100 times 100 equals 10,000 percent (Matthew 19:28-30)!

Those who operate these properties will enjoy the fruit of their labors. They will suffer no losses from unscrupulous landlords, underworld payoffs, and political harassments (Isaiah 65:21; Amos 9:14; Jeremiah 32:15; 39:10; Ezekiel 28:26).

The persons in mortal bodies during the Millennium, and later in the New Heaven, will live on properties owned and managed by the Royal Redeemed.

Diversities of talents to be given to the Royal Redeemed at the Judgment Seat of Christ will require no formal education. This will enable the Royal Redeemed to effectively direct the reconstruction operations performed by natural people during the Millennium and in the habitations of the New Heaven.

Those individuals who show no compassion for the Tribulation saints and Israel during the reign of Antichrist will be removed from society forever under the Judgment of the Nations. Only the righteous citizens of nations will continue to exist in the Millennium and beyond: The "...meek shall inherit the earth" (Matthew 5:5).

A RIGHTEOUS GOVERNMENT

The Millennium not only will begin on the foundation of a righteous government but also on the basis of a people who survive the test of the Judgment of the Nations. During the Tribulation days many unconverted people, at the risk of their lives, aid the Tribulation saints and shelter them from the wrath of Antichrist. They will be included within the righteous nations even though they are unredeemed. However, they must accept Christ as their Saviour before they reach the age of 100 years or they will die natural deaths (Isaiah 65:20). When they die, these unredeemed will appear at the White Throne Judgment for the final disposition of their cases. If, in Christ's opinion, they deserve the right to exist in the New Heaven, the Royal Redeemed sitting enthroned in the royal jury box will indicate their approbation of the decision of the Supreme Justice. The good deeds of the unredeemed appear to have certain influences on this final judgment day. Jesus said, "He that receiveth you receiveth me, and he that receiveth me receiveth him that sent me, He that receiveth a prophet in the name of a prophet shall receive a prophet's reward; and he that receiveth a righteous man in the name of a righteous man shall receive a righteous man's reward" (Matthew 10:40,41). They must, however, at some point make their decision for Christ or be lost forever as decreed by Jesus Christ.

WEEP NO MORE

The Tribulation of the last days brings much sorrow and loss of life. As at the tragedy of the Battle of Armageddon, carcasses are strewn over the entire globe: "And the slain of the LORD shall be at that day from one end of the earth even unto the other end of the earth . . ." (Jeremiah 25:33).[1] When the Lord returns to earth His brightness will slay millions who will have taken the mark of the Beast" (2 Thessalonians 2:8; Matthew 24:27; Luke 17:24-37). So there will be great sorrow among the survivors over the loss of loved ones, friends, and property.

Isaiah 25:8,9 affirms: ". . . the Lord GOD will wipe away tears from off all faces ; and the rebuke of his people shall he take away from off all the earth: for the LORD hath spoken it. And it shall be said in that day, Lo, this is our God; we have waited for him, and he will save us: this is the LORD; we have waited for him, we will be glad and rejoice in his salvation." "For the people shall dwell in Zion at Jerusalem: thou shalt weep no more for he will be very gracious unto thee at the voice of thy cry; when he shall hear it, he will answer thee" (Isaiah 30:19). As the Psalmist said, "They that sow in tears shall reap in joy" (Psalm 126:5). "In that day shall this song be sung in the land of Judah; We have a strong city . . ." (Isaiah 26:1). ". . . The inhabitants of the world will learn righteousness . . . Lord, thou wilt ordain peace for us" (Isaiah 26:9,12). "Therefore the redeemed of the Lord shall return, and come with singing unto Zion; and everlasting joy shall be upon their head; they shall obtain gladness and joy; and sorrow and mourning shall flee away" (Isaiah 51:11).[2]

THE RACIAL ISSUE

The human species come in an artist's palette of colors—yellow, reddish-tan, deep brown, light tan, creamy white, white, pale pink, and black. There are differences other than skin color, hair texture, and facial features; there are dissimilarities in the shape of bones, consistency of ear wax, and body chemistry. Many of these variations are the result of migration into new environmental situations. Some anthropologists and genetecists believe that there are two or three dozen races. Some insist upon five pure races as our origin.

Before 1500 A.D. there was little migration and almost no mixing. Since travel between nations and continents of the world has become rapid and commonplace, differences have been diminished greatly. During the Millennium it can be expected that national identities will be retained, but that racial differences practically will be eliminated. For one reason, the Pangaea will contribute to intermingling of the races as continents will be closer together and not separated by vast bodies of water. Racial differences, which have been the cause of many wars, will now cease to exist.

However, this is in great contrast to the last days when racial problems will escalate, as predicted by our Lord Jesus Christ when He said, "For nation shall rise against nation . . ." (Matthew 24:7). The word "nation" here means race (ethnos). This was not a problem in Christ's day. Roman citizenship did not rely on race. Paul the Apostle, a Jew, was a Roman citizen. Africa until recently was a dark continent.

MILLENNIAL OFFICIALDOM

KINGS:

The Royal Redeemed will reign with Christ in their apprenticeship during the Millennium; then go on to rule in the New Heaven; and continue to govern in the Father's Kingdom called the "New Heavens."

PRIESTS:

Theirs is a spiritual ministry to reach all of the unredeemed in natural bodies whose names are in the book of life. This will involve both the Manifested Kingdom of Christ during the Millennium and the New Heaven. They will not serve as "priests" in the Father's Kingdom since all who survive will be redeemed and in immortal bodies by that time.

JUDGES:

They not only will sit as jurors in the royal jury box during the White Throne Judgment when Christ officiates as the All-Supreme

Court Justice, but they also will serve as judges during the Manifested Kingdom of Christ throughout the Millennium and on into the New Heaven. This will discontinue in the Father's Kingdom because by then all ungodliness will have ended.

LORDS:

They are to "inherit the earth" and "all things." Besides reigning with Christ, the Royal Redeemed will administer all of the real property in the Millennium and the New Heaven under Jesus Christ, the "Lord of lords;" but they will govern only as "kings" in the Father's Kingdom (Revelation 22:5).

PROPHETS:

The ministry of the prophets will come to an end with the termination of the final two prophets (or witnesses) upon the earth during the Tribulation period: ". . . I will cause the prophets . . . to pass out of the land" (Zechariah 13:2). This position will be missing from the portfolio of the Royal Redeemed.

REDEEMED:

When all of the redeemed come into the family of God, their spiritual ministry as priests will be concluded because there are no more people to be redeemed. The only enduring honor bestowed upon the Royal Redeemed will be that of "kings" over the redeemed who are not in the corporate Wife of Christ: ". . . they shall reign for ever and ever" (Revelation 22:5).

[1] See illustration "The Sword of the Lord".

[2] There have been over 15,000 major wars throughout recorded history.

80

Millennial Blueprint

One proof that the Millennial Earth will also become the New Earth is that Jerusalem and other cities in their final restored beauty will not be destroyed.

While the Manifested Kingdom of the Lord in its inception (the Millennium) will have a perfect government, it will not be populated by perfect people in physical bodies. The offspring born unto them will have the old Adamic nature and the tendency to live out of harmony with the divine laws. Many will not know Christ as their personal Saviour (Revelation 20:7-10). This also will be the case in the New Heaven where the redemptive plan of God will be consummated at its dispensational end (Revelation 21:27, 22:11,14,15).

Eventually there will be no more unredeemed. At that point of time in the New Heaven, death will be destroyed and all of the redeemed will then possess eternal immortal bodies: ". . . when he (Christ) shall have put down all rule and all authority and power (that is, those in rebellion living in mortal bodies). For he must reign, till he hath put all enemies under his feet" (I Corinthians 15:24,25).

MILLIONS OF MISSIONARIES

The Holy Spirit will prevail throughout the world during the Millennium (Joel 2:28-32; Ezekiel 36:25-27), and this will be the first time in history when there will be a universal knowledge (not salvation) of the Lord (Isaiah 11:9; Habakkuk 2:14; Zechariah 8:22,23). The missionaries in natural physical bodies all will be Israelis (Isaiah 2:2,3; 40:9; 52:7; 66:18-21; Micah 4:2,3; Zechariah 8:23). It will be the first time that the Lord's commission to evangelize every creature will be fulfilled to the letter: "And he said unto them, Go ye into all the world, and preach the gospel *to every creature*" (Mark 16:15). The only religion permitted to exist will be the gospel exalting Jesus Christ according to Malachi 1:11.

Getting the gospel to every *nation* is another essential. This is the mission of the Church and when it is fulfilled it will serve as a sign of the approaching pre-Tribulation Rapture. When Jesus was asked by the disciples: ". . . and what shall be the *sign* of thy coming?" (Matthew 24:3), the response was this: "And this gospel of the kingdom shall be preached in all the world *for a witness unto all nations:* and then shall the end come" (Matthew 24:14). The word "end" refers to the time that the post-Tribulation Rapture takes place. This will terminate the period of the Church Age spanning from Pentecost to the sounding of the "last" trumpet (Revelation 14:14-16; I Corinthians 15:51-54; Daniel 12:1).

RESTORATION OF JERUSALEM

Reconstruction of the city of Jerusalem will be the first project on the agenda. The fantastic size of this Millennial city is described in Zechariah 2:1-3; the beauty of the Millennial Temple is portrayed in Ezekiel, chapters 41 through 48.

The city of Jerusalem will be an enclave of small rural-type communities: "Jerusalem shall be inhabited as *towns* without walls for the multitude of men and *cattle* therein" (Zechariah 2:4). There will be brond rivers and streams serving the intercity communities (Isaiah 33:21). Although navigable, the rivers will offer no threat of invasion by hostile naval forces.

It appears that each of the Royal Redeemed will reign over individual rural-style communities. Some will have greater rule, according to Luke 19:17-19, while other members of the Royal

Family will have charge over lesser community functions (Matthew 25:21-23). Some may even be teachers to assist the populace in learning the new universal language (Zephaniah 3:9): "... Yet shall not thy teachers be removed into a corner any more, but thine eyes shall see the teachers (highly esteemed)" (Isaiah 30:20). Why? To understand the words of the Lord: "And thine ears shall hear a word behind thee, saying, This is the way, walk ye in it..." (verse 21). The Holy Spirit will aid as the Divine Tutor and Counsellor (John 14:26; 16:13), assisting both the student and teacher (Isaiah 32:15; Habakkuk 2:14). So indeed, there is yet to be a more perfect language for all of the people on the earth. One proof that the Millennial Earth will become the New Earth is that Jerusalem and other cities in their final restored beauty will not be destroyed: "... thine eyes shall see Jerusalem a quiet habitation... *that shall not be taken down... (nor) ever be removed"* (Isaiah 33:20).

THE MILLENNIAL TEMPLE

The Millennial Temple will have to be built on Mount Moriah, which is the old Temple site, known in the Bible as Mount Zion. Moriah means "God provided." This Temple must be distinguished from the Tribulation Temple, which Antichrist pollutes, and to which Christ returns when He comes to the earth the second time: "... The BRANCH (Jesus Christ)... he shall build the temple of the LORD... and shall sit and rule upon his throne... and they that are afar off shall come and build in the Temple of the LORD..." (Zechariah 6:12-15).

The Millennial Temple will be constructed under the personal direction and supervision of the Lord Jesus Christ according to the specifications outlined by the prophet Ezekiel. Workmen will arrive from all over the world with their families, bringing with them their gold, silver, and precious stones in order to build and embellish the magnificent structure (Zechariah 6:12-15; 14:14; Haggai 2:8,9; Isaiah 16:5; 60:1-22). Much of the cache of wealth will come from the stockpiles left by Antichrist, as well as from the gold and silver idols and icons cast into the streets (Ezekiel 7:19).

When Jesus Christ alights on Mount Olivet, the cherubim, with the glory overhead, will descend upon Jesus as He moves toward Jerusalem. The shekinah (visible) glory will hover over the entire city as He enters the Temple and then will settle over the Temple like a gigantic coronet (Ezekiel 11:22-25; 43:1-5; Isaiah 4:4,5).[1] The eastern gate, through which Christ once entered triumphantly on a little colt, has been closed since 810 A.D. by the Turks because of their concern that the Messiah, Jesus Christ, would come and enter through this gate and take possession of the Temple Mount. It will not be reopened by divine decree until Christ Jesus is about to come back to earth again (Ezekiel 44:2).[2] At this wailing wall Israelis pray for their Messiah to enter the gate, unaware that it will be Jesus Christ (Revelation 1:7).

The Temple to which Christ will return is called "of all nations a house of prayer" (Mark 11:17), and He will sit on the throne of King David (now one of the Royal Redeemed) during the reconstruction of one thousand years. The fact that Christ will sit upon the throne at Jerusalem in the Millennial Temple is clearly predicted in I Chronicles 17:12; 2 Chronicles 3:1; Luke 1:32,33; Psalm 132:11; Joel 3:17. When the earth has been refurbished by Christ, it will become the New Earth and will be transported at the end of the Millennium into the second heaven ("New Heaven"). At that time the New Jerusalem, the Holy City, will descend upon it. Christ will then relinquish the throne to King David and reassume His position beside His Heavenly Father (Hosea 3:5; Jereminh 30:9; Ezekiel 37:24,25). During a considerable portion of this time in the New Heaven, Christ will be sitting as the Magistrate upon the bench at the White Throne Judgment.

The Lord's Manifested Kingdom will be worldwide (Zechariah 14:9; Daniel 7:13,14; Revelation 11:15; Zechariah 9:10; Psalm 72:8-11). The curse of multiple languages, which came as a result of man's rebellion against God, finally will be removed in the Lord's Manifested Kingdom. A perfect universal and international language will be introduced, one free from the vulgarity and corruption that have tainted all languages of the world: "For then will I turn to the people a *pure language* ('clean speech,' *Moffatt*), that they may *all* call upon the name of the LORD, to serve him with *one consent*" (Zephaniah 3:9). There also will be an international flag depicting a huge crown with twelve beautiful gems (Isaiah 5:26; 11:1-12; 2:5; Zechariah 9:16).

MILLENNIAL HOLIDAYS

"These are the feasts of the LORD . . . which ye shall proclaim in their seasons . . . it shall be a statute *for ever* . . . " (Leviticus 23:4,14). The five festivals to be retained are those of Unleavened Bread, Pentecost, Trumpets, Atonement, and Tabernacles, and Jesus fulfilled them all: *"Blotting out the handwriting of ordinances* that was against us, which was contrary to us, and took it out of the way, nailing it to the cross . . . Let no man therefore judge you in meat, or in drink, or in respect of an holy day, or of the new moon, or of the sabbath days: *which are a shadow of things to come,* but the body is of Christ" (Colossians 2:14-17).

The word "for ever" shows these feasts will always be observed, but only as holidays. Christ is the "body" of the ancient feasts and all that remains is the shell or shadow of them. Their full and true meaning was kept only by Jesus Christ: "Think not that I am come to destroy the law, or the prophets: I am not come to destroy, but to *fulfill*" (Matthew 5:17). "Christ *hath redeemed us from* the curse of the law, being made a curse for us . . . " (Galatians 3:13). "Christ is become of no effect unto you, whosoever of you are justified by the law; ye are fallen from grace" (Galatians 5:4). "In burnt offerings and sacrifices for sin thou hast had no pleasure . . . Then said he (Christ Jesus), Lo I come to do thy will, O God. *He taketh away the first,* that he may establish the second (Grace). By the which we are sanctified through the offering of the *body* of Jesus Christ once for all" (Hebrews 10:6,9,10).

PILGRIMAGES TO JERUSALEM

Not everyone will be required to make pilgrimages to Jerusalem on a regular basis. The word "many" in Isaiah 2:2-4 shows this will be on a voluntary basis. However, there are exceptions as seen in Zechariah

14:16-19. Nations in history which have invaded and fought against Jerusalem, and which have survived the Battle of Armageddon and the Judgment of the Nations, will be required to make annual pilgrimages. If they, as nations, fail to comply they will suffer severe consequences, such as lack of moisture and the presence of plagues. The Egyptians and Assyrians will be required to make annual pilgrimages (Zechariah 14:17-19). Assyria will represent the Arabs who accepted Christ as their personal Saviour after the "first resurrection" in the final Rapture of the Tribulation saints.

Egypt will survive because it will have provided sanctuaries for Israel upon its flight from Antichrist (Isaiah 19:18). Egypt, however, will become desolate and uninhabitable for the first forty years of the Millennium as a punishment for its cruelty to Israel in ancient history. Because Egypt held Israel in bondage for 400 years at the time of Moses, God will punish Egypt one year for each ten years of Israel's bondage. How merciful of God!

Egypt will be scattered among the nations in the same way Israel suffered dispersion when it rejected its Messiah (Ezekiel 29:9-12). At the end of forty years Christ will regather the Egyptians into their own land, and here they will remain "a base nation" throughout the Millennium (verses 12-15). Israel will occupy all of the Arab lands as a result of Islam's persecution of Israel and the Christians (Obadiah 17).

A ROYAL RECEPTION

Pilgrims arriving in Jerusalem to worship the Lord Jesus Christ in His Temple will receive a royal welcome. There will be elaborate banquet settings: ". . . the pots in the Lord's house shall . . . be holiness unto the Lord of hosts: and all they that sacrifice shall come and take them . . ." (Zechariah 14:20,21).

CITY TRAFFIC

Since the city will be composed of many rural-style communities with each resident possessing a vineyard, farm, or ranch, it is easily understood why there will be many outstanding, gorgeous horses moving about its streets. These horses presumably will be the white steeds that descended with Christ and the Royal Redeemed to the earth: "In that day there shall be upon the bells of the horses, HOLINESS UNTO THE LORD . . . " (Zechariah 14:20). The metabolism of these heavenly steeds will be such that it will preclude excrementitious litter on the streets. According to prophecy, most, if not all, of the *mortal* horses will be destroyed because they have been a symbol of military aggression throughout history. They will be replaced by the descending white horses: "And I will overthrow the throne of kingdoms, and I will destroy the strength of the kingdoms of the (Gentiles) . . . and the horses and their riders shall come down, every one by the sword of his brother" (Haggai 2:22). "That ye (the vultures) may eat the flesh . . . of horses . . . " (Revelation 19:18).[3]

SAFE CITIES

"In that day, saith the LORD of hosts, shall ye call every man his neighbour under the vine and under the fig tree" (Zechariah 3:10).

There will be great interaction and socializing in the Millennium. This will be due partly to the layout of the cities which will take on the characteristics of a smaller community: "Woe unto them that join house to house, that lay field to field, till there is no place alone in the midst of the earth" (Isaiah 5:8). It is strange but true that one can be lonely in a crowd. Loneliness seems to increase when living quarters are too congested; also crime becomes more prevalent. Pollution and infectious diseases spread more rampantly. Because it creates so many social problems, the apartment-style or condominium-type of living was never God's intent for mankind.

Today there is flight from the cities because they are unsafe and crime-ridden; also, a more healthful life is found in the rural atmosphere.

Of the Millennium we read: ". . . they shall dwell safely, and none shall make them afraid" (Ezekiel 34:28). One reason that the Millennial cities will be safe is that they will have low density populations, thereby eliminating ghettos, slums, or blighted sections. Every family will have a homestead outside of the cities because, in the Millennium, land will be available that is now unsuitable and unproductive for an agrarian society.[4]

"Thus saith the Lord of hosts, There shall yet old men and old women dwell in the streets of Jerusalem (a model of all cities), and every man with his staff in his hand for very age" (Zechariah 8:4). This text also shows that there are natural people during the Millennium.

NO CONGESTION

In God's coming world, the superabundance of land, coupled with the incredible general prosperity (Isaiah 65:21; Amos 8:14), will allow cities to have wide boulevards with an abundance of foliage, immaculate hedges, and shrubbery. There will be no weeds or obnoxious pests because the curse from the ground will have been lifted.

God has always opposed congested areas. This is why the Bible reveals that cities of the future will be places where even children will play *in the streets* (Zechariah 8:5), a clear evidence that there will be areas of open space between structures and between communities within the cities. Moffatt translates it: "open spaces."

TIME MACHINES

There are two kinds of vehicles involved in the Battle of Armageddon. Both are described as chariots—those used by natural men (Nahum 2:1-13) and others used by holy angels (Joel 2:1-11,15; Matthew 25:31).

In Revelation, chapter 19, Christ is on the white charger leading the Royal parade to earth. He is followed by the Royal Redeemed on their white horses: "... let the bridegroom (Christ) go forth of his chamber (first), and (then) the bride (Royal Redeemed) out of her closet" (Joel 2:16). Christ has the "sword" and He will "make war." It is His vesture, and not His Bride's garment, that is stained with blood (Revelation 19:13). She will not be engaged in this conflict for she is neither armed nor in battle gear; she is "clothed in fine linen, white and clean." After the successful conclusion of the Armageddon conflict we read: "... the year of my redeemed is come" (Isaiah 63:4). Then Christ extends His hand of forgiveness to all survivors: "For he said, Surely they are my people, children that will not lie: so he was their *Saviour*" (Isaiah 63:8).

When the Royal Redeemed and the holy angels return to earth with Christ, they will bring with them countless royal horses and chariots. One reason that the Mount of Olives will divide is to provide a great landing strip for the innumerable royal horses and chariots which will descend to earth (Zechariah 14:4). What will happen to all of these royal steeds and chariots? They will comprise the future transportation system during the Millennium and the New Heaven because all of the previous modes of transportation will have been destroyed, including the mortal horses: "And I will cut off . . . the horse from Jerusalem . . . " (Zechariah 9:10). In Zechariah 12:4 we read: ". . . I will smite *every horse* with astonishment . . . " (see also Psalm 76:6,9; Haggai 2:21,22; Micah 5:10). Yet we read of horses in this city in Zechariah 14:20.

In Joel 2:1-11 we see victory for the holy angels in their chariots, as opposed to the wicked forces of Antichrist and the False Prophet. Every war machine will be destroyed and all wicked men will be pursued and ferreted out of their hiding places. These war machines and mobile units will be converted into farming implements and useful products during the Millennium (Isaiah 2:4; Micah 4:3). God predicted: ". . . I will *burn her chariots* into smoke, and the sword (of the Lord) shall devour . . . and the voice of thy messengers (of Antichrist) shall no more be heard" (Nahum 2:13). It will be a bloody battle, as foretold in Nahum 2:3, where Nineveh served as a prophetic example of the last days: "The shield of his mighty men is made red, the valiant men are in scarlet . . . . "

These royal horses and chariots of the Royal Redeemed offer fast, efficient (fuelless), reliable, safe, and comfortable transportation on the ground and in the air. No wonder we read that children and senior citizens in the streets have no fear of being trampled upon and killed.[5]

THE DOVE TRANSIT LINES

The *Royal Redeemed* will have an unusual mode of transportation which exceeds the speed of light and instantly brings "time" to the diminishing point. Speed reduces time, thus beyond the third dimension one's travel is so rapid that time is irrelevant.

The natural people in the Millennium will have the challenge of accelerating their modes of transportation: "Who are these that fly as a cloud, and as the *doves* to their windows?" (Isaiah 60:8). The dove is a symbol of peaceful usage of the system. This is a prophecy of man's ability to produce transportation of superior speed during the Millennium. However, a major breakthrough must be made.

There are four fundamental forces in the universe. They are gravitation, electromagnetism, and weak and strong nuclear interaction forces. Albert Einstein, in his Unified Field Theory, sought to connect all four forces and reduce them into an understandable equation or formula. He was unsuccessful in achieving this breakthrough before he died. Perhaps God delayed this for the time of the Millennium.

One reason that Einstein knew that the Unified Field Theory was attainable is that it exists in hyperspace. It converts electromagnetic energy to gravitational force, which is similar to converting matter into energy. This breakthrough in the Unified Field Theory could achieve a new frontier in space travel to other planets. Such a propulsion device would have speeds of up to 24,000 miles per second, which is thirteen percent of the speed of light. Mass cannot retain its structure safely beyond this speed. This vehicle would receive its energy by osmosis directly from the environment in space.

In the Millennium there will be a blending of royal transportation and human transportation because there will be the incentive for faster and better modes of travel. High-speed land trains and super-speed private cars, such as magni-levitation ones, may conceivably become commonplace. Under Christ's reign we can expect to see this accomplished, together with many other wonderful advanced breakthroughs in science. The Millennium will be the preparation of Planet Earth to become the New Earth in order that it may occupy strategic place in the New Heaven. Another objective of the Millennium is the completion of the apprenticeship of the Royal Redeemed prior to their expanded services as kings, priests, lords, and judges in the New Heaven.

OLD GIVES WAY TO THE NEW

We read in Nahum 3:3 that the holy angels in their royal chariots will soar up to Christ as He clutches a "sword" in one hand and a "spear" in the other. They will hold up His hands in a gesture of (V)ictory over all of the enemies: ". . . the bright sword and the glittering spear . . . ."

The results will be astounding: ". . . the sword shall cut thee (the enemies) off . . . " (verse 15).[6] The "sword" is used in close combat, but the "spear" is used for distance. Like streaks of lightning, the destructive power of the spear will penetrate from great distances into hidden places where the ungodly with the mark of the Beast have fled from the revealed Christ (Revelation 6:12-17; Luke17:30-37).

Everywhere that the holy angels go they will employ a scorched earth policy: "A fire devoureth before them; and behind them a flame burneth: the land is as the garden of Eden before them, and behind them a desolate wilderness; yea, *and nothing shall escape them*" (Joel 2:3). The terrible searing of the flames is described: "Before their faces (of the holy angels) the people shall be much pained: *all faces shall gather blackness*" (verse 6). In Isaiah 24:6: ". . . the inhabitants of the earth are *burned,* and few men left."

When the battle for the possession of the earth is over and the Judgment of the Nations is completed, the Millennium will begin. What a change will take place: "And I will multiply upon you man and beast (large numbers on ranches); and they shall increase and bring fruit (great orchards): and I will settle you after your estates, and will do better unto you than at your beginnings (addressed to the primordial earth): and ye shall know that I am the Lord" (Ezekiel 36:11).

LAND GRANTS

One of the liabilities of young people today is their inability to purchase homes, farms, orchards, or businesses to get their start in life, unless someone is capable and willing to financially assist them. What a delight it would be if every ambitious young person could get this boost! In former days, in many countries, it was possible for the younger generation to assume its parent's farm, business, and home, but in our modern society it has often become unfeasible. Under Christ Jesus as the "Lord of lords," and His Royal Redeemed as "lords," every person in the Millennium will be given an opportunity to get his start under the laws of righteousness.

IRON HAND OF RIGHTEOUS LAWS

Everyone will obey Christ's laws, which will issue from Jerusalem, the world's new governmental and ecclesiastical center. Knowledge of these laws will become universal and those determined to violate them will receive prompt administrative justice under the Royal Redeemed who are serving as "judges" to enforce the laws (I Corinthians 6:2; Revelation 2:26,27).

Those who remain unconverted to Christ will not survive beyond the age of one hundred years (Isaiah 65:20). In Ezekiel 36:25-27 we see how the newly redeemed persons will be changed: "A new heart also will I give you, and a new spirit will I put within you: and I will take away this stony heart out of your flesh, and I will put my spirit within you, and cause you to walk in my statutes, and ye shall keep my judgments, and do them."

Under the dispensation of human government, before the law of Moses, we read: "And surely your blood of your lives will I require: at the hand of every beast will I require it (animals even incited to violence), and at the hand of man; at the hand of every man's brother will I require the life of man. Whoso sheddeth man's blood by man shall his blood be shed: for in the image of God made he man" (Genesis 9:5,6). This same principle will be enforced during the Millennium: "Violence shall no more be heard in thy land, wasting nor destruction within thy borders..." (Isaiah 60:18).

SOCIAL LIFE

"...My cities through prosperity shall yet be spread abroad... (Zechariah 1:17). Cities are absolutely necessary for a good cultural life. Concert halls, museums, art galleries, sports arenas, restaurants, playhouses, clubs, and libraries all require centralization for easy accessibility.

These Millennial cities will be constructed worldwide, beginning with the rebuilding of the cities of Judah, which will have been largely destroyed during the judgments of God and Armageddon. The construction crews will be natural people under the complete guidance and supervision of the Lord and His Royal Redeemed, who are "lords" under Him. These builders will construct edifices in which God's concern for beauty, harmony, utility, and efficiency will prevail; they will be as sparkling jewels and they will continue to exist into the New Heaven. No less than eighty-seven percent of the population today lives in the cities. Perhaps in the Millennial agrarian society, the percentage will be reversed.

At the beginning of the Millennium the world's population will be proportionately small. There will have been much loss of both human and animal life, but this is now "the time of the restitution of *all* things" (Acts 3:21). A rapid population explosion is attributed to the longevity of life because of a decrease of premature deaths caused by abortions, wars, accidents, diseases, famines, natural disasters, and violence.

Although the earth is currently inhabited by an estimated 1.4 million species of animals and 250 thousand species of plants, they represent only one percent of those which it is believed have ever existed on the earth. The rest are extinct. This also confirms that the earth had a primordial existence, reaching far beyond the general consensus of only a few thousand years.

An "increase" certainly includes restoration of *many* extinct animal and plant species which will constructively fit into the new environment. In the New Heaven we can expect to see *all* of the destroyed creatures and plants restored into habitable places suitable for their required environments. These areas no doubt will serve as interesting sights for the inhabitants to observe and appreciate the wonders of all of God's creations (Isaiah 11:7; 65:25).

[1] The first Dome of the Rock (Moslem Mosque) was erected on the Temple Mount in 688 A.D.

[2] This eastern gate probably will be reopened by the time the Tribulation Temple is dedicated.
[3] Natural transportation, according to Isaiah 34:9,10, is unlikely to require oil for mobility during the Millennium. See footnote on page 284.
[4] In 1982 a team of scientists found grain and barley in Egypt's Nile Valley. Radiocarbon dating has established them to be 4,800 years old. This area is believed to be the origin of agriculture. See Genesis 3:17-19; 4:2.
[5] ". . . I will cause thee to *ride upon the high places of the earth . . .*" (Isaiah 58:14).
[6] See page 613.

81
Great Leap Forward

The Millennium is the first generation in the Lord's Manifested Kingdom which will extend for one thousand generations.

Man's creative abilities distinguish him from all other creatures: "And God said, Let us make man in *our image* after *our likeness* . . . So God created man *in his own image,* in the image of *God* created he him; male and female created he them" (Genesis 1:26,27). "But there is a God in heaven that revealeth secrets . . . " (Daniel 2:28).

Man's astounding abilities to create identifies him with his Creator. This was first apparent at the tower of Babel: "And the Lord said, Behold . . . now nothing will be restrained from them, which they have imagined to do" (Genesis 11:6). Moffatt translates it, " . . . if this is what they do, to start with, nothing that they ever undertake will prove too hard for them."

This is echoed in Job 32:8: "But there is a spirit in man and the inspiration of the Almighty giveth them understanding." Man, besides being soul and body, is also "spirit" like his Creator (I Thessalonians 5:23; Acts 7:59; John 4:24). This is the "spark" from the Almighty that gives man an intelligence that is superior to that of other creatures on earth. For instance, man is capable of inventions enabling him to reach into heaven; to wit, he has now landed on the moon.

During the Antediluvian period from Adam to the Flood, which comprises about one thousand years, mankind established a highly developed society and culture. He made great strides in agriculture and the arts and improved his standard of living (Genesis 4:20-22). Unfortunately, he omitted God and became an inventor of evil and destructive devices (Romans 1:30; Proverbs 8:12; Ecclesiastes 7:29; Psalm 106:39). One reason that the Flood came upon mankind was to prevent the destruction of the entire human race, which would have abolished God's plan of redemption, much to Satan's delight.

FROM ADAM TO ATOM

The machine age, the industrial revolution, and the nuclear, space, and electronic ages are evidences of man's recent ingenious inventions. The Millennium will bring man's creativity and development into even more complete harmony with nature under peaceful rules established by Jesus Christ, the world's final Ruler. The Millennium will lay the final groundwork for the earth to later fulfill its role as the New Earth within the New Heaven.

Man is presently at the stage in which he has the capacity of destroying the human race. The foregoing events probably could have happened shortly after Babel if God had not intervened to slow man's progress. Babel was the door that could have proven to be the breakthrough into science and technology.

God retarded the progress of man from about ninety percent to a snail's pace. This was accomplished by creating multiple languages, races, continents, and islands, thereby confusing and scattering man upon the earth. God again will intervene in man's progress, but this time after mankind has been almost annihilated (Matthew 24:22).

As of this writing, ninety percent of all scientists who have existed are alive.[1] There has been a proliferation of scientists, scientific laboratories, and resultant new products flooding the marketplaces. It has been approximately three hundred seventy years (1611) since Galileo pierced the heavens with his six-power telescope and opened the door to modern astronomy. Before Thomas Edison's day there was serious concern about closing the United States Patent Office in the belief that there was nothing new to invent! Since then, science has advanced at lightning speed, with the Patent Office being deluged with new ideas.

Imagine how much greater our progress would have been if all scientists were engaged in working for the benefit of humanity. Instead, approximately fifty percent have been involved in warfare products and the deployment of destructive devices. One-half trillion dollars have been expended in the development of warfare throughout the world. If all of these resources and energies could have been channeled to solve man's problems, it is reasonable to assume there would now be cures for most, if not all, diseases. Man also could have eliminated pollution and dangerous contaminants. He could be living in a veritable paradise. However, man's prejudices, greed, and evil inventions have crippled his full potential for the good of mankind (Ecclesiastes 7:29; Romans 1:30). This does not take into account the loss of lives and ideas because of wars and superstitions of the past.

Think of the horrendous waste of financial resources because of the military demands. At this writing, the world is spending 100 billion dollars annually just for stockpiling nuclear weapons. They have produced more nuclear warheads (over 60,000) than they have targets. The United States, the Soviet Union, and China lead the world in military spending. A nuclear war would terminate an estimated 2 to 3 billion lives.

THE MIRACLE MICROCHIPS

Microchips are thin squares of silicon, about a quarter of an inch in size, consisting of circuit mazes crammed with thousands of bits of information. Most chips fall into two categories—memory and random-logic functions. Their integrated circuits simulate aerial views of cities with streets and skyscrapers, including a few structures in a state of conflagration. Others suggest industrial areas and parks.

Microchips form the thinking and remembering sections of the electronic brain. A single two-dimensional micro-processor chip today contains over 100,000 transistors and circuit elements with multiple bits of information. A gallion arsenide chip can now be etched with features down to 0.2 microns, which is thinner than the wavelength of light itself.

Plans call for a three-dimensional chip that can manage over 10 million binary transactions. A new hybrid electro-optical chip can contain more bits of data and operate at greater speed than is possible with the silicon semi-conductor chip. It operates at speeds in the trillionth of a second, but this is not fast enough for the future. The latest super-computer called SX-2 can do 1.3 billon calculations in a second. Plans are under development for chips from compounds of arsenic, gallium, indium, and antimony inasmuch as electrons travel faster in these materials.[2]

REDESIGNED ECOSYSTEMS

Efforts are now underway to shrink compounds below the size of most living cells. Parts presently in use are as little as two microns (thousandth of a millimeter) in width. Experimental compounds are expected to slice a micron in half. Microlithographers deal in measurements called angstroms. A thousand angstroms are one-tenth of a micron and a micron is a quarter of the thinkness of human hair.

Integrated circuits, a fraction of an inch square, contain tens of thousands of transistors which are etched on a single wafer. A chip capable of carrying over one million bits of information is now a reality. Today's circuit patterns are etched on chips by photolithography. Eventually engineers will use the x-ray etching techniques.

The tiny mini-micro memory chip is about to be placed into a slimline card just 1/30,000 of an inch thick so it can be handled and inserted into future terminals and other proposed sophisticated devices. Technicians envision the day when millions of memory facts will be placed in a single mini-microchip, thereby terminating massive paper work, taping, and microfilming. Every facet of human life will be affected. Libraries simply will be a repository of microchips inserted into a plastic holder for convenience of handling and for positioning into readout machines or duplicating devices for the reader's use.

The electronic microminiaturization revolution that is emerging will be incorporated during the reign of Christ on earth. He no doubt will adapt and utilize all scientific technological advances to improve society's conditions for the natural people as well as those in the New Heaven. This will demonstrate the full capability that God has implanted within man and why the Scriptures have called man a "god" (Psalm 82:6; John 10:34).

MICROBIOLOGY

There is now on the planning board a biochip that can be inserted into the human brain for the purpose of restoring sight, hearing, and damaged spinal nerves. It also will aid memory. These organic circuits will have a measurement less than one-hundredth of a micron in width, compared with a human hair that is about sixty-two microns wide. These biochips will be used with a biocomputer.

The human brain weighs three pounds, burns electricity at a rate of about twenty watts, and occupies a volume of one-tenth of a cubic foot. Within this small space is held a memory of ten billion items of information.

By 1995 the silicon brain should cross the line of human capacity—ten billion facts in a briefcase, operating on twenty watts. Computers even then will lack the creative inspiration and intuition that characterize the higher realms of human thought imparted by the Creator and that are likely never to find their way into a computer.

Every human brain cell is connected directly to many other brain cells; in some cases, as many as 100 thousand. The makeup of a computer is quite different. The memories are like a set of pigeonholes stacked against a wall, with no connection from one pigeonhole to another.

During the Millennium all restraints and hindrances will be removed and man's scientific mind will be released for inventions that are to bring solutions to many problems that now plague mankind.

[1] About one million scientists are alive today.

[2] Only mass cannot exceed the speed of light.

82

The Ages to Come

The joy of living is accepting challenges and seeing the fruit of accomplishments. Idleness is the route to boredom and a sense of uselessness.[1]

"That in the ages to come he (God) might shew the exceeding riches of his grace in his kindness toward us through Jesus Christ" (Ephesians 2:7).

". . . Knowledge shall be increased" (Daniel 12:4).

Unlimited changes will take place during the Millennium. Satan will be bound, so there will be no tempter (Revelation 20:2) to inflame the passions of mankind into violence. Sin will be in the human heart but will be more passive and subdued. Universal peace will prevail, thereby eliminating all military forces and armaments (Isaiah 2:4; 9:6,7; Micah 4:3,4).

Social interaction will be peaceful. There will be full justice of the law (Isaiah 11:3-5; 57:16). There will be worldwide prosperity (Isaiah 65:22; Micah 4:4,5). One common language will be taught and spoken (Zephaniah 3:9); and there will be but one flag (Isaiah 11:12).

Human existence will be transformed in various ways: life will be prolonged (Isaiah 65:20-22; Zechariah 8:4); no one will be sick (Isaiah 33:24); there will be no blindness, deafness, or disabilities

(Isaiah 35:5,6). There will be changes in the animal kingdom as well (Isaiah 11:6-8; 65:25).

As an agrarian society, all nations will be able to supply their own food during the Millennium. At *present* one hundred countries depend solely upon the United States and Canada for grain shipments. Thirty-four percent of American crop land, however, is declining because of excessive loss of topsoil. Losses in productivity have been marked by the increased use of fertilizers which has risen from 14 million tons in 1950 to 133 million tons in 1980. During the *Millennium,* deserts (which represent a land area equal to the size of the Western Hemisphere), will become productive (Deuteronomy 11:13-15; 28:11,12; Joel 2:21-26; Amos 9:13-15; Isaiah 35:1,2; Ezekiel 36:8-11; Zechariah 8:12). This will be as a result of the water canopy above the earth, which also will increase the amount of light (Isaiah 49:10; 30:26; 60:19-22; Zechariah 14:7; Ecclesiastes 12:2).

A MILLENNIUM OF RECONSTRUCTION

There will be one thousand years of reconstruction and development in preparing the world for its role as the New Earth. The cities of the future will be decent, joyful places in which to live and work because of the absence of blight, slums, crime, and social unrest. There will be magnificent architecture, involving floating megastructures of many intriguing designs such as tetrahedral (pyramid style), sphere, arcosanti, geodesic, and biomorphic biosphere (whose huge footings touch the ground). There will be educational opportunities, cultural events, sports, and inspiring entertainment. Christ will accomplish all of this and much more during His administration.

Our world is already making rapid scientific changes, but the greatest progress lies in the Millennium, as knowledge feeds upon itself. The Pangaea will eliminate the need of intercontinental travel across oceans, so travel time will be greatly reduced. There will be asteroid colonies, with regular shuttle service to the space stations as floaters, where scientific research will make astounding breakthroughs. When the Planet Earth becomes the New Earth located in the New Heaven, there also will be transplanetary means of communication and interplanetary travel between the habitable planets. Think of the possibilities in the scientific field when those who embrace Christ as their Saviour will be able to pursue their research beyond the present limitations of life's existence. They will live on through the Millennium and into the New Heaven to continue their accomplishments. The redeemed will live in physical bodies for one million years through the Millennium and the entire dispensation of the New Heaven until death is destroyed, when they will receive their bodies of immortality: "... the people that dwell there shall be forgiven their iniquity" (Isaiah 33:24b). "*All* the inhabitants shall not say, I am sick ... " (Isaiah 33:24a). "For God giveth to a man that is good in his sight wisdom, and knowledge, and joy ... " (Ecclesiastes 2:26).

THE MILLENNIAL TRANSFORMATIONS

Three major changes will involve Planet Earth as it prepares to become the New Earth: 1) *redemption* of the earth (Romans 8:19-22), including its climate, vegetation, and topography; 2) *regeneration* (Matthew 19:28), establishing society under a new government; 3) *restitution* (Acts 3:21), consisting of rebuilding of structures, beautifying of the earth's surface, reviving of many extinct creatures and plants that will complement the new environment, and changing the natural habits of all living creatures.

It will take one thousand years to prepare the planet for its primary role as the New Earth. It will then leave its present orbit and vacate the planetary heaven before the cosmos is destroyed, to be relocated in the New Heaven where it will serve as the central capital of future universes.

During the Millennium the Royal Redeemed will serve their apprenticeship as kings, lords, priests, and judges. It will be a time of forgiveness of the penitent (Isaiah 33:24). Their services will not be confined to that of kings, and their territory will be expanded throughout the New Heavens in the *future* Kingdom of the Heavenly Father beyond the New Heaven.

A MILLION YEARS FROM NOW

The Millennium, which consists of one thousand years during which major changes will occur, is also called a "generation" (Psalm 22:30). It is only one generation in a series involving the occupation of the earth: "The Lord shall reign for ever . . . unto all generations" (Psalm 146:10).

". . . His kingdom is an everlasting kingdom . . . from generation to generation" (Daniel 4:3,34).

The Millennium is the first generation in the Lord's Manifested Kingdom which will extend for one thousand such generations (Deuteronomy 7:9; Psalm 105:8; I Chronicles 16:15). It will take that long to finalize the accomplishments under the plan of redemption: "Then cometh the end (for all of those registered in the book of life to be redeemed and to come to the end of their mortality) . . . " (I Corinthians 15:24). Christ then will have successfully "put all enemies under his feet" (verse 25). Death will be destroyed.

SUPER-CIVILIZATIONS

The redeemed natural people, who will emerge from the Millennium, will not be able to live comfortably on the New Earth in their natural bodies where there is no night (Revelation 21:25). Their mortal bodies will require rest, so they will be transferred into habitable regions in the far reaches of the New Heaven under solar systems compatible for them to live normal, productive, and happy lives. Here there will be super-civilizations making fantastic scientific progress in the interplanetary systems of the New Heaven. It is obvious that such an explosive population, over a period of a million years, never could be accomodated on the New Earth. That is the reason for the New Heaven. The New Earth will be adequate only for the Royal Redeemed and the visitors who will be making pilgrimages to the Holy City, Jerusalem.

The Royal Redeemed, as "royal priests" (I Peter 2:5,9) in their bodies of immortality, will minister spiritually in the far-flung habitable reaches of the New Heaven. This ministry will be terminated when the million years conclude and the final person is redeemed. The next transfer will be to the Heavenly Father's Manifested Kingdom (I Corinthians 15:24). At that time death will be destroyed and all of the redeemed will have immortal bodies to fit into the dispensational environment in the heavenly Father's Manifested Eternal Kingdom called the "New Heavens."

The human beings who are living in natural bodies in their respective solar systems in the New Heaven will have the capacity of transforming their places into veritable paradises as a result of their God-given talents: "But there is a spirit in man: and the inspiration of the Almighty giveth them understanding" (Job 32:8).

The purpose of the Millennium, as the first generation of the projected one thousand such generations (one million years), is to transform the Planet Earth into the New Earth under the reign of Jesus Christ. Once it becomes the New Earth, and is transferred near God's throne as His footstool, further development will continue into the extended solar systems in the remote regions of the New Heaven.

Imagine the tremendous leap forward in scientific progress! Our wildest imagination can never fathom that which will be unfolding under the peaceful conditions of one thousand generations with a longevity of one thousand years each. The spark of man's intuitiveness and the ability to achieve his visions is illimitable.

The joy of living is accepting challenges and seeing the fruit of accomplishments. Idleness is the route to boredom and a sense of uselessness.

[1] See chart on page 632.

83
The Flight of Spacecraft Earth

All of the inhabitants, who are redeemed but in natural bodies, will be the forebears of natural people in the New Heaven.

Peter first made reference prophetically to the Antediluvian earth in this manner: "... the earth standing *out* of the water ..." (2 Peter 3:5), and compared "... the longsuffering of God (that) waited in the days of Noah ..." (I Peter 3:20) to the longsuffering of the Lord: "... longsuffering to usward, not willing that any should perish, but that all should come to repentance" (2 Peter 3:9).

Second, Peter referred to earth as being "... *in* the water ... (that) perished" (2 Peter 3:6). It disappeared under the Flood but later emerged cleansed of the works of men in a changed planet. On this premise Peter predicted that the earth will disappear again, but this time from its place in the universe, and will emerge as the "New Earth" in the "New Heaven" (verse 13).

As in the days of Noah when those within the ark were sealed in by God, likewise, one day He will protect Spacecraft Earth and its passengers as they take off on their long journey to the second or New Heaven.

THE SCARLET SKY

"... When it is evening, ye say, it will be fair weather: for the sky is red" (Matthew 16:2).

As the wicked perished in the Flood, so at the end of the Millennial period, the wicked who are led by Satan in a revolt against Jerusalem and its inhabitants, will also perish: "... and fire came down from God out of heaven, and devoured them" (Revelation 20:9). Their ashes will lie on the ground to be walked upon by the saints (Malachi 4:1-3). This final destruction of the wicked upon the Planet Earth is called "the day of God." All of the habitations previously occupied by Satan and his angels in the planetary heavens also will be destroyed (2 Peter 3:12). This "day of God" must be distinguished from "the day of the Lord" which refers to the time of the final judgments upon the last Gentile kingdom under Antichrist—Babylon.

EARTH'S SPACESUIT

"... The windows from on high are open, and the foundations of the earth do shake... the earth is moved exceedingly... and shall be removed like a cottage..." (Isaiah 24:18-20).

Prior to the take-off, on its long journey through the universe, Spacecraft Earth will be sealed hermetically in a huge bubble by the Glory of God to protect it and its natural inhabitants from the dangerous environment through which it must pass, and to survive travel at and beyond the speed of light; otherwise, according to Einstein's theory of physics stating $E=mc^2$, mass would be converted into pure energy. The Spacecraft Earth must have divine protection.

The trajectory of the spacecraft will be toward the second or New Heaven. All of the inhabitants, who are redeemed but in natural bodies, will be the forebears of natural people in the New Heaven. The safe passage of these natural passengers has its precedence in Enoch, Elijah, the Hebrew children in the fiery furnace, and even

Philip's physical transportation from one place to another (Acts 8:39).

THE COSMIC CORRIDOR

The spacecraft will take off in the direction of the North Pole (Psalm 48:1-3; Job 26:7; Ezekiel 1:4), which is toward the Throne of God (The universe rotates on its axis with its north pointed toward God's Throne as do the stars and planets). When the spacecraft arrives at its destination in the New Heaven, the Throne of God in the Holy City will come down directly from the third heaven onto the earth.

Since we do not know the exact route the spacecraft will traverse to its new site, we shall present an imaginary one. This will help us to better understand some facts in our study of God's Word.

All is in readiness. Christ is on Spacecraft Earth at the controls of His throne room in Jerusalem. The Heavenly Father is at the command post of His throne room in the New Jerusalem. The final countdown commences.

Excitement mounts as lift-off time approaches. All is A-OK! The powerful thrusts of the Holy Spirit's firepower cause the huge spacecraft to vibrate. There is a sense of incredible power as the spacecraft slowly moves upward and gains momentum. It makes its final elliptical orbit around the sun, and then as a new surge of Spirit-power is felt, it heads into deep space. Jesus Christ, as Captain, is at the helm. He wants His passengers to catch final glimpses of that which lies south of the cosmos in the depth of eternity.

JOURNEY FROM DARKNESS

"For the LORD God is a *sun and a shield*..." (Psalm 84:11).

The spacecraft is like a domed planet brilliantly illuminated because of the presence of the Lord Jesus Christ (Revelation 21:23). It speeds on through the cosmos. When it veers away from the radiance of the sun, the blackness of space surrounds the magnificent ship.

As the spacecraft leaves the black sphere of the cosmos, it enters eternity's expanse where the original light is its normal environment. Space scientists are amazed that the superluminal motion of quasars appears to be faster than the speed of light coming into our cosmos. Does this suggest that the light which comes into our universe creating these quasars is the original light from the habitation of eternity beyond our universe, and not photon particles?

The voyagers begin to see in the far distance something that appears as a vast, dark, ink-like blot in the brightness of eternity's spread. As they approach, they see that this blackness serves as massive prison walls for incarcerated beings.

EDGE OF DOOM[1]

Christ informs His passengers that this is the place called "outer darkness," where the disobedient children of the Kingdom of Heaven are incarcerated as a disciplinary necessity. After they have completed their sentences, they will go to the White Throne Judgment. At that time they will be reunited with their physical bodies in a resurrection to appear before the Lord Jesus Christ (John 5:27-29; Daniel 12:2).

Since the names of those in "outer darkness" still remain in the book of life (I Corinthians 3:15; Jude 23), they will go on to the New Heaven in their mortal bodies (Matthew 10:28) to await the time when death will be destroyed and they, too, like the Royal Redeemed, will receive bodies of immortality.

Unfortunately, they missed the "first resurrection" since they were not clad in wedding garments (Matthew 22:11-14; I Corinthians 5:1-11; Revelation 16:15). Therefore, they cannot qualify for a body of immortality or be in the corporate Wife of Christ. Their loss, after their rehabilitation, is enormous and is permanent throughout eternity; that is, they never can be in the corporate Wife of Christ, nor ever reign as kings, priests, lords, and judges, and they already have missed the Millennium.

These children of the kingdom in "outer darkness" will be filled with deep remorse, as evidenced by "weeping and gnashing of teeth" (Matthew 8:12; 22:13; 25:30). "Thou feedest them with the bread of tears; and givest them tears to drink in great measure" (Psalm 80:5). But after their rehabilitation we read: "And God shall wipe away all tears from their eyes; and there shall be no more . . . sorrow, nor crying . . . " (Revelation 21:4). This must be distinguished from Revelation 7:14 which refers to the last day Tribulation saints.

But simply to wipe away tears is inadequate. The haunting memories of what they have endured and of what they have missed not only could spoil eternity for them, but also it could become an unbearable burden. Therefore, God says: " . . . the former things are passed away" (Revelation 21:4). And in Isaiah 65:17: " . . . and the former shall not be remembered, nor come into mind." What tragedy if God were not so merciful!

BOTTOMLESS PIT[2]

The spacecraft continues further into the depths of eternity's bright expanse. In the distance can be seen an ominous sight which appears to be a huge cylinder suspended in an upright position with a top and bottom that fade away into endlessness. Christ informs the passengers that this is the "bottomless pit" where fallen angels are being held until the time of their judgment, and that the Royal Redeemed will serve some day in the royal jury box in judgment over these angels (I Corinthians 6:3).

BELOW OUTER DARKNESS

The location of the "bottomless pit" is below the place called "outer darkness," as is evident from Jude 6: "And the angels which . . . left their own habitation, he (God) hath reserved in everlasting chains *under darkness* . . . ." This place cannot be located in cosmos for that is scheduled for conflagration before the White Throne Judgment is set up in the globe of darkness left vacant by all of its galaxies (Revelation 20:11).

What kind of place is this Stygian darkness? 2 Peter 2:17 reads: " . . . to whom the *mist of darkness* is reserved for ever." Moffatt translates it: " . . . for whom the *nether gloom of darkness* is eternally reserved." The word "for ever" tells us this place will remain throughout the future as an eternal reminder and warning to all holy angels. In Colossians 2:15: " . . . when he (God) cut away the angelic Rulers and Powers from us . . . " (*Moffatt*). As a result, they must appear at the White Throne Judgment before the Royal Redeemed who are the royal jurors (I Corinthians 6:3).

TERMINAL ISLAND[3]

Spacecraft Earth makes an even deeper penetration into the expanse of eternity. Soon the travelers are greeted by a strange and eerie sight, which appears like a field of raging flames shooting upward in the light of eternity. Scripture describes it as a "fire not blown" and its occupants are "wailing and gnashing their teeth." These expressions only can be an *attempt* at conveying the tragedies of those incarcerated there. Obviously, fire *per se* would destroy the flesh and bones (including teeth). Whatever it is can best be described in the Scripture which defines what Satan will endure in this place: ". . . a violent death shall you die, there *by the deep* . . . you shall die *cruel death* . . . your fate is awful" (Ezekiel 28:8, *Moffatt*). The word "death" in Scripture means not annihilation, but separation.

The word "death" conveys the idea that there is separation, not only *from* one's "spirit" being, which is that closest affinity to God; but also a separation *from* what might have been; *from* any hope that conditions will improve, or that there is a way out. What is the ultimate? "Death" comes from the Greek word meaning "to lose, mar, or perish," which is the origin of the idea of a "lost" soul. Everything has been obliterated; the spirit has returned to God so there is now *no* God-consciousness: there are *no* high ideals but there is a total degeneration of the person himself as the sentence is imposed. Death is being totally cut off from the past and future. If there is any suffering beyond this, one never would want to find out for himself.

BACKDOOR OF HEAVEN

The aforementioned three places of incarceration (outer darkness, bottomless pit, and the lake of fire) cannot be located within the universe for several reasons:

First, Satan, together with the fallen angels, is held in the bottomless pit throughout the period of the Millennium. He then is loosed for a season and finally cast into the lake of fire *forever*. Since the cosmos will be destroyed, leaving only emptiness, the lake of fire will have to be located elsewhere.

Second, *no judgment* will have been made concerning those incarcerated in the bottomless pit during the time of Satan's confinement, so the judgment of those detained there would have to take place after the destruction of the planetary heavens.

Third, Christ and His saints are to *sit in judgment* over the fallen angels (I Corinthians 6:3; Matthew 8:29). This possibly cannot occur until after the Millennium and the destruction of the universe.

Fourth, according to Jude 6, the bottomless pit will remain *forever*. Therefore, it cannot be destroyed during the conflagration of the interplanetary systems when all that remains is a vacuum.

Fifth, *no other judgment* is indicated in Scripture except the White Throne Judgment, which is to be set up in the vacant area left by the destroyed cosmos (Revelation 20:11).

Sixth, the earth must be returned near God's Throne as His footstool. This requires that the earth make a *successful escape* from its present location before the destruction of the universe.

Seventh, the sun cannot be the eternal lake of fire as some theologians would have us believe because *it, too, will be destroyed* along with the rest of the cosmos.

Eighth, the bottomless pit could *not be bottomless* if it were located within the earth. It would have limitations and it then would be all bottom.

Ninth, hell within the earth must remain *the holding place* for the ungodly human beings until the time of the White Throne Judgment. Therefore, earth must escape the conflagration of the interstellar heavens *if those in its lower regions are to survive* to be judged at the White Throne Judgment.

Tenth, the sea no longer will hide those to be resurrected; they will be revealed for their resurrection (Revelation 21:1), as all of the water will be evaporated. This will occur *after* Planet Earth has become the New Earth: "And the sea gave up the dead which were in it . . ." (Revelation 20:13). The word "sea" (singular) shows that the

Pangaea already has occured. It is the "sea," not "seas," that shall give up the dead. It is at the time of the evaporation of the sea on the New Earth that the day and night cycle ends: "He hath compassed the waters with bounds, *until the day and night come to an end*" (Job 26:10).

The marshy basin where the sea once was located will be filled by the River of Life which flows from the Throne of God onto the New Earth (Ezekiel 47:1-12). The River of Life already exists in the third heaven where God's Throne is located on Mount Sion: "Praise him, ye heavens of heavens, and *ye waters that be above the heavens*" (Psalm 148:4).

[1] See chart U.

[2] From the Greek word, Tartarus.

[3] Gehenna, the valley of Hinnom (Isaiah 66:24) became the synonym or symbol of the lake of fire (Mark 9:44,46,48).

84
Return to Cosmos

The resurrection of the unbeliever will be a reuniting of the soul to a mortal *body, but . . . of the believer . . . a reuniting of the soul to an* immortal *body.*

Enough of these deplorable scenes in the wide expanse of eternity! However, they definitely will serve as reminders to all mortal passengers of the importance of loyalty to God, and this experience will be impressed upon their posterity when they arrive in the New Heaven.

The eternal security of the Royal Redeemed is assured (Revelation 3:5; I John 3:2). The lessons they learn from this detour will be an aid to them in their evangelization as royal "priests" throughout the expanse and dispensation of the New Heaven.

Spacecraft Earth now swings away from these hideous sights and heads towards cosmos, this time taking a different route. It moves toward an immense eddy with a massive black hole at its center. Within this hyperspace there is neither time nor light, but calmness, as in the eye of a hurricane. The passengers observe the black hole as a huge chasm and collector of matter which acts as a smeltering plant to transform the aged and dying systems into molten ore and finally into pure energy.

As the spacecraft enters the white hole[1] beyond the black hole, the passengers see the cosmic laws of physics and chemistry at work producing new matter from the pure energy received from the black hole. These are the cosmic blast furnaces in the circle of cosmos: "His (God's) going forth from the *end of the heaven* (apogee), and his *circuit* upon the ends of it: and there is nothing hid from *the heat thereof*" (Psalm 19:6). The passengers scrutinize the genesis of new elements and products, the embryo of new stars, planets, and other celestial objects. What a sight as they enter and emerge from one black and white hole after another, witnessing the production of new solar and galactic systems! Oh, the wonders of these space factories on such a grandiose scale!

As the spacecraft emerges from one of the white holes;[2] in the universe,the passengers are aware of the contrast of heavenly "things" with those which they knew on Planet Earth before the voyage was launched (Philippians 2:10).

As the passengers pass through one of the white holes (which astrophysicists describe as like the "eye of a hurricane"), they are greeted by a myriad of holy angels who are excitedly and enthusiastically waving to welcome them to their home in the New Heaven. For many voyagers, this is their first glimpse of angels. What a grand and glorious scene!

Some of the stations outside of the white holes have evidences of previous occupation by fallen angels. The Scriptures speak of these as dominions, habitations, and principalities (Ephesians 1:21), but obviously the evil beings are in absentia as they have been evicted by holy angels from the astropremises (Revelation 12:7-9). There is great rejoicing among the holy angels as they proclaim the celestial quarters as liberated: "Therefore rejoice, ye heavens, and ye that dwell in them" (verse 12).

THE LOST WORLD

The "dry places" to which Jesus Christ exorcised evil spirits are the cosmic areas which are unsuitable for survival of human beings, but which are adequate for spirit beings (Matthew 12:43; Luke 11:24). "The heaven, even the heavens, are the LORD'S: *but the earth hath he given to the children of man*" (Psalm 115:16). There are no other living creatures in the universe except angelic beings.

Satan, once the masterpiece of God's creation, was very intelligent, but his wisdom became warped: ". . . Thou sealest up the sum, full of wisdom, and perfect in beauty . . . *thou hast corrupted thy wisdom* . . . Thou hast *defiled thy sanctuaries* . . ." (Ezekiel 28:12, 17,18).

Satan became sovereign over evil spirits. He organized serpentine gods and made them "rulers of the darkness of this world" (compare Daniel 10:13,20). ". . . Spiritual wickedness in high places" (Ephesians 6:12). Moffatt translates it: "For we have to struggle *not* with blood and flesh but with the angelic Rulers, the angelic Authorities, the Potentates of the dark present, the spirit force of evil in the heavenly sphere."

The passengers can clearly identify Satan's corrupt concepts of architectural beauty. While the structures are ingeniously constructed from the materials thrust forth from the white holes, there is a diabolical nature about the design that the graffiti-like scrawlings and caricatures. They are obviously the brainchild of extraterrestrial brobingnagians, evil creatures with distorted mentalities. They depict violence, terror, and utmost wickedness. No wonder Job (15:15), by divine inspiration said: ". . . Yea, the heavens are not clean in his (God's) sight." These labyrinth structures in Satan's hinterlands are condemned and will be crushed out of existence in the fiery death pangs of the final massive black hole of the cosmos.

ASTROPLANES?

In the distance there appears to be a landing field. Could this be the launching and landing strip for Unidentified Flying Objects (UFOs) in fulfillment of the prophecy about "wonders" in the heaven above (Acts 2:19)? The word "wonders" refers to intangibles; the word "signs" attests to tangibles. Some believe Zechariah 5:1-8 is a prophecy of UFOs and that their mission is described as "wickedness." Because Satan knows the astrophysical laws of the various worlds, his ingenious, inventive, and clever mind may have designed and produced these UFOs (if they really exist) for personal gratifica tion to enable him to occasionally send his wicked scouts to wander about over the earth to ascertain what is taking place. Since fallen angels cannot possess physical bodies without encountering the worst of God's wrath (Jude 6), these devious machines may serve as alternatives. We know from the Scriptures that only holy angels have the authority and capacity to materialize without divine rebuke (Genesis 18:1,2,16,22; 19:1; Luke 24:4; Acts 12:7). Materializing, however, is not their normal function. It is reserved for special occasions. The evil angels who violated this divine law are imprisoned in the bottomless pit, thus enabling God to control their activities. They may make appearances, but they are not to materialize as fallen angels, and no angels are to inhabit physical bodies. This is the exclusive right of the Holy Spirit.

SATAN'S CHANNEL INTO THE SOUL

Scriptures denote two redemptions affecting the human race, which place mankind in an enviable position above angels. The first is the redemption of the soul; the second is the salvation or deliverance of the physical body into a tangible immortal body (I Peter 1:9; Colossians 1:14; Hebrews 10:39; James 5:20. Compare Hebrews 9:28; Romans 8:8,23; 13:11). An immortal body is superior to that of an angelic one, because it has the combination of both instant mobility and physical-like tangibility.

This unique position of mankind means that Satan makes vicious attacks on the body and soul in order to corrupt and demoralize the human race. Through the body, via the mind, he seeks to infiltrate the soul. By preventing salvation (redemption) from rescuing the soul, Satan can halt the second salvation (redemption) which is the immortality of the body. Our defense lies, not only in accepting Christ as our personal Saviour, but also in allowing *the atoning blood of Christ,* the *eternal Written Word,* and the *Holy Spirit* to keep washing our minds from Satan's contamination. In Scriptures it is called "sanctification"[3] (2 Thessalonians 2:13; I Timothy 4:5; John 17:17; Hebrews 10:10; 9:14; Titus 3:5; Ephesians 5:26). Salvation will assure not only immortality of the soul, but also immortality of the body (Ezekiel 18:4,20; Matthew 10:28; John 5:24; I John 3:14; JameS 5:20); however, the "spirit" is eternally immortal (Ecclesiastes 3:21; 12:7). The first death does not mean cessation, but separation. Example: When one dies, his soul and spirit are separated from the body. However is spirit entity can be further separated from his soul (Ecclesiastes 12:7) if he is unredeemed. the resurrection of the unbeliever will be a reuniting of the soul to a *mortal* body (Matthew 10:28; Revelation 20:14), but the resurrection of the believer will be a reuniting of the soul to an *immortal* body (I Corinthians 15:35-37; Romans 8:11,23).

NEW SENSES

As the voyagers behold the vivid novel colors and hues in these astrodominions, they acquire an unusual awareness. They wonder whether there will be new strains of music and instruments existing in different dimensions from those to which they are accustomed. Will there be more than eight notes to an octave? Will there be an extension of the six basic colors? Paul the Apostle said, "Eye hath not seen, nor ear heard, neither have entered into the heart of man, the *things* which God hath prepared for them that love him" (I Corinthians 2:9).

Paul logically arrived at this conclusion as the result of an experience when he was caught up into Paradise. He was totally unprepared for what he saw and heard and thus was unable to convey his impressions into words. All he said was that he "heard unspeakable words, which it is not lawful for a man to utter" (2 Corinthians 12:4). Moffatt translates it this way: "...and heard sacred secrets which no human lips can repeat."

The spacecraft glides through one hyperspace after another, each unfolding new wonders and beauties. It makes incredible time through the subterranean cosmic caverns. The occasional sight of fiendish structures and malevolent and impious etchings along the way at various stations makes them realize the real character and degradation of Satan and his fallen subjects. How grateful they are to be escaping his influences forever!

[1] Discovered by what is called "back-end" signal analysis.
[2] White holes are viewed by astrophysicists as supergigantic clouds that encase the pure energy which is the source of new stars and other celestial objects.
[3] Sanctification means to be separated: not only *from* the ungodliness of the world, but also separated *unto* God.

85

Out of This World

In the glorious light of eternity's endless expanse...there is no calibration of time.

As Spacecraft Earth approaches the boundary of the cosmic globe, the passengers become conscious of a glorious pathway of light penetrating into our universe.[1] The spacecraft heads toward this brilliant beam which is pouring through a vast doorway or portal from outside the cosmos. This particular pathway of light is coming from the Throne of God, so immediately angels line its sides to form a wide corridor through which the spacecraft will pass. It is the expressway upon which millions of deceased redeemed and innocent have been escorted by angels into the Heavenly Paradise. This thoroughfare has been in use ever since those in lower Paradise were transferred by Jesus Christ after His resurrection.

Quasars, cosmologists tell us, are the best evidence for a noncosmological red shift. Previous to 1981 astronomers discovered more than 1500 quasars, or pathways of light, which enter our cosmos and converge like the point of a "V" with tremendous energy. Although a quasar emits one hundred times as much luminosity as does all of the Milky Way, its light comes from a compact area trillions of times smaller than the galaxy.

"Where is the *way* (literally, *home*) where light dwelleth? and as for darkness, where is the place thereof (edge of the universe), that thou shouldest know the paths (pathways of light through these cosmic openings) to the house thereof (the Heavenly Father's house being the habitation of eternity)?" (Job 38:19,20).

ALIEN WORLDS

As the spacecraft emerges from the cosmos, it is surrounded by billions of minute, bright objects reaching into the vast distances in the habitation of eternity. Christ informs the passengers that these cosmic neighbors are the "Heavens of Heavens," the residences of holy angels.

As the travelers look back upon the universe which they just left, they realize they have departed from the multiple clocks of time and calendars in the cosmic vault. Now they are in the glorious light of eternity's endless expanse where there is no calibration of time. As they observe the immense sphere of the blackness of the universe, it seems suspended in the open territory of eternity. Inside the sphere are billions of galaxies and myriads of streaking comets and other celestial wanderers which are barely discernible.

God's creations, they realize, are responsible for the beginnings and calibrations of "times." As a creature of "time," man is prone to theorize that there must be a beginning because he lives in a dimension with limitations.

COSMONOVA

The origin of the universe began with pure energy forming the first and lightest elements, hydrogen and helium.[2] These fused to start the chain of heavier elements in the numbers we now see in our universe, until they reached the heaviest element, uranium. A sign of the approaching last days of cosmos and man's government on earth was predicted in Scriptures as beginning with man's fissioning of uranium to release its enormous power in the first nuclear explosion (Luke 21:26).

When a star collapses and pulls its galaxy into a black hole, the star's nuclear fuel is spent, and an explosion occurs when only the lighter element, hydrogen remains.[3] Similarly, there will be a time when the synchronization of all energy throughout the universe will end in one titanic explosion. It will end where it all began—with helium, hydrogen, and finally pure energy. Time is running out in the cosmos but it will not end until after the Millennium when Planet Earth makes its successful escape to become the New Earth in the New Heaven: "But the heavens and the earth, which are now, by the same word (that preserved the earth during the Flood) are *kept in store, reserved* unto fire against the day of judgment and perdition of ungodly men" (2 Peter 3:7).

DANCE OF DEATH

Suddenly there is a magnitudinous explosion. The concussion rocks the spacecraft to and fro like a drunkard (Isaiah 24:20). The cosmic globe, like one gigantic hydrogen bomb, detonates into a colossal fireball. The searing heat, as from a blast furnace, strikes the spacecraft at full force. But with Christ in their midst, as when He stood with the three Hebrews in the fiery furnace, the spacecraft and its voyagers are shielded from the heat, the gamma rays, and other dangerous elements coming from the blast. They suffer no harm. Everything remains intact. All are safe.

As the passengers continue to look back from a safe distance in their spacecraft, they see multitudinous galaxies sending out supergigantic mushroom clouds in all directions in interlacing patterns. As they reach their zenith, they begin to recede toward a central point, as if drawn by powerful suction toward an open drain. The implosion that follows pulls in these eddy-like clouds at an incredible speed. There appear among these clouds faint flickers of light, like pulsars, as the galaxies are brought together in their final gasps of life into one massive grave-like black hole. The smaller galaxies seem to linger a little longer as dwarf galaxies, but soon all of these mushroom clouds merge and converge into one massive black hole. Grudgingly and violently these galaxies give up the remaining matter by first expanding like a balloon, shuddering for a moment, then in a blast of gamma rays equivalent to the detonation of many billions of megaton hydrogen bombs, finally succumb to death and a grave in space.

COSMIC REQUIEM

Space scientists believe that when galaxies collide with great velocity and powerful gravity everything compresses with such tremendous density and pressure that it creates temperatures of one hundred trillion watts. This is sufficient to produce fusion *ignition that will set off* explosions at temperatures ranging up to one hundred million trillion degrees kelvin.[4] Astrophysicists believe that matter cannot exist even at a temperature of one trillion degrees kelvin. Such radiation (ten million times that of matter) would destroy atoms and disintegrate their nuclei. Astronomers believe that in the distant future the cosmos will contract and become denser until it is finally reduced to less than a handful of pure energy: ". . . in thine hand is there not power and might . . . ?" (2 Chronicles 20:6).

If the earth were compressed into a black hole its mass would be less than one inch in diameter. To go into pure energy it would be reduced even further. The radiation at first would evaporate slowly, then accelerate until it ultimately would disappear into pure invisible energy.

The universe is nearly empty, consisting mostly of space and cold, and its volume of matter can be condensed to the size of the earth. If it were not for the white holes, all stars eventually would consume their hydrogen and helium and burn out. This would leave only a dark remnant being crushed out of existence by its own gravity.

The death knell of the universe is the process which the passengers on the spacecraft are witnessing as the Almighty sets off the trigger mechanism that produces the cosmic explosion: ". . . the heavens . . . shall perish . . . they shall all wax old like a garment; and as a vesture shalt thou fold them up . . . " (Hebrews 1:10-12). ". . . The heavens shall pass away with a great noise, and the elements shall melt with fervent heat . . . " (2 Peter 3:10). The words "pass away" is a redemptive term: ". . . old things are passed away; behold, all things are become new" (2 Corinthians 5:17).

Before the eyes of the passengers everything soon disappears into a single powerful bright object of pure energy like a coal of fire. The nuclei heart stops beating. The Creator reaches forth His hand and plucks the pure energy chestnut unto Himself.[5]

VIEW FROM THE EDGE OF DARKNESS

How can the Creator collect unto Himself the searing hot ember of pure energy? It is because our God is "a consuming fire" (Deuteronomy 4:24; Hebrews 12:29; I Timothy 6:16). All created light also will return to pure energy for it, too, consists of particles called photons.

Only by divine appointment and by proper vestment can anyone hope to approach the Almighty. The Royal Redeemed will be vested in immortality like Christ Jesus (Matthew 5:8; I John 3:2; Philippians 3:20,21; Revelation 4:4; 5:9,10) so that they can approach God at any time without dire consequences.

COSMIC VACANCY REFURBISHED

Had the Creator desired, He could have allowed the pure energy to remain in the singular black hole to be reactivated by the Holy Spirit into new galaxies, as occured in the primordial creations. However, the Creator had other plans for the vacant sphere of blackness left by the disappearing cosmos. Within this huge, black orb God is to set up His Great White Throne Tribunal. "And I saw a great white throne, and him that sat on it, from whose face the earth and the heaven *fled away;* and there was found *no place for them"* (Revelation 20:11).

All persons not tried previously, including angels and those in "outer darkness," must appear in resurrected mortal bodies at the final tribunal (John 5:22-30; Matthew 8:29). The wicked, who infiltrated the Kingdom of God (Heaven) as intruders or who did not appear at the Judgment Seat of Christ, must now appear at this final court: ". . . at the end of this world (age)" (Matthew 13:40). The holy angels will escort all incarcerated beings to this White Throne Judgment. The indictment rendered will determine the length of their fate in the lake of fire.

RELIGIOUS VIOLENCE[6]

That which occurs in the name of religion is sometimes appalling, unconscionable, vicious, and unmerciful, especially when clothed in political garb. It appears from Matthew 13:38 that the "wicked" *in* the Kingdom of Heaven (God) already will have sealed their doom since their names have been deleted from the book of life. They must wait for the final tribunal to review their deeds to determine the duration of time they are to spend in the lake of fire under the second death: ". . . and burned in the fire . . . in the end of this world (age) . . . And shall cast them into a furnace of fire . . . . " The time is indicated as to when this court session will have its final hearing: "Then shall the righteous shine forth as the sun *in the kingdom of the Father* . . . " (Matthew 13:43). The tribunal must conclude its final session at the dispensational end of the New Heaven, just before the Manifested Kingdom of the Heavenly Father when all who will be redeemed "shall shine as the sun" in their bodies of immortality.

BRIGHT LIGHT AHEAD

Like a train approaching a station, Spacecraft Earth slowly and gently comes to a stop in the second heaven, and as it does, it becomes the New Earth.

Christ now directs the attention of the voyagers toward an extremely bright object in the distance. As the spacecraft hovers closer, it is obvious that a brilliant corona is surrounding the huge cubical object. Below the luminous aureole an iridescent light exhibits vivid colors streaming downward like shimmering, sparkling water in a magnificent display of grandeur and beauty. It is the Holy City and it descends from the larger planet, Paradise, upon the New Earth as if lowered by a sky hook from a giant out-of-sight crane.

ODYSSEY ENDS

The Royal Redeemed are escorted toward one of the enormous gates of pearl into the Holy City. Those in natural bodies follow at the rear of the parade into the city and immediately are transferred into huge galleries that extend high above Mount Sion where they can observe the scenes and activities below. Here they are adequately shielded from the glory of God, like Moses in ancient times.

The Royal Redeemed appear before the Throne for special ceremonial introductions by Christ to the Heavenly Father. After that event, those in natural bodies receive a tour of the Holy City, and are accompanied to their respective habitations within the New Heaven where they are to reside until they receive their immortal bodies when death will be destroyed forever. The Holy City is to become the interuniversal capital city of all the habitations in the New Heaven.

The top of Mount Sion is a massive plateau, smooth and glistening like highly polished crystal (Revelation 4:6; 15:2). "The Lord hath prepared his throne in the heavens . . . " (Psalm 103:19).

[1] See illustration III, page 605.
[2] Hydrogen is lighter than helium.
[3] The hydrogen bomb is more powerful than an atom bomb.
[4] A kelvin is 459.67 degrees Farenheit.
[5] Space mathematicians call it a giant singularity. There would be no wormhole (tunnel through hypersapce) and no white hole.
[6] See page 578, "Two Narrow Roads."

86

Oasis in the New Heaven

The Heavenly Father's Kingdom will be an entirely new dimension created only for those in bodies of immortality.

"But will God indeed dwell on the earth?" (I Kings 8:27). As the redeemed earth ascends with its precious cargo from the Millennium toward its port in the New Heaven, the Holy City simultaneously descends with Mount Sion, upon which God is enthroned. Those remaining in Paradise are summoned to appear at the White Throne Judgment before the Magistrate, the Lord Jesus Christ.

We know that no others will enter Paradise after this because the trees of life will have been transferred from there into the Holy City now on the New Earth. The trees of life have served their purpose well, enabling those in paradisiacal bodies to survive until the final tribunal. Anyone once having partaken of them will find it no longer necessary to have further access to them. This is an advantage to those residing great distances from the New Earth.

It is highly probable, at the initiation of the Heavenly Father's Manifested Kingdom, that the Holy City will reascend to its original site in the third heaven. This will be after the final person whose name is in the book of life has been redeemed in the New Heaven, and the final tribunal has ended its sessions.

On the New Earth we now see four enthroned authoritative powers all converging to finalize the plan of redemption. David is placed upon his throne in Jerusalem to aid and direct the evangelization of Israel's future generations. The Father, the Son, and the Royal Redeemed are seated on their thrones (compare Revelation 3:21) upon Mount Sion to reach the descendants of the Gentiles (compare Revelation 21:3,24-27). This occurs during the dispensation of the New Heaven up to the time death is destroyed. Then all who finally are redeemed are transferred into immortal bodies and enter the Father's Kingdom. The fact that death, as the last enemy, must be destroyed, shows that mortality reigns in the New Heaven and that immortality eventually will become the new lifestyle after all are redeemed.

The New Earth cannot accommodate the expanding population of natural people who emerge from the Millennium—or their posterity. This will be true until the last person is redeemed in the New Heaven.[1] The time allocated and required under God's foreknowledge to complete the redemptive objective is one million years. Therefore, there will be universes in the New Heaven to facilitate the physical requirements of these natural people because they will be unable to live comfortably under the environment and the brilliance of the undiminished Glory of God (also called "Son-light," Revelation 21:23, and the "Sun of Righteousness," Malachi 4:2. Compare Daniel 10:5-8; Acts 9:3,4) which will flood the entire New Earth where God has now established His throne.

The Royal Redeemed will reign under Christ, the "King of kings," and they also will serve as "*priests.*" In addition, they will escort the distant residents who must make their required pilgrimages to the Holy City on the New Earth (Revelation 21:24-27; 22:14). The New Earth will be the capital site for all those living in the far-flung habitations of the New Heaven.

At the end of the dispensation of the New Heaven, the redeemed who are residing in the outpost habitations will receive their immortal bodies because death will be destroyed. Then they will join Christ and the Royal Redeemed to be transported to the Heavenly Father's

Manifested Kingdom (I Corinthians 15:24). At that point, God's Family will be complete and *imputed* with His righteousness. The door of redemption will then be closed forever since no one will remain to be redeemed. The Royal Redeemed will reign as kings forever over all of the redeemed who will have come in after the first resurrection (Revelation 22:5); however their positions as priests, judges, and lords will be eliminated because such services no longer will be required. In the same way, the prophets' ministries will cease when the Millennium begins because it is the time of fulfillment of all remaining prophecies.

The Heavenly Father's Manifested Kingdom will be an entirely new dimension created exclusively for those in bodies of immortality with their unique kinds of atomic properties: "And he that sat on the throne said, Behold, I make all things new . . . " (Revelation 21:5).

NEW FRONTIERS

The mortal redeemed, who are located in the New Heaven, will be placed within solar systems to match their interests and desires. Planets will be geared to specific flora and fauna (Isaiah 65:25). There will be a variety of landforms, and climates, with areas set aside for many extinct and prehistoric creatures that have been reactivated and restored for the delight and pleasure of all inhabitants. Perhaps even new creatures will be created.

Life will be stimulating because of one's God-given spirit entity. There will be peace and the materials required to bring into fruition that which God originally intended when He created our first parents. There will be time to achieve the quality of life in the areas of work, recreation, adventure, the arts, and science.

ROYAL ARISTOCRACY

As the Royal Redeemed and those emerging from the Millennium enter the eastern gate of the Holy City, they encounter many fascinating sights, similar to an Alice-in-Wonderland fantasy. Here, at the foothills of Mount Sion, rise broad boulevards winding gracefully up the mountain terrain amid beautiful landscapes. Lined on both sides, as well as through the center of the boulevards, is an abundance of trees of life (Revelation 22:2).

The Royal Redeemed, who are in their immortal bodies, will make an appearance before the Heavenly Father on the huge dais which resembles a sea of glass (Revelation 4:6). The redeemed in mortal bodies will be transported instantly, like Philip in Acts 8:39, to massive observation galleries. Like the 144,000 in mortal bodies, who are sealed by God in the final Tribulation on earth, and like the post-Tribulation redeemed mortals who are protected from the brightness of the Lord Jesus when He returns to the earth, these redeemed are protected from the brightness of the Glory of God emanating from the Father's majestic Throne.

TWO VACANT THRONES ON MOUNT SION

Everywhere one looks in this huge hippodrome-like structure there are myriads of holy angels. They appear to be standing on invisible balconies surrounding the interior of the great city. It is obvious that beside the Heavenly Father there are two vacant thrones. The first throne adjacent to the Heavenly Father's right side is the Throne of the Lord Jesus Christ. This is the one He occupied as Mediator during the pilgrimage of the Christians on Planet Earth. Why as Mediator? All prayers, in order to receive the attention of the Heavenly Father, must go through His Son. And did He not say, "I pray *not* for the world, *but for them* which thou hast given me . . . " (John 17:9). Mediation for the unconverted is the responsibility of the believers for their own kin and friends. It also inspires them to witness.

The second vacant throne to the right of Christ's throne has been reserved a long time for the Royal Redeemed, to share, as Christ promised: "To him that overcometh will I grant to sit with me in my throne, *even* as I also overcame, and am set down with my Father in his throne" (Revelation 3:21).

The radiant splendor of the cherubim and seraphim casts its brightness against the huge rainbow which envelops the thrones that shimmer in unbelievable beauty (Ezekiel 1:1-26). These heavenly creatures are constantly worshipping and exalting the majesty of God (Isaiah 6:1-8).

ROYAL INTRODUCTIONS

"Blessed are the pure in heart: for they shall see God" (Matthew 5:8). The people in mortal bodies could never survive this scene if they were not shielded like Moses before the presence of God (Exodus 33:18-23). They are unable to stand before the Heavenly Father as can the Royal Redeemed, thus they are assembled in a protective area where they can observe all that is to occur.

Now there is a very proud and dramatic moment for Jesus as He stands with His assembled corporate Royal Wife. He addresses the Father and makes individual introductions of His Royal Redeemed to the Father. Perhaps brief biographies of each are presented, highlighting how they were introduced to Christ as their personal Saviour, the services they have rendered, and any special hardships they may have endured as believers. Christ then turns to the holy angels and expresses His exuberance in having each of the Royal Redeemed as His Royal Wife (Luke 12:8; Revelation 3:5). Each of the Royal Redeemed echoes his sentiments upon being presented to the Father and the holy angels, then makes a brief emotional response of his or her love for Jesus Christ: ". . . thousand thousands ministered unto him . . . (for many) came to the Ancient of days (the Heavenly Father), and they brought him (Jesus Christ) near before him (the Father)" (Daniel 7:10,13).

As the introductions are completed and the responses made, the Father hands to each member of the Wife of Christ a sparkling white stone. It contains the Father's personal recognition and confirmation of their rechristening with a new name given by the Heavenly Father (Revelation 2:17). As they enter this new relationship in the Family of God, everything of the past is erased.

They are informed that after the ceremonies there will be a great royal banquet prepared especially for this occasion (verse 17). They also are presented "the morning star," which is the Star of Jesus Christ, a signet which will signify forever to all creatures everywhere that they are the Wife and Queen of Christ and members of the Royal Family. This is not the "star of David" (Revelation 2:28; 22:16) which will go to the redeemed of Israel during the reign of King David after

Christ relinquishes the throne to sit beside His Heavenly Father and His Royal Wife.

"... And his name shall be in their foreheads" (Revelation 22:4). This beauty mark will identify the Wife or Queen of Christ wherever she may travel and will serve in lieu of a crown. Are the servants in Revelation 22:3 those who were rehabilitated from "outer darkness" and who must assume some menial obligation? If not, who can they be? (Compare Matthew 25:1-30).

The scene is so overwhelming and emotional that, without exception, all of the Royal Redeemed weep copiously in sheer ecstasy. The Heavenly Father descends from His Throne and personally wipes away all tears from their eyes. When He dries the tears He simultaneously removes from each person all sad memories and scars of the past (Revelation 21:4).

A similar experience will occur at the end of the Kingdom of Christ before He turns everything over to the Father and to the newly redeemed in the New Heaven. Sad memories will be abolished and the redeemed will be unaware of kin and friends who have failed to embrace Jesus Christ as personal Redeemer and Lord of their lives: "For behold, I create new heaven*s* (plural, a different dimension), and *a new earth:* and the former (New Heaven) shall not be remembered, nor come into mind" (Isaiah 65:17).

"He (God) withdraweth not his eyes from the righteous: but with *kings* (Royal Redeemed) are they on his throne: yea, he doth establish them *for ever and they are exalted"* (Job 36:7).

The redeemed in mortal bodies, who are located in the observation galleries, now will be escorted by the Royal Redeemed through the Holy City to receive information about partaking of both the fruit and leaves from the tree of life. This will be for permanent assurance of continued health and life until death is exterminated. Then they will receive their immortal bodies, along with all of the future redeemed in the New Heaven, and be transported to the newly created Heavenly Father's Manifested Eternal Kingdom.

[1]There are yet five great spiritual awakenings in our future with these impressive results: 1) Tribulation saints; 2) Righteous Nations; 3) Remnants of Gentiles and Israel; 4) Millennial harvest; 5) New Heaven bonanza.

87

Redemption of the Earth[1]

The Kingdoms of Christ and His Father will merge as the New Heavens.

"Glorious things are spoken of thee, O city of God, Selah" (Psalm 87:3). Because of its vast size this city is also called a "heavenly country" (Hebrews 11:16). It is fifteen hundred miles long on each of its four sides, as well as in height (Revelation 21:16). It is shaped like a cube, and contains 2,250,000 square miles or 2,375,000,000 cubic miles. The wall surrounding the city is two hundred fifty feet high and of jasper, a precious stone reflecting a mixture of colors—white, red, yellow, brown, and black. The city itself is of pure gold with twelve enormous foundations erected in tiers, and constructed of massive and extremely rare and precious stones which are the originals of God's creations, not earth's copies and synthetics. These foundation stones *resemble* sapphire, emerald, jasper, chalcedony, sardonyx, sardius, chrysolyte, beryl, topaz, chrysoprasus, jacinth, and amethyst. The twelve foundations serve as huge ramps rising from the earth and leading up toward the twelve gates of pearl, which are separated by the crystal clear river of life flowing between the trees of life which line its banks.[2]

THE HOLY CITY'S FACADE

The twelve tributaries of the river of life create gorgeous waterfalls cascading past the gates of pearl down the terraced slopes of the twelve foundations. Their colorful gems form a background reflecting through the waterfalls under the light of the Glory of God. These tributaries, as they flow out over the land onto the New Earth, teem with all types of fish life (Ezekiel 47:10).

The moment of true excitement mounts as the newly redeemed progeny from the New Heaven make their pilgrimages through the gates of pearl and catch their first glimpse inside the city. Before them lie the foothills of Mount Sion which gradually increase in elevation to peaks looming at heights of almost fifteen hundred miles. The beauty is unrivaled anywhere as they observe the golden boulevards winding their way up the magnificent mountain ranges of Sion.

THE WEDDING RING OF THE ROYAL REDEEMED

God's majestic Throne will be seated atop Mount Sion and can be viewed above the wall surrounding the Holy City for a distance of at least two hundred and fifty miles: ". . . come hither, I will shew thee the bride, the Lamb's wife. And he carried me away in the spirit to a great and high mountain, and shewed me that great city, the holy Jerusalem, descending out of heaven from God, Having the glory of God: and *her light* (the Holy City) was like unto *a stone most precious,* even *like* a jasper stone (original), clear as crystal" (Revelation 21:9-11). These facts emerge: 1) the light that will illuminate the New Earth and serve as a beacon to travellers from distant habitations in the New Heaven will be the city itself; 2) John was viewing the scene from a mountain high above the Holy City on some distant planet or satellite; 3) when the city descended God was not located on Mount Sion ("descended . . . *from* God"); 4) the vision was designed to let John see the Royal Redeemed, not God ("I will shew thee the bride, the Lamb's wife"); 5) according to Hebrews 12:22,23 God will not sit on Mount Sion within the Holy City until *after* the Royal Redeemed have been located within the Holy City: "But *ye* (the

Royal Redeemed) *are come unto Mount Sion*... To the general assembly and church of the firstborn (Christ)... *and to God*...."[3]

When the Father, the Son, and the Royal Redeemed are all gathered together in the Holy City, on the New Earth, we find that: "... *the glory of God did lighten it,* and the Lamb (Jesus Christ) is the light thereof" (Revelation 21:23). Thus, the Holy City itself has tremendous luminosity.

The light from God's Throne will complement the light of the Holy City. The light inside of the Holy City undoubtedly will have an unusual effect on the inside walls, which are of original pure diamond. They will have billions of facets, perfectly cut into their sides, with light striking them from God's Throne which will reflect shimmering colors from innumerable prisms. These walls of original diamond serve as the Lord's wedding ring and envelop the body of the Royal Redeemed who are now His *one* Royal Wife.

When Planet Earth comes to rest in the second heaven as the New Earth, the Holy City descends (from the third heaven) upon it at the precise center of the Pangaea. Upon cue, the Royal Redeemed who surround the walls will begin to converge upon the twelve gates of *pearl* and walk on streets of pure gold. As they enter the Holy City, they will find themselves beholding the brilliance and radiance of the billion prisms of glorious colors flooding the vast interior of the city, which is 6,000 miles in perimeter.

A HUGE BEACON LIGHT

The largest beacon light ever seen by mankind is foretold in Revelation 21:23: "... the glory of God did lighten it (New Earth), and the Lamb is the light thereof." Christ is called "... the SUN of righteousness" (Malachi 4:2).

The interuniversal wayfarers making their pilgrimages to the Holy City will see this bright light atop Mount Sion. Its radiance will brilliantly illuminate all of the Pangaea on the New Earth leaving no shadows, since it does not consist of photon particles.

DISTANT UNIVERSES

The Royal Redeemed will escort the newly redeemed mortals between the distant locations in the hinterlands of the New Heaven to the Holy City via interuniversal transit lines traveling beyond the speed of created light. After their tour through the Holy City and over the New Earth, the newly redeemed will return to their places of residence to continue to live for Christ, and to develop and transform those distant planets into veritable Gardens of Eden with ideal climates.

When the Royal Redeemed approach the Holy City with their loads of newly redeemed guests from distant habitations, the sentry angels standing guard at each of the city's gates (Revelation 21:24-27) will approve their entry because the royal priests are their escorts. As the newly redeemed enter the Holy City, they will look across the river of life to the boulevard leading out from the city in growing anticipation of what they are about to witness. As they pass the gate of pearl and begin to mount the foothills, they will see great ravines, valleys, plains, plateaus, mountain peaks, ridges, rolling hills, wooded sections, palatial gardens, and blue lagoons in a lush rural setting all within the Holy City (Hebrews 11:15,16; Isaiah 11:9; 65:25).

These excursionists notice the river of life, with its ripples and waterfalls, winding its way down the slopes of Mount Sion. It begins at the summit, near the Throne of God, like a massive spring of water gushing and bubbling up similar to an artesian well. As the river approaches the foothills, it fans out into twelve main streams, each headed toward one of the gates of the city; thus all of the land on the New Earth is adequately watered, as was Paradise, which once surrounded the city. There are no longer great bodies of water separating the land into continents.

CHORALES

The caravans of pilgrims are made aware of beautiful angels silently soaring about like graceful, white birds with feathers. They hear music and singing coming from afar and symphonies of worship to the Almighty God. The Royal Redeemed escorts inform the group that there are innumerable areas where the spirit of worship to Deity is constantly in progress, thus accounting for the warm atmosphere of adoration to God. Some of the participants are the 144,000 sealed Israelites singing and playing in unexcelled harmony. There are other chorales composed of the Royal Redeemed.

As the redeemed approach the summit, the melodies of the seraphim and cherubim that are clustered about the Thrones worshipping the Heavenly Father and His Son, Jesus Christ, command their attention and reverence, and swell their hearts with pride. Since delegates of the Royal Redeemed share a throne beside the thrones of the Father and His Son, these beautiful lyrics must be included in the music and accompaniment directed toward the members of the Royal Wife and Oueen of Christ: "To (the Royal Redeemed) will I grant to sit with me in my throne, even as I also overcame, and am set down with my Father in his throne" (Revelation 3:21).

MUSEUM OF REDEMPTION

The newly redeemed guests are accompanied to a museum of unbelievable charm, elegance, and size. Within this magnificent edifice they see on display for the first time the incorruptible and precious atoning blood of Jesus Christ, which flowed in His veins, and was shed on a cross for the redemption of all, then assumed by the Holy Spirit for Jesus Christ to present to His Father (Hebrews 9:12).

They move on to view the original Ark of the Covenant (Testimony) which served as a pattern for the one that God had Moses construct in the wilderness of Sinai.

Among the invaluable memorabilia in the scheme of redemption are the book of life and the divine writing instrument which God used when all of the names in the eternal past were inscribed; also there is the book of the seven seals which is now open. On display in one section of the museum are the millions of crowns that the Royal Redeemed placed at the feet of Jesus after the pre-Tribulation Rapture. Gabriel's trumpet and all instruments of special significance in connection with the process of redemption will be on display, together with the millions of royal palm fronds which the Tribulation saints received at their post-Tribulation Rapture.

Many invaluable Old Testament mementos are in evidence, such as the Ten Commandments given to Moses, Aaron's rod that budded, and the manna that retained its freshness within the Ark of the Covenant.

TOUR OF THE KING'S PALACE

In Psalm 45:15 we read: "... they shall enter into the King's palace." What unbelievable beauty and wealth is evident as their guides escort them through this palace which was brought into existence by the Heavenly Father for His Son in the pre-dawn of the creation of all heavens.

The tourists are escorted to the antechambers of the King's palace called the "ivory palaces." Here are well-organized chamber rooms with an unbelievable profusion of the most lavish and beautiful clothing ever imagined. Each of the ivory palaces has its own vestments for specialized occasions.

PHYSICAL TRANSFORMATIONS

As the guests prepare to leave the Holy City, they partake of the trees of life with a fruit of their choice. Each month there is a new variety of fruit and the supply is inexhaustible. Each fruit titillates the tastebuds, has perfect texture, and is free of flaws and blight. As the fruit is digested, new sensations permeate the bodies of the newly redeemed. They are making the wonderful and interesting discovery that their fatigued bodies are being rejuvenated. The gerontological problems for the redeemed have been permanently resolved.

At the request of the Royal Redeemed, the visitors reach for leaves from the trees of life and in doing so, they discover their physical blemishes disappear, the aging processes retreat, and their youthful appearances return minus all infirmities. Their bodies revert to that period of life when they were the most beautiful and attractive. The ones who are in their prime of life and at their greatest point of attractiveness find their aging processes halted (Revelation

22:2; Ezekiel 47:12). The trees of life contain the antibodies necessary to destroy the death cells and put them into remission until the Holy Spirit transforms their bodies into immortality.

CLONED COLONIES

The travelers are informed by their guides that since they have accepted Christ Jesus as their Saviour and Lord, they no longer need fear death or be concerned about their future health. They will remain healthy, vigorous, and youthful. They will retain their desired attractiveness until death is finally and eternally destroyed, at which point they will experience a much greater transformation into bodies of immortality. No longer will they be confined to the habitable places in the New Heaven which require extrinsic modes of transportation; they will use intrinsic ones. They will be capable of living in the environment of the New Earth where there is constant light, as well as in the new dimension of the Heavenly Father's Manifested Kingdom. They will be able to behold the Glory of God face to face and not require a protective shield (Matthew 5:8; I Corinthians 13:12; John 17:24; Jude 24; 2 Corinthians 3:18). They also will have the opportunity of exploring the vastness of Paradise in the third heaven, including the royal ranches with their millions of royal white steeds and chariots.

ROYAL TRANSIT SHUTTLE LINES

The newly redeemed must again board the Royal Chariot Transit Lines to return to their distant residences in the New Heaven. There they will remain and enjoy themselves as normal persons in their physical bodies until they, too, are changed into immortality. They will be required to make occasional pilgrimages to the New Jerusalem, as was the case in the Millennium.

In the meantime, they are urged to assist the Royal Redeemed as royal priests in evangelizing their friends and loved ones to accept Jesus Christ as Lord and Saviour. In this manner, they, too, will be able to avert the aging processes that will bring certain death (Isaiah 65:20-25).

CHERUBIC INNKEEPERS

"Are they (angels) not *all ministering spirits,* sent to *minister for them who shall be heirs* of salvation?" (Hebrews 1:14).

The Psalmist (Psalm 45:8; 48:3,13) states that within the Holy City there are "palaces," presumably for the accommodation of all newly redeemed tourists. Never have human beings enjoyed such luxurious elegance. What splendor, comfort, and services! They are hosted by angels for a time and given the royal treatment in the palatial inns.

PRIMA DONNAS IN THE NEW HEAVEN

While the latest redeemed are not kings and queens, they are honored with the title "princes" (Psalm 47:9) and described as the "kings' daughters" because they are the spiritual children of the Royal Redeemed who are serving as royal "priests." The Royal Redeemed will continually reach out to the expanding population of the many habitations in the wide expanse of the New Heaven (Luke 1:33; Revelation 1:6).

"Kings' daughters" are among the honorable women: upon thy right hand (Revelation 3:21) did stand the queen (Royal Redeemed) in gold of Orphir" (Psalm 45:9). These "kings' daughters," therefore, have to be God's grandchildren through redemption. God's only daughter (the Royal Redeemed) through redemption, who is the Wife of His only Son, is "adopted," as well as are His spiritual grandchildren (Romans 8:21; 9:8,26; Galatians 3:26; Ephesians 1:5; I John 3:10; 5:2; Matthew 5:9).

FROM "TIME" TO "ETERNITY"

When all who can be redeemed will have accepted Christ as their personal Saviour during the one-million-year dispensation in the New Heaven, the curtain will be lowered on this period of "time"; then it will rise for the final scene in the dispensation of timeless eternity in the Father's Kingdom. All of the redeemed will have received their immortal bodies because death is destroyed, and they all will be transported into the Father's Manifested Kingdom (I Corinthians 15:24), where there will be the merging of two kingdoms into one (2 Peter 1:11; Matthew 26:29; Luke 12:32; Daniel 7:27). It will be known as the "New Heaven*s*" (Isaiah 65:17; 66:22).[4]

This merging event will be celebrated in the Father's Kingdom with Christ Jesus sharing in a Holy Communion around the greatest Lord's Supper ever held. We may presume that the twelve apostles and the elect saints, with the redeemed of Israel, will serve this vast multitude (Matthew 26:29). Thus, the redeemed will be "toasted," as it were, into the Heavenly Father's new dispensational and dimensional Manifested Kingdom.

Just as the kingdom of Antichrist and the False Prophet merge as "Babylon," to expand their kingdom as one, and to take possession of the earth, so the Kingdom of Christ and His Father will merge as the "New Heaven*s*" to take possession of all creations: past, present, and future that are to remain forever (I Corinthians 15:24; Isaiah 65:16-18; 66:21,22).[5]

ENORMOUS SALT BASINS

"And *the sea* gave up the dead which were in *it* . . . (Revelation 20:13).

The words, "the sea," show that the Pangaea has taken place and that the water still remains prior to the White Throne Judgment. This indicates the Planet Earth apparently loses little or none of the water in "the sea" while enroute to its destination as the New Earth. As soon as the White Throne Judgment nears its sessions, the waters within "the sea" dissipate under the Glory of God leaving only "salt." Those who have perished in it are now destined to be resurrected unto mortal bodies and united with their souls so they can appear at the final tribunal. This excludes the Royal Redeemed, who were in the "first resurrection," as well as the survivors from the Millennium.

"And I saw a new heaven and a new earth . . . and there was *no more sea*" (Revelation 21:1).

The words, "no more sea," inform us that this evaporation takes place after the earth is redeemed to become the New Earth. According to Ezekiel, chapter 47, the river of life will begin to fill specific areas, such as the massive salt basins (Ezekiel 47:11). Certain salt marshes will be preserved as reminders of the history of Planet Earth. Some of these water-filled basins will contain a great variety of fish (verses 9,10), presumably reserved for the pleasure of the Royal Redeemed, the sole residents on the New Earth. This tells us that they are very real tangible persons who enjoy moments of recreation. The Royal Redeemed will not spend their entire time in administrative services, but will enjoy a balanced life, which includes relaxation.

THEN COMETH THE END

"Then cometh the end (no more to be redeemed), when he (Christ) shall have delivered up the kingdom *to* God, even the Father, when he shall have put down all rule and all authority and power" (I Corinthians 15:24).

Persons to be evangelized by the royal "priests" include the masses of children who were cut off from life on Planet Earth and held in Paradise, together with billions of innocent adults who have never heard the Gospel of Jesus Christ, but whose lives justify that God retain their names in the book of life. They eventually must accept Christ as their personal Saviour or they will be eternally lost souls, since there are no exceptions in the plan of redemption. Innocence has no atoning value, nor does martyrdom, even for good causes. Shedding one's blood has no atoning value (Matthew 19:14; 2 Peter 3:9; Acts 17:30; Luke 13:3,5; Matthew 12:32).

Multitudes will be born of natural parents in the New Heaven (survivors from the Millennium) and they also must receive Christ as their own Redeemer. The Royal Redeemed, as "priests," will assist in reaching them with the message of the Gospel and in urging them to accept Christ into their lives.[6]

TOO BIG TO BAG

Some believers hold the pathetic concept that only their exclusive group will inherit the glories of the future heavens. How utterly inconsistent for their limited populations to be rambling about in the vastness of God's future creations! One's vision must justifiably include untold numbers of redeemed for the following reasons:

First, God has placed a great price upon souls. One soul is worth more than the whole earth, which He plans to redeem. This value far exceeds that of angels because they never have been provided a Redeemer.

Second, Christ died for all who will be redeemed, making them exclusive in the eyes of God.

Third, those redeemed in the "first resurrection" will become the corporate Wife of Christ, the Queen in all of God's creation.

Fourth, the Royal Redeemed are to inherit *all* things in God's creation.

Fifth, the Royal Redeemed will judge angels.

Sixth, since angels are innumerable, why shouldn't all of the redeemed at least equal or exceed their number, since they are greater in value in God's estimation?

Seventh, since the huge gates into the Holy City are twelve in all, they must have been designed to accommodate and handle abundant traffic in and out of the New Jerusalem.

Eighth, the vast numbers of trees of life at each of the gates must be there to serve a great multitude of redeemed.

Ninth, the Tribulation saints, alone, are referred to as "a great multitude, which no man could number, of all nations, and kindreds, and people, and tongues . . . " (Revelation 7:9). They represent but a small portion of all of the redeemed.

Tenth, Daniel (7:10) foretold of "thousand thousands (that) ministered unto him . . . ."

[1] Redemption is the process of restoring to its original state something that has degenerated.

[2] This shows there is more than one tree of life.

[3] Since the Heavenly Father is remaining on the Planet Paradise, He probably is creating a new water system for irrigation.

[3] See chart X.

[4] See chart X.

[5] So long as mortality reigns, the Gospel of Jesus Christ will go forth, for the ultimate goal is immortality when the last enemy, death, is destroyed.

88
The Best Kept Secret

This final thrust of evangelism will be under the Royal Redeemed who will be reigning with Christ as His royal "priests."

". . . The mystery (of redemption) *which from the beginning* of the world hath been *hid* in God, who created all things by Jesus Christ" (Ephesians 3:9). Paul, in Colossians 1:26, distinguishes the ancient ages (during which the revelation of redemption was hid) from the generations of mankind on the earth: "Even the mystery (of redemption) which hath been *hid from ages and from generations* ('from angels and men,' *Amplified New Testament*), but now is made manifest to the saints."[1]

The angels who existed during those ages and generations revealed their intense interest in the manifestation of the gospel of God's grace under redemption to mankind: "of which salvation . . . the angels desire to look into" (I Peter 1:10-12). *Living Bible* translates it: ". . . it is all so strange and wonderful that even the angels in heaven would give a great deal to know more about it."

". . . The *generations* . . . of the earth . . . in the *day* (primordial creations in *one dispensational* day) that the LORD God made the earth . . . " (Genesis 2:4). "Behold, the Lord maketh the earth *empty,* and maketh it *waste,* and *turneth it upside down,* and scattereth abroad the inhabitants thereof" (Isaiah 24:1). ". . . Which *removeth* the mountains, and *overturneth* them in his anger" (Job 9:5). ". . . The everlasting mountains were *scattered* . . ." (Habakkuk 3:6). "Behold . . . the waters . . . also he sendeth them out, and they *overturn* the earth" (Job 12:15).

Geologists say that they have confirmed evidence that the earth has experienced at least seven colossal ice (glacial) ages. The earth literally has rolled over at times because of the weight of the glaciers cascading from pole to pole, slicing through, tearing up, and overturning mountains at their "foundations," as it were. The mountains of ice that swept through the oceans sent enormous tidal waves inundating the land, changing the topography of the earth, creating new seas and pockets of lakes: ". . . which *divided* the sea when the waves thereof roar . . ." (Jeremiah 31:35). ". . . All the foundations of the earth are out of course" (Psalm 82:5).

Scripture shows that the prehistoric inhabitants during those cyclical ages were *scattered* (not necessarily all destroyed). The prophet Isaiah used this ancient historical record as a classical example to prophetically warn what can happen when ". . . the earth (is) defiled under the inhabitants thereof . . . Therefore *hath* the curse *devoured the earth* . . ." (verses 5,6).

"Dead things are *formed* (progressively developed) from under the waters (such as oil and other minerals), and the inhabitants thereof" (Job 26:5). Inhabitants are the results of procreation and, therefore, are progressive. Only the first pairs needed to be created but the succeeding generations of posterity take time. The aeons of ancient ages, which made possible such treasures as lie within the earth, are compatible with God's chosen will when we recognize that "generations" of all creatures develop progressively through procreation.

THE ORIGIN OF LANDFORMS

The uniformitarianist position of the evolution shows that the landscapes and deep-sea topography are the results of catastrophic and evolutionary processes, as follows: 1) During the ancient ages Planet

Earth occasionally was subjected to inversion caused by the excessive imbalance of polar ice and snow, which sent gigantic glaciers across the land and through the ocean. These altered the earth's surface by carving out paleoplains, slicing through the deep-sea floor, scalpelling out underwater ravines, valleys, submarine canyons, and sending mountainous tidal waves sheering into the coastlines. Normally, sedimentary rock on the bottom is older than that on top. But in many sites, this is inverted; there is a repositioning of strata. In juxtapositions often many "ages" are missing; 2) The Pangaea broke up into continents and islands. Geostatic pressure perhaps may have been capable of floating the formation into relocation. Gravity alone appears to be hydraulically impossible. Earthquakes and volcanic actions seem more feasible; 3) and, finally, erosion.

THE FOUR DISPENSATIONAL AGES[2]

When God created the "world without end" He divided its future history into four main ages, dispensations, or times: ". . . throughout *all* ages. World without end. Amen" (Ephesians 3:21). First, ". . . *from the beginning* of the world" (Ephesians 3:9). Paul, in Colossians 1:26, divided the ancient ages before man appeared on the scene from those after his appearance; second, he separated man's appearance on the earth from the revelation of redemption. Paul divided that period into ages as well: "Which in other ages was not made known unto *the sons of man,* as it is *now* revealed unto his holy apostles and prophets by the Spirit" (Ephesians 3:5). The third major division is the period of the revelation of redemption, the gospel of God's grace to mankind: "if ye have heard of *the dispensation of the grace of God* which is given me to you-ward" (Ephesians 3:2). ". . . A *dispensation of the gospel* is commited to me" (I Corinthians 9:17). Finally, there are the ages following redemption[3] for those who respond to accept their eternal inheritance: "That in *the ages to come* he might shew the exceeding riches of his grace in his kindness toward us through Christ Jesus" (Ephesians 2:7). "That in the dispensation of the *fulness of times* (ages) he might *gather together* in one all things in Christ both which are in *heaven,* and which are on *earth,* even in him . . . In whom also we have obtained an inheritance . . ." (Ephesians 1:10,11).

THE LAST IS BEST

God's plan of redemption, which was hidden for an infinite length of time, will continue its restitutive effectiveness after Calvary, through the Millennium, and beyond into the New Heaven—a period of some one million years—one thousand generations with each generation one thousand years (Deuteronomy 7:9; I Chronicles 16:15; Psalm 22:30,31; 90:4; 102:18; 105:8; 2 Peter 3:8).

This revelation of redemption will continue under "the dispensation of grace" and "the dispensation of the gospel" until "death" is destroyed (I Corinthians 15:26); then all of the redeemed (not just the Royal Redeemed) will receive immortal bodies since no more unredeemed remain to be salvaged. This final thrust of evangelism will be under the Royal Redeemed who will be reigning with Christ as His royal "priests" (Revelation 1:6; 5:10; 20:6; I Peter 2:5,9).

[1] The Millenium of one thousand years, called a "generation" (Psalm 102:16-18), serves as an example of what delineates or separates the many generations of heaven or of the earth. Each generation ends in the judgments of God and begins afresh with new occupants. The last one thousandth such generation, affecting the redemptive plan of God toward humanity, will end with the finality of all sin and rebellion against God. It will consummate into the Heavenly Father's perfect eternal Kingdom as indicated in I Corinthians 15:24-28; Ephesians 1:10.

[2] See chart on "Four Dispensational Ages."

[3] The Millennium, New Heaven, and the New Heavens.

89
The Highest Royal Tribunal

In God's redemption is mercy; and in His judgment there is justice.

The Judge—The Lord Jesus Christ.

The Jury—The Royal Redeemed.

The Executioner—The Second Death.

". . . And the dead were judged out of those things which were written in the *books,* according to their works (deeds)" (Revelation 20:12).

". . . There is nothing covered, that shall not be revealed; and hid, that shall not be known" (Matthew 10:26).

"Do ye not know that the saints shall judge the *world?* . . . know ye not that we shall judge *angels* . . ." (I Corinthians 6:2,3).

In God's redemption there is mercy; and in his judgments there is justice. This final judgment of the litigants will be based exclusively upon "works" (deeds). The Scripture, ". . . after death is the judgment" (Matthew 9:27), no way precludes a sentence of guilt or that all appearing there will receive equal punishment. the word "works" denotes varying degrees of indictments.

It is clear that at this judgment there are names in the book of life (note word "whosoever" in Revelation 20:15). Why will these people not have been in the first resurrection with the Royal Redeemed? Since the Royal Redeemed, who will have been in that first resurrection, are serving on the royal jury panel, it shows a distinction between the Royal Redeemed, who are judging, and the redeemed who are being judged. Obviously, those on trial whose names are still in the book of life are not lost, neither are they serving as jurors, so what is their destiny? They must be those who will go on into the New Heaven: "And the nations of them which are *saved* shall walk in the light of it (the Holy City): and the *kings* of the earth (the Royal Redeemed serving as kings) do bring their glory and honour into it" (Revelation 21:24).

This final judgment in no way implies that those with good works, who are generally law-abiding, decent people, will be saved by their works. If they reject Christ after receiving the knowledge of the truth (I Timothy 2:4); Romans 10:14) they must receive their sentences, but they will be tempered on the basis of their past performances.

THE COURT'S IRREVOCABLE DECISIONS

The Highest Supreme Court Justice will render his verdict based exclusively on "works." Jesus, the Magistrate, sitting on His bench will render a fair and impartial decision: ". . . as I hear, I judge; and my judgment is just . . . " (John 5:30). "O the depth of the riches both of the wisdom and knowledge of God! how unsearchable are his judgments, and his ways past finding out!" (Romans 11:33).

TWO NARROW ROADS

Jesus Christ referred to two narrow roads: one is the way of *salvation* to eternal life: "Because strait is the gate, and narrow is the way, which leadeth unto *life,* and few there be that find it" (Matthew 7:14). The other is the path of *religion* that leads to a closed door: "Then said one unto him, Lord, are there few that be *saved?* And he said unto them, Strive to enter in at the strait gate: for many, I say unto you, will seek to enter in, and shall not be able. When once the master of the house is risen up, and *hath shut to the door,* and ye begin to stand without . . . saying, *Lord, Lord, open unto us;* and he shall answer and say unto you, *I know you not* whence ye are: Then shall ye begin to say, We have eaten and drunk *in thy presence,* and thou hast taught in *our streets.* But he shall say, I tell you, I *know you not...*" (Luke 13:23-27).

All on the broad road are not irretrievably lost. They can repent and turn from their unredeemed ways. Regardless of which three roads are selected, "All have sinned..." (Romans 3:23). The word "many" on the broad way shows that not "all" are on that road, but that some have been in a religious indoctrination of rituals and compliance to church creed all of their lives: "There is a way which *seemeth right* unto a man, but the end thereof are the ways of death" (Proverbs 14:12; 16:25).

The road to eternal life is the way of "grace" (Ephesians 2:8-10), which is the unmerited favor of God. This is for all who will come to Him for forgiveness without making human effort to improve themselves to merit acceptance. The narrow road of religion leads to a closed door. Jesus said, "I am the door..." (John 10:9).

ROYAL JURORS ON THRONES

"And I saw thrones, and they sat upon them, and judgment was given unto them..." (Revelation 20:4). "And I appoint unto you a kingdom, as my Father hath appointed unto me: That ye may eat and drink at my table in *my* kingdom, and *sit on thrones judging the twelve tribes of Israel*" (Luke 22:29,30). Not only will Israel make an appearance at the judgment but so will the Gentiles and the angels.

Presumably all of the Royal Redeemed will participate in making important decisions as members of the royal jury at some point during the High Court's session. Undoubtedly, if some of these jurors must serve as witnesses, they will be dismissed from jury duty in order to assure that a fair and impartial decision will be rendered.

One can almost visualize the stream of witnesses who will appear before the Justice and the royal jurors. The evidence will be clearly conclusive. The inexcusable conduct, atrociousness, and vicious wickedness of the accused will be so totally unmasked that the decision rendered by the High Court will be incontestable. Christ, the Magistrate, will render the final verdict. Because of this fact, all cases will encounter a prompt and final decision.

OH MY GOD, WHAT HAVE I DONE!

When the victims of the accused appear as witnesses to relate their pathetic, emotional, and sometimes gruesome stories, the charges of abuses and atrocities suffered by the plaintiffs undoubtedly will come to haunt the defendants. In some cases defendants will have destroyed the lives of persons who otherwise would have accepted Christ as their Saviour, and effectively served Him, had they not been cut off from life prematurely and denied that privilege. In situations where children have been abused and killed, their angels will come forth to testify as witnesses on their behalf (Matthew 18:10).

These dispassionate and hardened criminals, including rulers who have acceded to corruption and vice, will come to realize fully the magnitude of their acts. In the presence of the haunting ghosts of the past as witnesses, they will break down in agonizing anguish and cry out, "Oh, my God, what have I done!" (Proverbs 1:24-32).

Divine justice is clearly spelled out in the Scriptures: "For with what judgment ye judge, ye shall be judged, and with what measure ye mete, it shall be measured to you again" (Matthew 7:2; Mark 4:24). "And if any mischief follow, then thou shalt give life for a life, Eye for an eye, tooth for a tooth, hand for a hand, foot for a foot, burning for burning, wound for wound, stripe for stripe" (Exodus 21:23:25). It is ridiculous to claim that they now are under the "Grace of God" for this is not the day of "grace" but of "judgment." "It were better for him that a millstone were hanged about his neck, and he were cast into the sea, than that he should offend one of these little ones" (Luke 17:2).

The guilty must go through this final tribunal individually in a reenactment of the emotional and physical suffering which they have imposed upon others. This simulation upon themselves will be equally traumatic with that which they have inflicted upon their victims. The reaction will be so dramatic that it will evoke them to call upon God to terminate their punishment. But God's reaction is plainly "spelled out" in Proverbs, chapter one: "When your fear cometh as desolation, and your destruction cometh as a whirlwind;

when distress and anguish cometh upon you, Then shall they call upon me, *but I will not answer* . . . Therefore they shall eat the fruit of their own way, and be filled with their own devices." From the *second death* these will be *ever aging,* in *pain* and *dying* in eerie darkness until justice is satisfied, because they will not possess immortal bodies. The redeemed will be *forever living* under the light of God's glory and remain *forever young.*

IS THIS ALL THERE IS TO LIFE?

Does the justice of God simply punish the guilty without restoring to the innocent victims that which they lost and suffered? That would be justice only to the guilty but certainly not to the innocent victims. The words, "justice" and "judgment," speak of variations of decision-making affecting both the guilty and the innocent.

". . . The *dead* shall hear the voice of the Son of God: and they that hear *shall live"* (John 5:25). But live for what—to be lost forever; to be denied again the fortunes of life; or to live again, but this time a normal, healthy, and happy life under favorable conditions in which they can determine their own equitable destinies? To physically and emotionally suffer and shed one's own blood as a victim offers no redemption. The victims must make that decision themselves in accepting and rejecting the atonement of Jesus Christ. Certainly not all of those who are spared at the final tribunal will go on to accept and serve the Lord in the New Heaven (Revelation 20:11-15; 21:27; 22:11,14). This also will be true of their offspring.

Those whose young lives have been terminated prematurely because of disease, war, famine, abuse, and other misfortunes conceivably could raise an accusing finger questioning whether this is all there is to life; whether this is justice; whether there are any equities in life. John Kennedy said, ". . . it is very hard . . . to assure complete equality. Life is unfair." Their hope, thank God, will be in the New Heaven when the scales will be tipped more favorably in their direction: "And shall come forth; they have done *good* (nothing here about a born-again experience), unto the resurrection of *life* (to live again in mortal bodies) . . . " (John 5:29). There is nothing here about receiving eternal life.

It is reasonable and most generous to conclude that a maximum of ten percent of humanity since Adam will have become born-again Christians by the time of the Millennium. This indicates that God's plan of redemption will have been less than successful, without the implementation of some supplementary rescue plan. It appears unjustifiable that God would proceed to create mankind or allow him to exist under these staggering statistics. To say glibly that the redeemed are of infinite worth implies that the lost are not. Where, then, is the sense of value? Does not the plan of redemption require maximum effort to reduce the number of unredeemed? Therefore, we conclude, redemption reaches into the New Heaven.

INSTANT RECALL

Science has discovered that the mind is like a memory computer that registers and holds everything intact in a person's life. God needs only to turn on this mental computer for the defendant to have instant recall. The full truth, emotions, hatred, verbal abuses, threats, unmerciful accusations and assaults, and motives will all come to light. The guilty will become his own incontestable accuser. Some theologians believe that the "books" at the final tribunal may very well be these memory computers.

DEGREES OF PUNISHMENT

Jesus Christ said, ". . . as I hear, I judge . . . " (John 5:30). That which the first death did to the being is analogous to the second death except that it will be on a more imposing and awesome scale. If the first death did not destroy the "soul's" consciousness, will the second death do so? The length of punishment will be determined by the "works" (acts or deeds) revealed at this final judgment. It would not be a judgment if all punishment were the same, and justice disregarded.

JUSTICE FOR ALL

Two questions are frequently asked: Is the justice of God ever satisfied at some point in eternity by calling for an end to the punishment or by decreasing the length of incarceration? What happens to the souls of those who have served out their terms of sentence?

That there are maximum sentences handed out is clearly defined in Scripture (Revelation 20:10; 14:9-11). The utmost penalty will be given to *all* fallen angels because they are distinctly "spirit" beings and a "spirit" cannot be destroyed for it as eternal as God Who is also "Spirit." It was in connection with fallen spirit beings that Jesus alluded to the awesome statement about the unpardonable sin (Matthew 12:24). The Pharisees were crediting the miracles of Jesus to Satanic power, which indicates that men can bring themselves into the same state as fallen angels when there is no inclination on their behalf to repent or to alter their evil behavior in the face of what they know about the Sovereignty of God and their impending punishment.

An example of the unpardonable sin is seen in Revelation 14:9-11, where in spite of adequate warnings, many still will take the mark of the Beast. They worship Satan as God and in their own deluded state, under God's judgment, they show no inclination to repent and turn to God: "And men were scorched with great heat, and blasphemed the name of God, which hath power over these plagues: and repented not to give him glory" (Revelation 16:9). "And for this cause God shall send them strong delusion, that they should believe a lie; That they all might be damned who believed not the truth . . ." (2 Thessalonians 2:11,12).

Apparently fallen angels all have committed this unpardonable sin and will receive maximum punishment because they had optimum light regarding what they were doing. Like the incorrigibles during the Tribulation, they have stiffened their wills against God to such a degree that they have hardened and made permanent this state of being. There is not even any desire to repent for the Holy Spirit has completely withdrawn Himself toward any redemption for their souls.

In regard to those under degrees of punishment, it is apparent that while the first death is limited to destroying the body, the second death can destroy *both* soul and body (Matthew 10:28).

Those who already have suffered time in the regions of the lost possibly may be accorded credit for the time of their incarceration and thereby have their terms of punishment shortened after the final tribunal. This certainly would be justice. The *properties of the soul* can be subject to punishment and annihilation as well as that of reward and survival. Death is *not* dying, but a merciful *end* to it. The second *death,* therefore, implies termination of dying.

THE LAST WALK

"... This is the second death" (Revelation 20:14).

"All darkness shall be hid in his secret places, a *fire not blown shall consume...*" (Job 20:26).

Wherever punishment is inflicted, it will be in the depths of darkness. The lake of fire is a strange kind of flame that does not light up the darkness but feeds upon itself. It can consume whatever God wills, whenever God wills, and for whatever duration He establishes at the final tribunal.

"For the wages of sin is *death,* but the gift of God is *eternal life* through Jesus Christ our Lord" (Romans 6:23). Note the word "eternal" is associated *only* with life and *not with death.* Does this imply that the second death has a termination of punishment under the justice of God, especially for those not receiving the maximum sentence? If so, then there is a time when the justice of God is satisfied regarding the degree and length of punishment; also, it indicates that sometime in the distant future the incarceration will terminate finally and forever for some.

There is no evidence in God's Word that man's "spirit" being appears at the White Throne Judgment, only his soul and body (John 5:25-30; I Peter 4:6). This is significant for the "spirit" cannot be destroyed. The "spirit" always returns to God (Ecclesiastes 3:21; 12:7).

It is evident that the lake of fire is eternal (Revelation 14:11). Now we must consider whether the justice of God is rendered fairly. The Scriptures are prolific about the fact that the Almighty Jehovah is a "just" God. Those who say that the Almighty can do that which

He pleases fail to consider that God will not act contrary to His character. This would give the illusion that He is a monster, not a "just" God. We must view God in this light if we are to appreciate and understand His true nature. Let us not create a misconception of God by our foolish innuendos. Placing Him in a wrong light can only destroy the facts and divert people from this God of great grace, love, and integrity.

EXTRADITIONS

Holy angels, as security guards, will escort humans, as well as fallen angels to make their appearance at the final tribunal.

They are witnesses to the tragic sufferings of those incarcerated: " . . . shall be tormented with fire and brimstone in the presence of the holy angels . . . " (Revelation 14:10). "The Son of man shall send forth his angels, and they . . . shall cast them (wicked) into a furnace of fire . . . " (Matthew 13:41,42). Thank God, the redeemed will be spared this tragic scene.

MISTAKES ONE CANNOT ERASE

Those in the Kingdom of Heaven who were redeemed but severely disciplined at the Judgment Seat of Christ, will be released from "outer darkness" to appear at the White Throne Judgment. Here the extent of their rehabilitation will be decided. Any restoration of whatever they lost by way of rewards and positions must be determined. There are, however, certain losses they never can recover, such as being in the corporate Wife of Christ as His Royal Queen and the high positions and honors conferred upon the Royal Redeemed. There certainly will be some restoration made to them but to a much lesser degree than the others. The Bible says: "And if the righteous scarcely be saved, where shall the ungodly *and* the sinner appear?" (I Peter 4:18). " . . . Shall be saved: yet so as by fire" (I Corinthians 3:15). "And now, little children, abide in him; that, when he shall appear, we may have confidence, and not be ashamed before him at his coming" (I John 2:28). "And others save with fear, pulling them out of the fire . . . " (Jude 23).

These rehabilitated saints, because of their past despicable conduct, will pay dearly for all eternity. No wonder God will have to blot out the past memories and comfort them in their demoted status: "And God shall wipe away *all tears* from their eyes . . . " (Revelation 21:4). Tears speak of emotions. They are visible evidence of the Soul. These are the ones Jesus referred to in His dissertations of the mysteries of the kingdom when He included the word "weeping" (Matthew 8:12; 22:13; 25:30). The word, "tears," also reveals to us the fact that those appearing at this final judgment will appear in flesh because tears are very tangible. Tears are not applied to the "wicked" because of their hardened state of being (Matthew 13:42,50).

When God wipes away tears He eradicates the causes by blotting out the memories that create such sadness, and furthermore, he helps the saints to joyfully accept their reduced rank throughout eternity.

THE EMPTYING OF PARADISE

"And I saw the dead, small and great, stand before God; and the books were opened: and *another book* (book of life) was opened . . . " (Revelation 20:12).

Some who appear at this final tribunal will have their names in the book of life, so obviously there must be a reason why they never made the "first resurrection." Therefore, they either died prematurely before they decided in favor of Christ; or they never were given the "knowledge of the truth" so they could make a proper decision; or their works were of a nature that required God to retain their names in the book of life: "How shall they call on him in whom they have not believed? and how shall they believe in him of whom they have not heard? . . . " (Romans 10:14). " . . . Sin is not imputed when there is no law" and " . . . where no law is, there is no transgression" (Romans 4:15; 5:13). This included children in Paradise who will be held there until the White Throne Judgment (Revelation 20:12). They must make their own decision. Innocence has no atoning value, nor does martyr's blood. Only the precious blood of Jesus Christ has atoning value.

The Royal Redeemed look forward to their resurrection as the time when they receive their bodies of immortality: "... I wait," said Job, *"till my change come"* (Job 14:14,15). On such a body, "... the second death hath no power" (Revelation 20:6).

There must be a mandatory resurrection to the final tribunal if Christ is to judge mankind in the "flesh" (I Peter 4:6). Therefore, the soul and body will be reunited in a resurrection (John 5:21-23; Daniel 12:2).

A "second death" can be possible only when these two are brought together. The word, "whosoever," in Revelation 20:15 is from the Greek *et tis* meaning "if anyone." It does not mean "all." Jesus cannot judge any human beings while they remain in "spirit" form; He must wait until their souls are reunited, via resurrection, with their bodies of "flesh."

REINCARNATION

Is it even remotely possible for someone after death to be reincarnated later into another body on earth? Besides being biologically impossible, there are these factors to consider: first, Scriptures only teach an incarnation at the resurrection into one's *own* body which is to go directly to judgment (John 5:22-31; Mark 5:1-17; 2 Corinthians 5:10). Jesus recognized that evil and lying spirits can possess a body and inject information into the mind when it is receptive to suggestions. This is not reincarnation but demon possession. It is Satanic!

The Bible teaches that the human "spirit" is never loose upon the earth after one's demise. The "spirit" goes directly to God (Ecclesiastes 12:7); the "soul" goes to its assigned destination in a conscious state (I Peter 3:19).

One final observation: Scriptures nowhere teach that a person will appear at the White Throne Judgment (or the Judgment Seat of Christ) in more than one body, as one is more than sufficient to be accountable for on that day.

FUTURE LOST

At this final tribunal there is no appearance at a higher court, no plea bargaining, no availability of bail bonds, no reduction in sentence, no

bribes, no plea of insanity, no "not guilty" plea, no pardon (He is Judge now, not the Savior), no stay of punishment, no commutation of sentence, no mistrial declared, no miscarriage of justice, no hung juries, no parole boards, no rescue squads, no "passing the buck," no defense attorneys, no excuses accepted, and no escape possible from the place of confinement. Tears will be ineffective. the decision is final.

There will be those standing before the Magistrate who have refused to heed the pleas for mercy from their victims. therefore, God will say to them: "When your fear cometh as desolation, and your destruction cometh as a whirlwind; when distress and anguish cometh upon you. *Then shall they call upon me, but I will not answer . . .* Therefore shall they eat of the fruit of their own way . . . " (Proverbs 1:27,28,31).

No wonder we read of them: ". . . there shall be wailing and gnashing of teeth" (Matthew 13:42).

SUMMATION

If there are no degrees of punishment:

1. The mild-mannered will suffer equally with the vicious.
2. Those who never have heard of the Gospel of Jesus Christ will suffer equally with those who have heard but rejected.

If there are degrees of punishment:

1. There will be terminal points of punishment.
2. Second death has the power to terminate the existence of both body and soul, according to Matthew 10:28.
3. Second *death* in the lake of fire is diametrically opposite of *eternal life.*

DEATH'S DOMINIONS

The first death has power to destroy only the body. In the same manner, the second death has the power to destroy both *body and soul.* The second death is the final, eternal, and absolute separation from God. The dominion of the soul is hell, a temporary place of incarceration for those not in the book of life. The dominion of the resurrected soul and body will be the lake of fire, a permanent place of incarceration for those not in the book of life.

Why is the destruction of death delayed until the dispensational end of the New Heaven when all of the redeemed receive their bodies of immortality to enter the Heavenly Father's Kingdom? It is to enable all of those who are incarcerated in the lake of fire under a terminal sentence to serve out their full punishment (see also the summary, "Why the New Heaven is Destroyed").[1] Those whose punishment extends beyond that point will suffer maximum interminable penalty without any hope of egress. There is eternal punishment, but not for all. *The soul is mortal only to the second death.*

The "last enemy" is the "second death" which has power over both body and soul. Until all of those in the book of life have been redeemed, it is essential that death remain to deal with the unredeemed. This also reveals to us that the White Throne Judgment will not conclude its sessions until the last unredeemed in the New Heaven have been sentenced and have served out their terminal sentence. "The last enemy that shall be destroyed is (the second) death" (I Corinthians 15:26).[2]

PHYSICAL FEATURES OF LOST SOULS

Man was created in the "image and likeness of God" but it is very unlikely that this image will remain with the lost. It is more liable to take on the distorted configurations of fallen angels. These disfigurations are reflected when demonic power possesses its victims (Revelation 9:1-11; 12:7-9).

[1] See page 596.
[2] See chart Y.

90

Genesis III—The Heavenly Father's Manifested Kingdom

In the Father's Kingdom, the New Heavens, there will be many realms ("dominions" - Daniel 7:27) over which the Royal Redeemed will reign forever.

"Behold, I make all things new . . . " (Revelation 21:5).

"For as the new heaven*s* and the new earth, *which I will make, shall remain* before me, saith the Lord . . . " (Isaiah 66:22). This infers that the new heaven (singular) shall *not* remain. " . . . In the heavenly places" (original, "heavenlies" Ephesians 1:20; 2:6).

The tree of life, the river of life, and the Glory of God from the Holy City will be removed from Paradise to be relocated on the New Earth in the second heaven when the New Jerusalem descends. This, together with the vacating of the population of Paradise in the third heaven, will bring regression and disarray, unless as previously stated, God makes other provisions before He descends into the Holy City where the Royal Redeemed are assembled (Revelation 21:2). Planet Earth, after the Millennium, will become a veritable Paradise as the New Earth: "And they shall say, This land (Planet Earth) that was desolate is become *like the garden of Eden* (and) inhabited" (Ezekiel 36:35).

According to Ezekiel, chapter 47, the river of life that once flowed on Planet Paradise, and which came with the descending

Holy City onto the New Earth, will begin to flow over the New Earth: " . . . These waters issue out toward the east country . . . and go into the sea (Pangaea has become a reality again) . . . and everything shall live whither the river cometh" (verses 8,9).

The next creation will be the New Heavens (universes) where the Royal Redeemed will have their respective realms ("dominions" Daniel 7:27) forever (Revelation 22:5). The wonders of the Heavenly Father's Kingdom, its size, and its composition excel all human description and all previous creations. It is an entirely *new* dimension, a new concept, in which there will be different laws of physics and chemistry, with atoms and molecules exceeding even their prototypes. These eternal "things" (2 Corinthians 4:18) are designed exclusively for all of the redeemed now in their eternal bodies of immortality.

The redeemed, in immortal yet tangible bodies, will require eternal substance if their lifestyle is to be happy, productive, and meaningful. They will create things from eternal materials of totally unique composition and arrangement which will not decay, rust, or deteriorate. Time will not be wasted in learning, preserving, and repairing. Serving under Christ in both the manifested kingdoms will be a self-fulfilling and rewarding opportunity. The redeemed will be majoring in majors and not majoring in minors.[1]

Within the heavens (universes or realms) it is possible there may be production centers (equivalent to white holes) that will pour forth ultra-matter in keeping with the new dispensational bodies of the redeemed. Thus, there will be a constant supply of eternal ultra-substances to utilize.

TIMETABLE FOR ETERNITY

Education will be gained, not by the difficult route of books, but by osmosis which will be faster, more complete, enduring, and proficient. Application of the knowledge will bring much fulfillment. The most informed person today can retain in his lifetime only less than one percent of the available material which is published. How much is missed!

With perfect enlightenment in the New Heavens, how will one occupy himself? Since the Heavenly Father, with His perfect knowledge, applies it to new creations, people obviously will translate their learning into achievements. Such ultimate knowledge will enrich and beautify the New Heaven*s* (realms) in His Kingdom. Inactivity, except for reflection and social interaction, is totally inconsistent with the Almighty. Passivity is due primarily to physical deficiencies and the denial of opportunities; however, the immortal body will be imbued with energy, activity, ability, and ambition to achieve.

". . . Knowledge, it shall vanish away. For we know in part . . . But when that which is perfect is come, then that which is in part shall be done away" (I Corinthians 13:8-10). "So also is the resurrection of the dead. It is sown in . . . weakness; it is raised in power . . . " (I Corinthians 15:42,43).

WONDERLANDS IN ETERNITY

The Heavenly Father's Manifested Kingdom is called the "New Heaven*s*." This plurality of "heaven*s*" gives credence to the concept there will be unnumbered redeemed just as there are innumerable angels that occupy the "Heavens of Heavens" (Hebrews 12:22; Revelation 5:11; Daniel 7:10) and they will have their own domains within the Father's Kingdom (Daniel 7:27). Note the words "dominions" and "heavens" in Isaiah 65:17; 66:22.

DISTORTING THE FACTS

The ideas that one will be roaming about in an enormous mansion, playing a harp, eternally sitting on a cloud, floating and whirling about as spirit beings, resting on the banks of the river of life, or staring constantly into the face of Jesus Christ is ludicrous, inconsistent, and unscriptural. Such narrow and ill-informed concepts involving only one's own religious clan, make heaven unreal, boring, and lonely, leaving one with no incentive to go there, except to escape eternal punishment because the alternative offers pain instead of lethargy. However, the reality is that the redeemed have an optimistic hope that is genuine: "Blessed be the God and Father of our Lord Jesus Christ, which according to his abundant mercy hath begotten us again *unto a lively hope* by the resurrection of Jesus Christ from the dead" (I Peter 1:3).

WORLD WITHOUT END

"Unto him be glory in the *church* by Christ Jesus throughout all ages, *world without end.* Amen" (Ephesians 3:21). "May he be given glory forever and ever through endless ages because of his master plan of salvation for the church through Jesus Christ" (Ephesians 3:21, *Living Bible*). "But *Israel* shall be saved in the Lord with an everlasting salvation: ye shall not be ashamed nor confounded *world without end."* (Isaiah 45:17).

The "Church" and "Israel," on Planet Earth are both involved in the eternal and ultimate destiny of God's primordial redemptive plan. Their survival is assured because they are the prime objectives in God's plan for the eternal future through redemption.

Someone has aptly said that frequently one cannot see the forest for the trees. Because we are part of the universe in which we live, we fail to comprehend God's plan beyond our cosmos. We readily accept the idea of the conflagration of our universe but fail to recognize that the polluted "New Heaven" likewise is scheduled for the same kind of inferno in its future. But again the earth, now the New Earth, will escape that fiery holocaust.

Once Planet Earth becomes the New Earth, not one unredeemed person will ever again set foot on its soil: "And the nations of them which are *saved* shall walk in the light of it . . . " (Revelation 21:24). "For without (in the New Heaven) are dogs, and soccerers, and whoremongers, and murderers, and idolaters, and whosoever loveth and maketh a lie" (Revelation 22:15).

Only the New Heaven*s* and New Earth (God's footstool) will remain forever, together with the "Heavens of Heavens," and possibly the Heavenly Paradise: "For as the new heaven*s* and the new earth, which I will make, *shall remain* before me, saith the Lord (Heavenly Father) . . . " (Isaiah 66:22). "Nevertheless we, according to his pro

mise, look for new heaven*s* (Father's Kingdom) and a new earth, wherein *dwelleth righteousness"* (2 Peter 3:13).
"... Ye heavens of heavens (angels' habitations)... also established *for ever and ever"* (Psalm 148:4-6).

Not only is our universe unclean in the eyes of God, but also the New Heaven: "...yea, the heaven*s* (planetary and New Heaven) being on *fire* shall be *dissolved,* and the elements shall melt with fervent heat" (2 Peter 3:12). "For behold, I create new heaven*s* and a new earth: and the former (new heaven) *shall not be remembered, nor come into mind"* (Isaiah 65:17).

The Planet Earth, because it has been redeemed together with Israel and the Church, will survive. The earth has a special place in God's plan of redemption: "Again, the kingdom of heaven is like unto treasure hid in a field; the which when a man hath found, he hideth, and for joy thereof goeth and selleth all that he hath, and *buyeth that field"* (Matthew 13:44).

SPACECRAFT EARTH'S THIRD FLIGHT

Earth's first flight was at the time Lucifer sinned, and as God's footstool, it was removed into an orb of blackness far from the Paradise of God, which is the location of God's Throne. Its second flight will be after the Millennium when it becomes the New Earth in the New Heaven. Its third and final flight will be back to its original position in the third heaven, the Heaven of Heavens. There, major changes will occur with the Planet Earth and it will revert to being the beautiful footstool of God. There is no indication that God will relinquish His position on the New Earth in favor of the Heavenly Paradise, because the former has become a veritable Paradise.

The destruction of death will coincide with the last execution by the second death. When the last unredeemed person is deceased and then judged at the White Throne Judgment, the sentence imposed will determine at what point his suffering will cease, unless he has been given a sentence of eternal punishment. At this point all of the redeemed will rejoice that death has been conquered, and they will view the cemeteries of the unredeemed in the New Heaven at the last

Memorial Day of all time: "And they shall go forth (into the many habitable places in the New Heaven to pay their final respects to their open gravesites), and *look* upon the carcasses of the men that have transgressed against me: for *their worm* (soul, Job 25:6; Psalm 22:6; Isaiah 14:11; Mark 9:44,46,48), shall not die, neither shall *their fire* be quenched (an eternal crematory); and they shall be an abhorring unto all flesh" (Isaiah 66:24).

Another reason that the New Heaven must be destroyed is that death has remained active over all of its inhabitants. This is with the exception of the redeemed who have been given a stay from death because of the tree of life; they will receive their immortal bodies at this time for death will be destroyed. The first death will end its dominion over all forms of life in the fiery blaze of the New Heaven. Mortality will be no more. This summarizes why the New Heaven will be destroyed:

1. Transgressors have occupied and polluted it with their sin (Isaiah 66:24; Revelation 21:27; 22:14,15).

2. All persons living are now redeemed (Isaiah 66:23).

3. All transgressors are deceased (Isaiah 66:24).

4. The New Heaven has served its purpose in God's plan of redemption (I Corinthians 15:24).

5. No Scriptures support its eternal existence, only the New Heaven*s* (Isaiah 66:22).

6. No Scriptures promise the redemption of the New Heaven polluted by sin.

7. The New Heaven will be turned into a veritable lake of fire for its own malfeasants (Isaiah 66:24).

8. Death will take its final gasp here in a fiery blaze forever (I Corinthians 15:25,26; Isaiah 66:24).

9. The Adamic nature of the redeemed will be replaced by the nature of the Second Adam, the Lord Jesus Christ (I Corinthians 15:22,45; Romans 4:11. Note the word "imputed").

10. The time has come for the Royal Redeemed to have their own realms to rule over in the Father's Manifested Kingdom, called the "New Heaven*s*" (Revelation 21:7; 22:5; Isaiah 66:22; Daniel 7:27; Matthew 13:43).

11. The roles of the Royal Redeemed as priests, lords, and judges have come to an end since there are no more unredeemed alive to require such services.

12. All of the redeemed from the Lord's Manifested Kingdoms (the Millennium and the New Heaven) will move their residences into something better—the eternal Father's Kingdom.

13. It is time for *"all* things" to become new (Revelation 21:5).

14. It is time for Christ Jesus to turn His Kingdom over to the Heavenly Father and His Kingdom (I Corinthians 15:24).

All of the redeemed in immortal bodies, who have been scattered throughout the New Heaven, will be assembled. Then the earth will become a spacecraft again, and with the Holy City riding upon it, it will make its final journey through a corridor of angels, to its original place in the third heaven, called the "Heaven of Heavens." As soon as the spacecraft, with its passengers, is at a safe distance, the New Heaven will become an inferno (2 Peter 3:12; Hebrews 1:10-12). The Holy City, including the Royal Redeemed as well, will become the epicenter of the Heavenly Father's New Kingdom with the Glory of God surrounding it. Now with the earth at its primordial location, the plan of redemption will have gone full circle, ending where it all began.

FUTURE FRONTIERS

The five sacred feasts of the Old Testament economy will be preserved and celebrated as holidays forever because they express the basics of redemption: atonement, Holy Spirit, fellowship, immortality, and inheritance. These special feasts in ancient times were observed as "a shadow of things *to come...*" (Colossians 2:16,17; Hebrews 8:5). The redeemed must never forget these major truths in the process of redemption and its consummate accomplishments.

Visualize the Planet Earth and our cosmos as spheres with their north poles in direct line toward the Throne of God (Job 26:7; Psalm 48:2; Isaiah 14:13). Surrounding God's Throne in the Heaven of Heavens (located in the third heaven) are the "Heaven*s* of Heavens" or abodes of the holy angels. Outside of these, or extending further north above the Throne of God and beyond the Heaven of Heavens is the area of God's Manifested Kingdom.[2]

In the Father's Kingdom, the New Heavens, there will be many realms ("dominions" Daniel 7:27) over which the Royal Redeemed will reign forever. The multitudes of rulers are consistent with God's Word: "And the kingdom . . . shall be given to the people (Royal Redeemed) of the saints (Redeemed) of the most High (Heavenly Father), whose kingdom is an *everlasting kingdom* and all *dominions* (realms) shall serve and obey him" (Daniel 7:27). "Then cometh the *end*, when he (Christ) shall have delivered up the kingdom *to* God, even the Father . . . For he (Christ) must reign, till he hath put all enemies under his feet. The last enemy that shall be destroyed is death" (I Corinthians 15:24-26). "Thy kingdom is an everlasting kingdom, and thy dominion endureth throughout *all* generations" (Psalm 145:13).

A KINGDOM OF KINGS

The redeemed may serve in high positions of great honor under the Royal Redeemed who are the Wife of Christ, the Queen of all God's remaining and future creations. They may hold honorable posts such as consorts, staff members in the royal palaces, ambassadors, or couriers between the many realms.

God said, "Behold, I make all things new." Who can conceive what the full extent of this will be? Can anyone imagine God going into *retirement?* The resourcefulness of His Kingdom is incomprehensible. Might the Creator not bring into existence other kinds of beings, as subjects, to dwell in these various realms under the Royal Redeemed as monarchs?

FESTIVALS FOR THE MASSES

Mass gatherings could take on a variety of distinctive flavors in each of the realms, according to interests of the reigning monarch. Some dominions may predominate in musical festivals, some in oratory, and others in art, but all will have worship and exaltation of the Creator, the Redeemed, and the Holy Spirit.

Nowhere will there be evangelistic appeals, prayers for the sick, bereaved, depressed, or the impoverished. These will have become non-existent long ago. Joy and worship will prevail. There will be no tears, unhappy faces, or sad memories; only endless bliss, well-being, and challenges of a different dimension.

"... These things saith the Amen... the *beginning* of the creation of God" (Revelation 3:14).

"... The former things are *passed away*... behold, I make all things *new*..." (Revelation 21:4,5).

"... And they shall reign *for ever and ever*" (Revelation 22:5).

Indeed the stakes are high for those who now faithfully serve God within the framework of the "mystery" Kingdom of God, the dispensation of the Church.

A PARTIAL CONNECTION

God tells us that "... he hath set the world in their hearts, (yet) so that no man can find out the work that God maketh from the beginning to the end" and "Then I beheld all the work of God... that is done under the sun: because though a man labour to seek it out, yet he shall not find it; yea farther; though a wise man think to know it, yet shall he not be able to find it" (Ecclesiastes 3:11; 8:17).

Obviously these are part of the "sacred things" which will be revealed in their time: "The secret things belong unto the Lord our God: but those things which are *revealed* belong unto us and to the children for ever, that we may do all the words of this law" (Deuteronomy 29:29). "But there is a God in heaven that *revealeth secrets*..." (Daniel 2:28). "And I heard, but *understood not*... And he said... the words are *sealed till the time of the end*" (Daniel 12:8,9).

Two facts are clear: first, there are secrets of God that he chooses to reveal in their time; second, they can be comprehended through divine ability.

SHADOWS ON SUBSTANCE

"Let no man therefore judge you in meat, or in drink, or in respect of an holyday, or of the new moon, or of the sabbath day:

Which are *a shadow of things to come;*but the body is of Christ" (Colossians 2:16,17).

All of the sacred feasts, even the Ten Commandments, under the Mosaic Law are but shadows of the real substance that will come to pass in the future because of redemption under Jesus Christ, the Redeemer. These all were zeroed in upon mortal beings who have a tendency to sin. They all point toward the true Substance, Jesus Christ.

When immortality of the body is realized for all of the redeemed, prefigured under the last of the sacred feasts called the feast of tabernacles, then redemption in every category under the Mosaic Law will have been completed. King David then will step down from his throne because the purposes for which he was enthroned will be finalized. Eternity, under the Law, applies to mortal beings and when mortality ends, the shadow vanishes and the Sun of Righteousness becomes the focal substance and celebration.

The Ten Commandments will have no further meaning since they are applicable only to mortal beings with the tendency to sin and disobey. But in immortality all such tendencies are gone: "But when that which is perfect is come, then that which is *in part shall be done away*" (I Corinthians 13:10).

THE FINAL CONNECTION

"God is faithful, by whom ye were called unto the fellowship of his Son Jesus Christ our Lord" (I Corinthians 1:9).

"For your fellowship in the gospel . . . " (Philippians 1:5).

In the distant past, long before God entered into His work of creation as we know it, there must have been unexcelled fellowship within the Godhead. The creation of "time" was designed to bring others eternally into that family of fellowship and to promote the achievement of everyone's fullest potential.

Throughout the eternal future, the redeemed not only will be occupied totally in creative activity, but also they will enjoy social interaction. They will then reflect and communicate the unending scope of their understanding regarding the wonders of God's creation. There are many with scientific interests in the Father's Kingdom who are in the enviable position of comprehending the total connection of the cosmic puzzle. Science, as we have shown in this book, subscribes to the existence of other dimensions beyond our third mortal cosmic dimension, but it is unable to describe them.

Everything that Satan has torn asunder will be restored, and all of the redeemed will be brought together under Christ, the Redeemer, so that they may forever bathe in the glories of the Father's Kingdom. Then they will be in a position to fully comprehend and eternally appreciate the total connection of God's redemptive plan. All of the undesirable flaws caused by sin, such as knowing in part, prophesying in part, and seeing things as through a dark glass (even with sophisticated scientific instruments), will be permanently removed. Now they will clearly understand it all (I Corinthians 13:9-12).

When both "time" and "eternity" merge under one final government, which is the Heavenly Father's Manifested Kingdom, then and only then, will all of the redeemed from every walk of earthly life and occupation observe, comprehend, and enjoy eternally

THE

TOTAL

CONNECTION!

[1]The spirit entity has the capacity to create third-dimensional properties. Therefore, it is consistent that the redeemed in the Father's kingdom will be able to instantly create from fourth dimensional properties any desired item without the laborious time-consuming task of construction.

[2]See chart on pages 632 and 634.

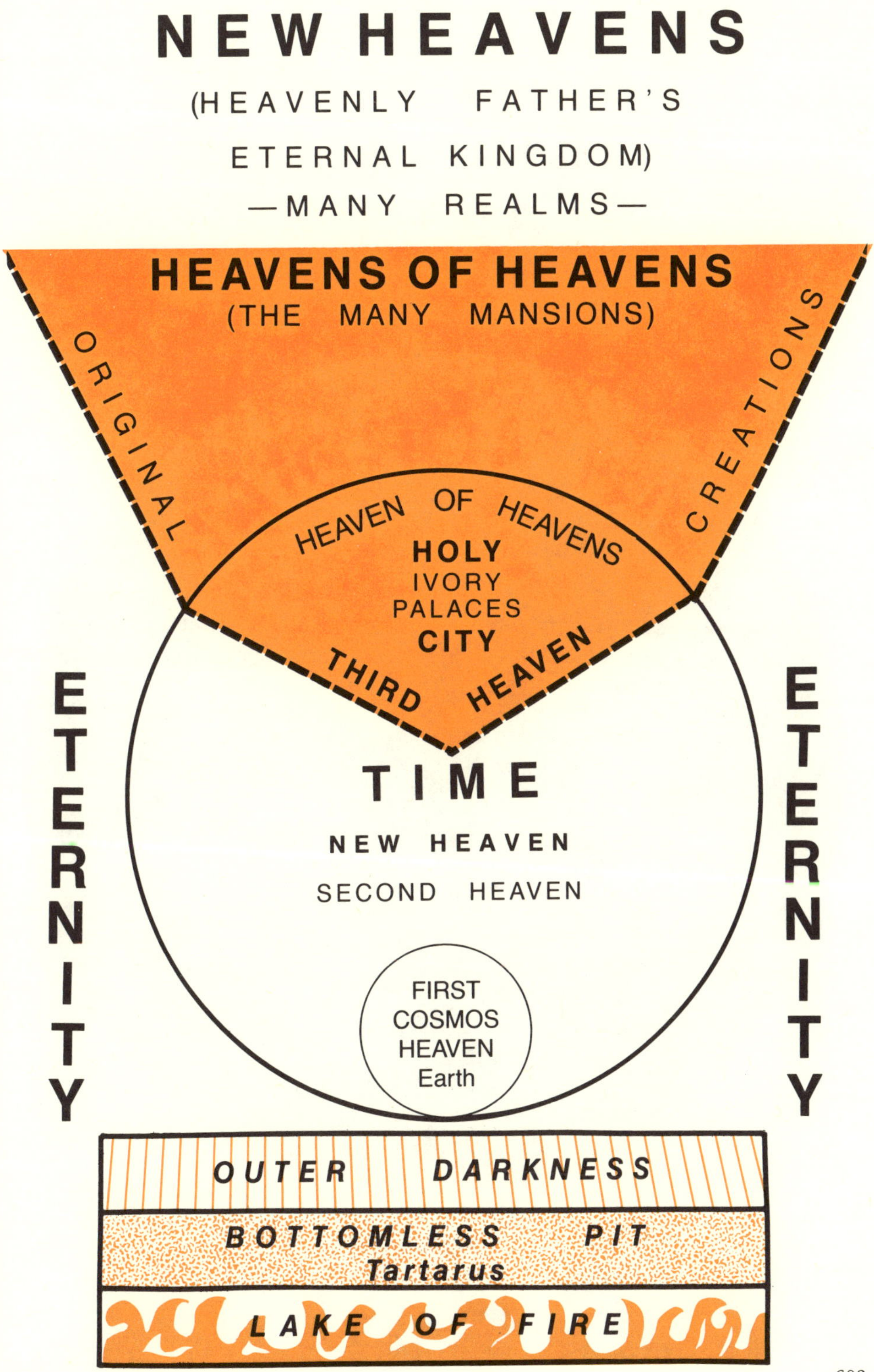
NEW HEAVENS
(HEAVENLY FATHER'S
ETERNAL KINGDOM)
—MANY REALMS—
HEAVENS OF HEAVENS
(THE MANY MANSIONS)
ORIGINAL
CREATIONS
HEAVEN OF HEAVENS
HOLY
IVORY
PALACES
CITY
THIRD HEAVEN
ETERNITY
ETERNITY
TIME
NEW HEAVEN
SECOND HEAVEN
FIRST
COSMOS
HEAVEN
Earth
OUTER DARKNESS
BOTTOMLESS PIT
Tartarus
LAKE OF FIRE

INTERSTELLAR PLANTINGS

Cosmic Seeds

LIGHT OF ETERNITY

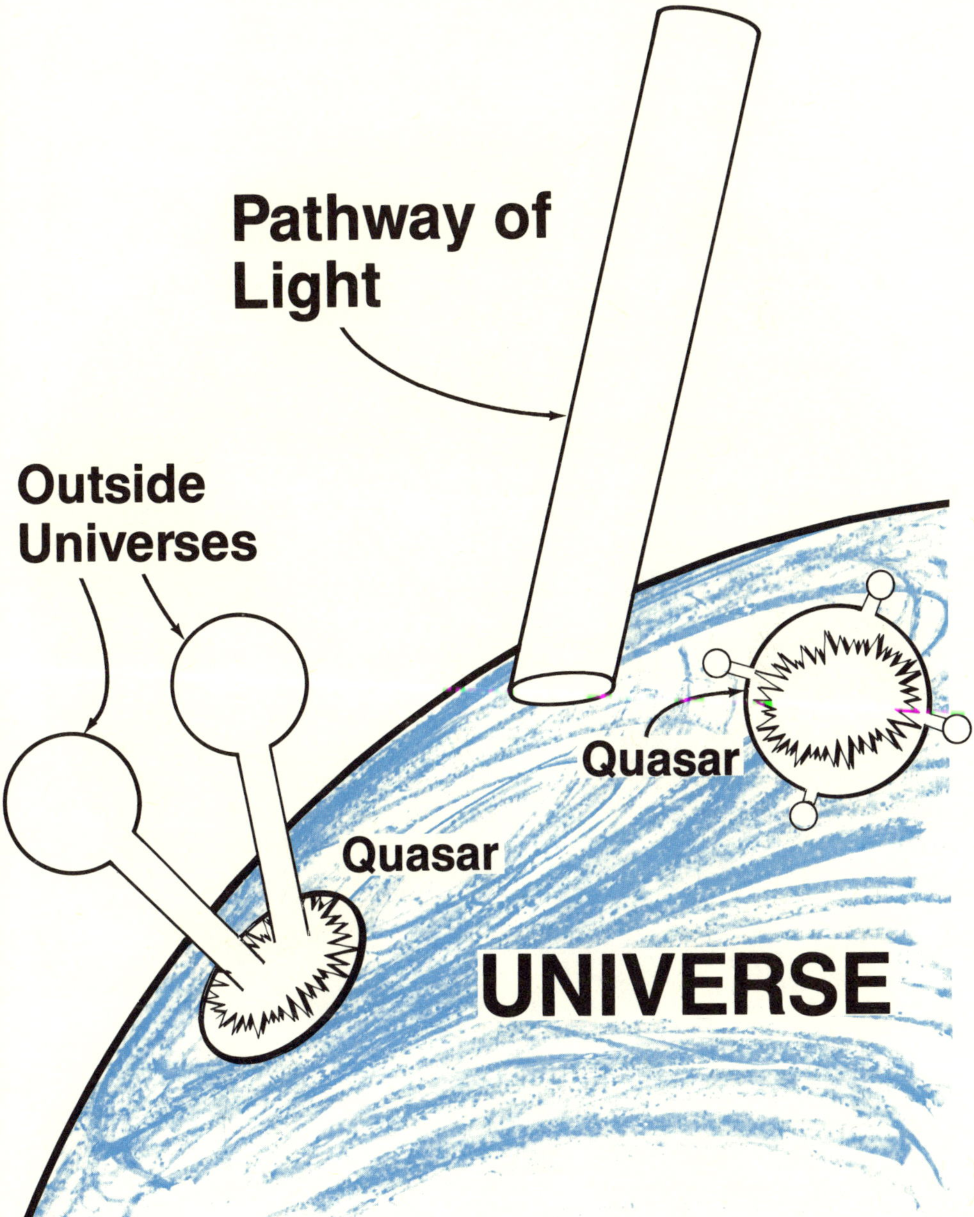

HOW INCREASED VELOCITY REDUCES TIME

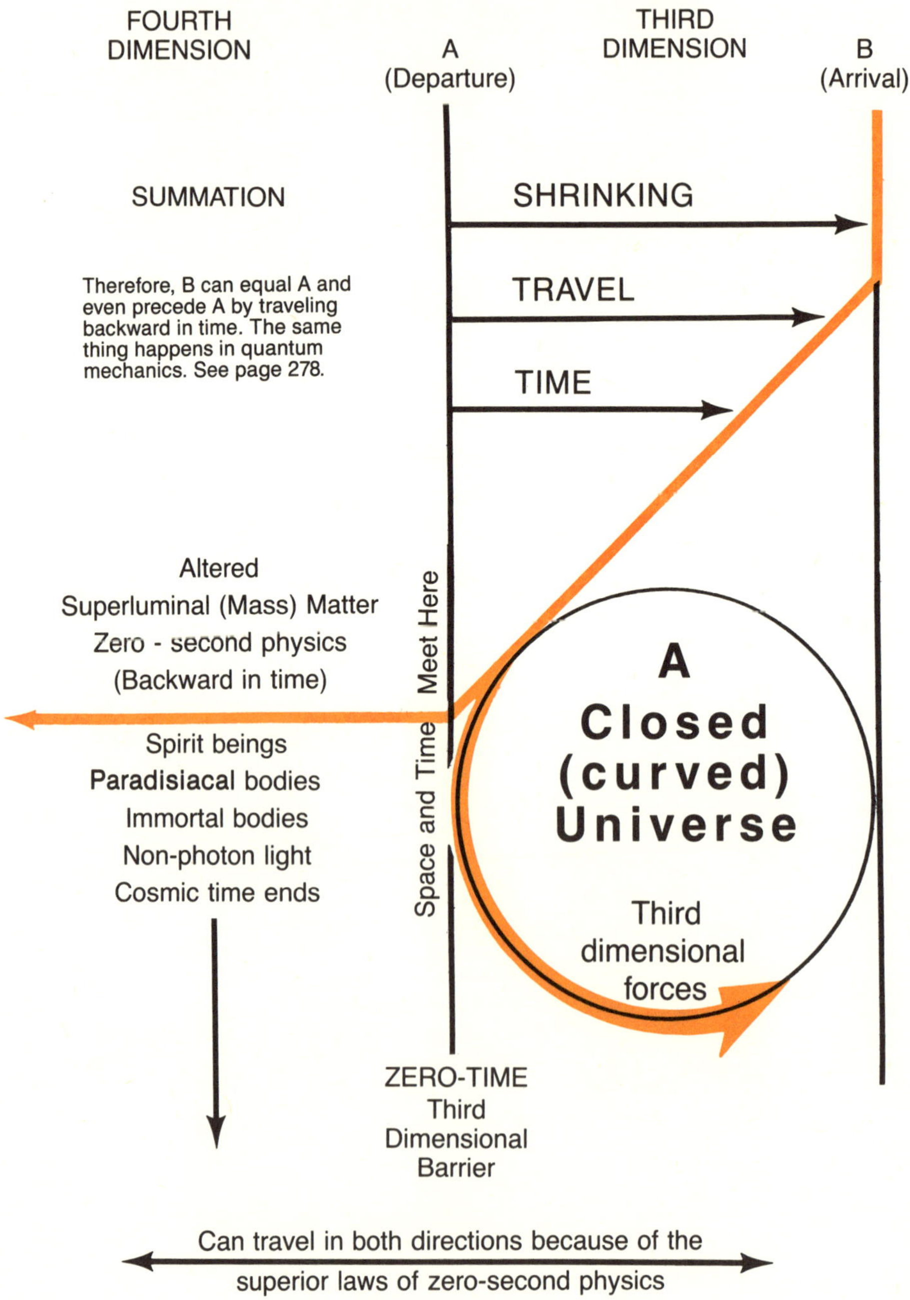

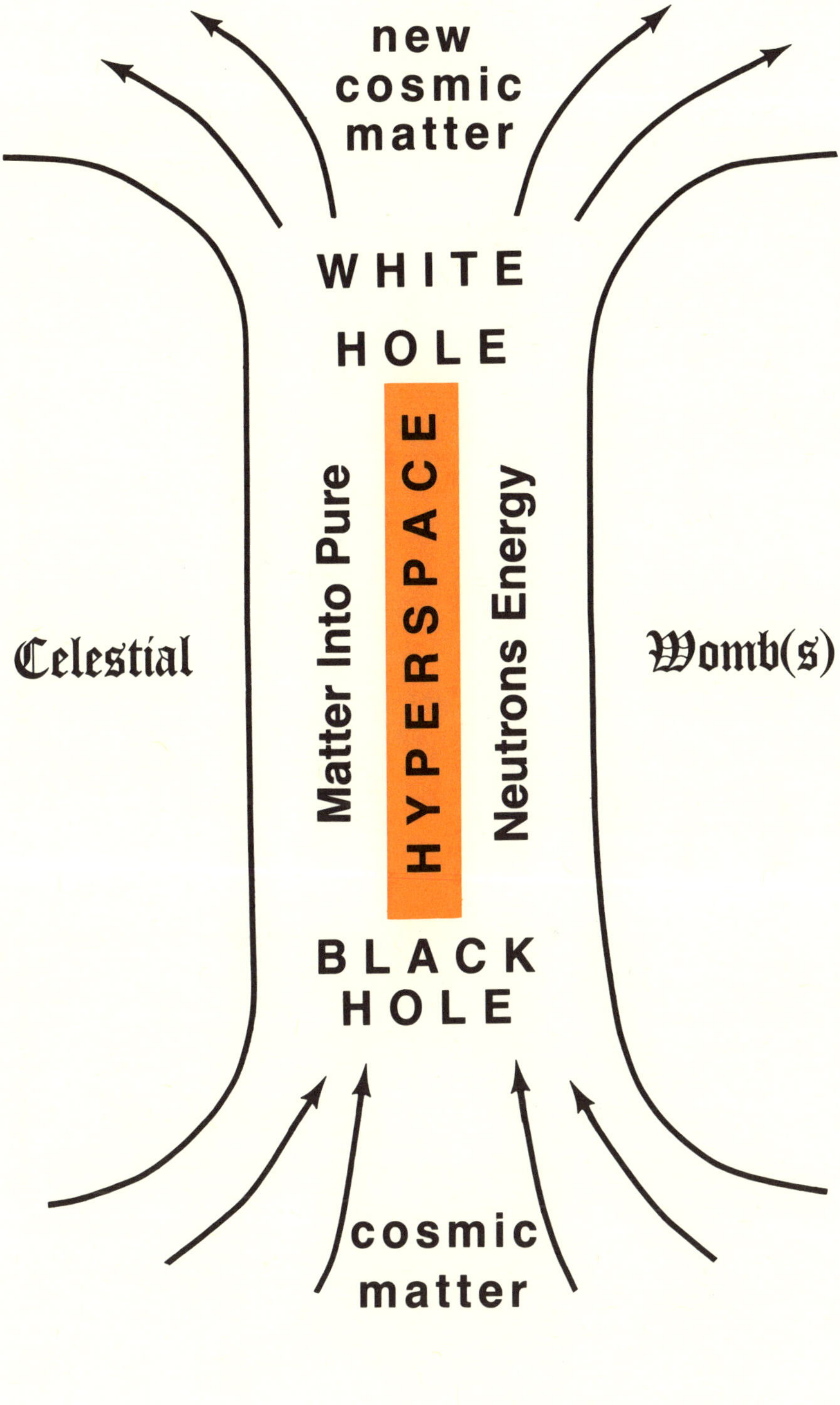
new
cosmic
matter
WHITE
HOLE
Matter Into Pure
HYPERSPACE
Neutrons Energy
Celestial
Womb(s)
BLACK
HOLE
cosmic
matter

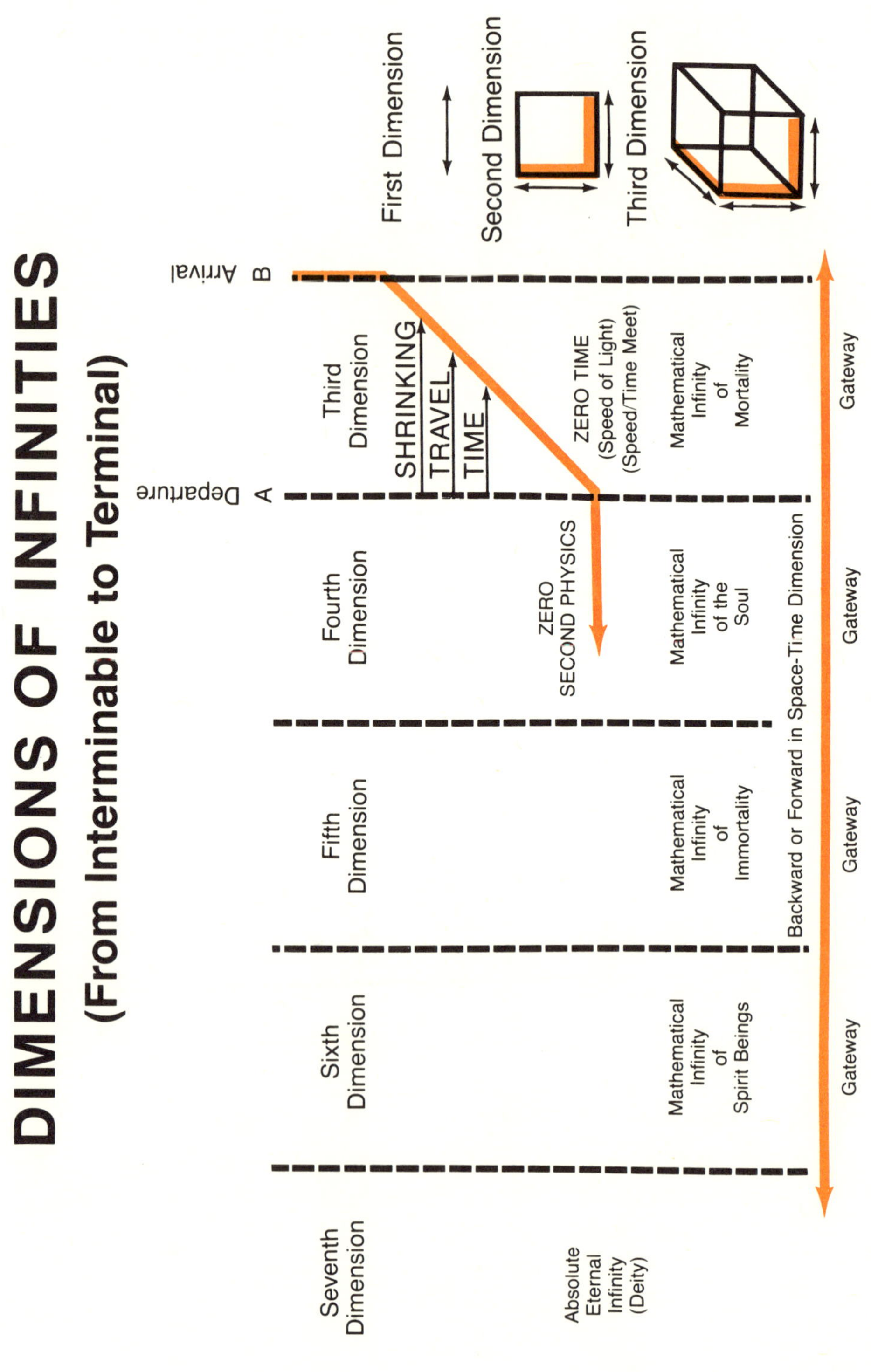
DIMENSIONS OF INFINITIES
(From Interminable to Terminal)
Departure
A
Arrival
B
First Dimension
Second Dimension
Third Dimension
Third Dimension
SHRINKING
TRAVEL
TIME
ZERO TIME
(Speed of Light)
(Speed/Time Meet)
Mathematical Infinity of Mortality
Gateway
Fourth Dimension
ZERO SECOND PHYSICS
Mathematical Infinity of the Soul
Gateway
Gateway
Fifth Dimension
Mathematical Infinity of Immortality
Gateway
Backward or Forward in Space-Time Dimension
Sixth Dimension
Mathematical Infinity of Spirit Beings
Gateway
Seventh Dimension
Absolute Eternal Infinity (Deity)

Visible & Invisible

Entities

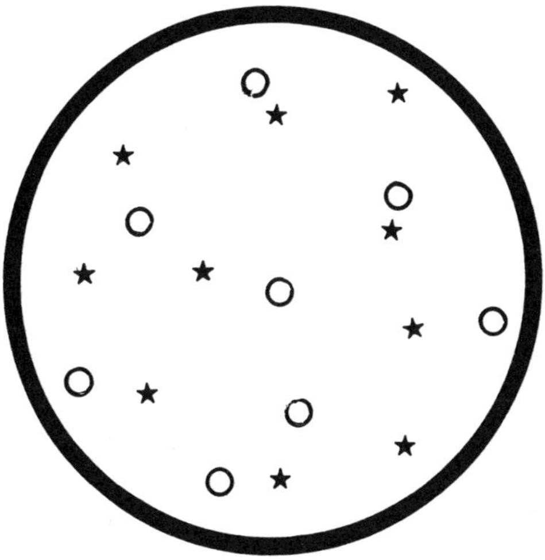

* VISIBLE ENTITIES: Known quantities

○ INVISIBLE ENTITIES: Unknown quantities

PHYSICISTS only accept entities that can be observed

THEORISTS subscribe to the possibility of existing entites yet unobserved

REALISTS recognize that beside visible entities there are invisible entities which scientists will never observe because they are fourth dimensional.

ILLUSTRATION VIII

DISPLACEMENT OF THE UNIVERSE

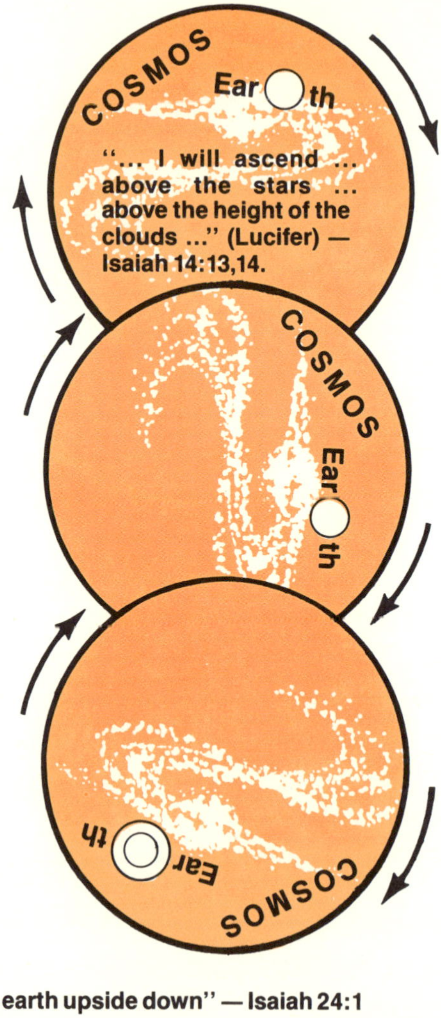

"The Lord turneth the earth upside down" — Isaiah 24:1

"And darkness was upon the face of the deep" — Genesis 1:2

"(He) sealeth up the stars" — Job 9:7

"The heavens ... had no light (the) heavens above be black" — Jeremiah 4:23,28

"I clothed the heavens with darkness, and I make sackcloth their covering" — Job 50:3

"... the foundations of heaven moved and shook, because (God) was wroth" — 2 Samuel 22:8

"He holdeth back the face of his throne ... the pillars of heaven tremble and are astonished at his reproof" — Job 26:9-11.

FLIGHT OF THE SOUL AFTER ONE IS DECEASED

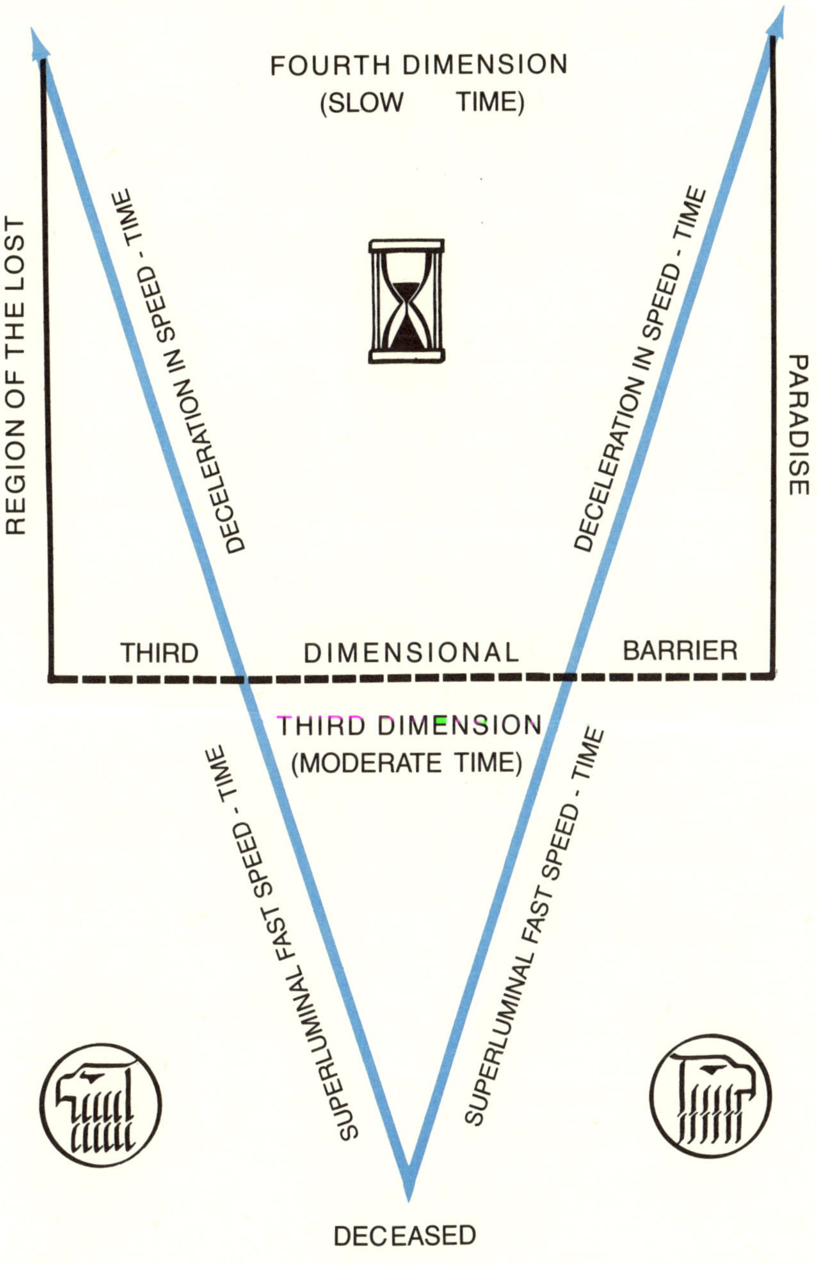

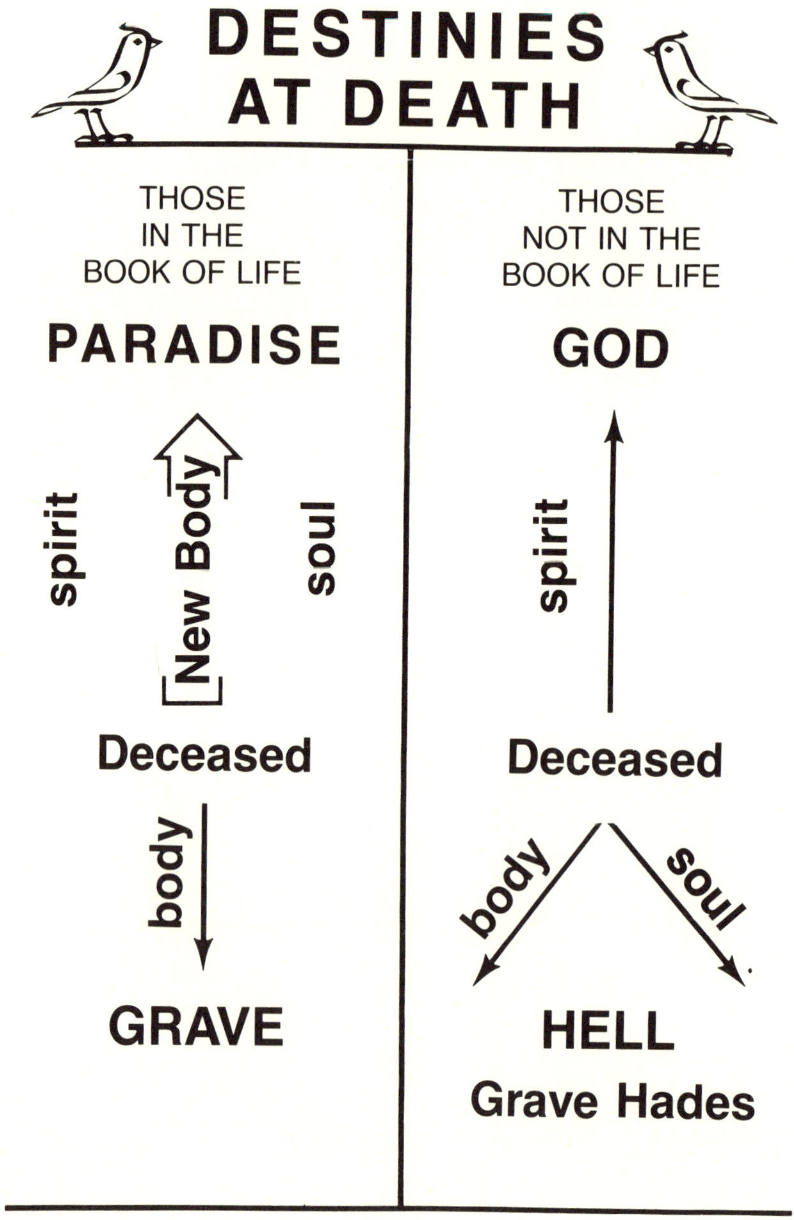
DESTINIES
AT DEATH
THOSE
IN THE
BOOK OF LIFE
PARADISE
spirit
New Body
soul
Deceased
body
GRAVE
THOSE
NOT IN THE
BOOK OF LIFE
GOD
spirit
Deceased
body
soul
HELL
Grave Hades

Sword of the Lord

REFERENCES:
HABAKKUK 3:11,12; JEREMIAH 25:3l;
REVELATION 19:15; MATTHEW 24:27;
LUKE 17:33-37

GENERATIONS OF THE EARTH AND THE PLAN OF REDEMPTION

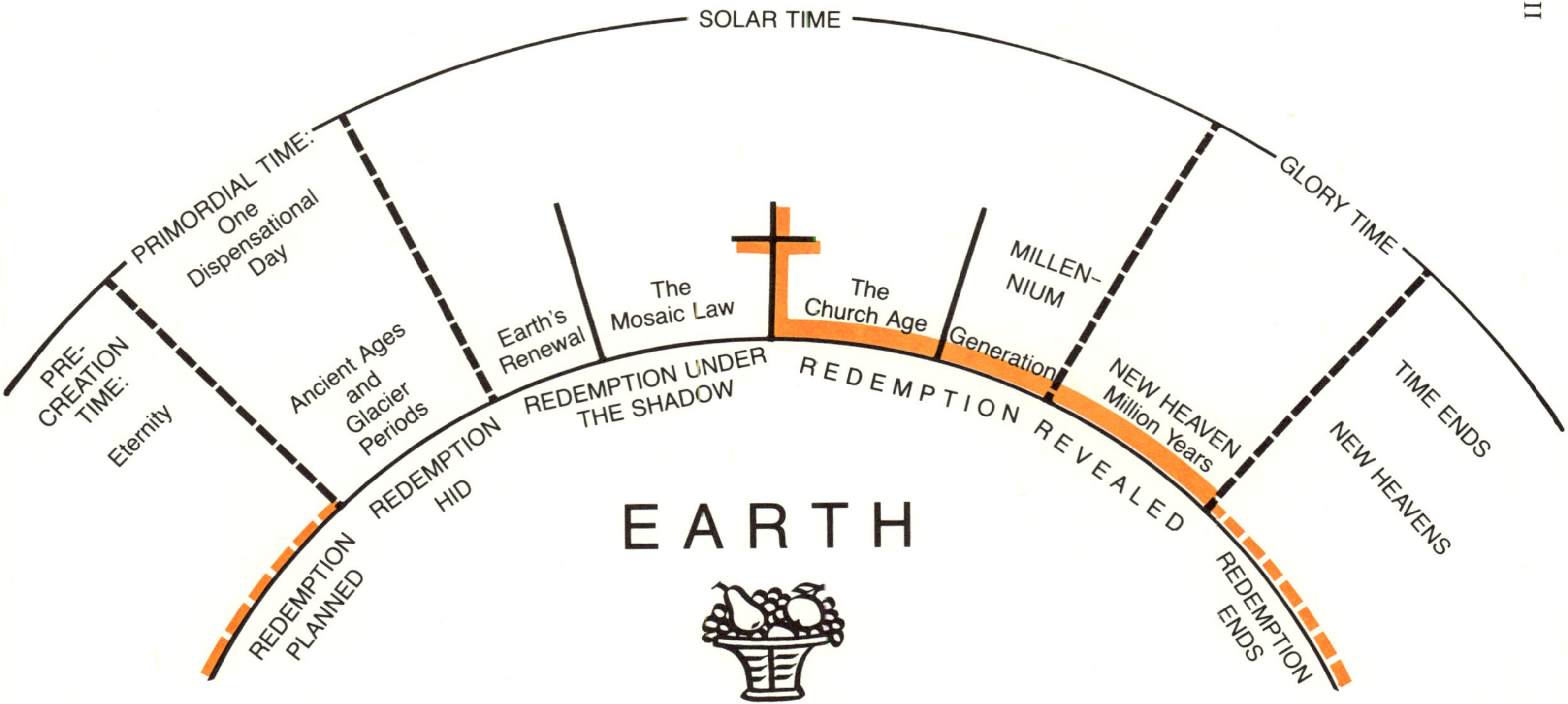

ROYAL MILLENNIAL HIGHWAY SYSTEM

ISAIAH 11:16; 19:23; 35:8; 49:11; OBADIAH 6-21

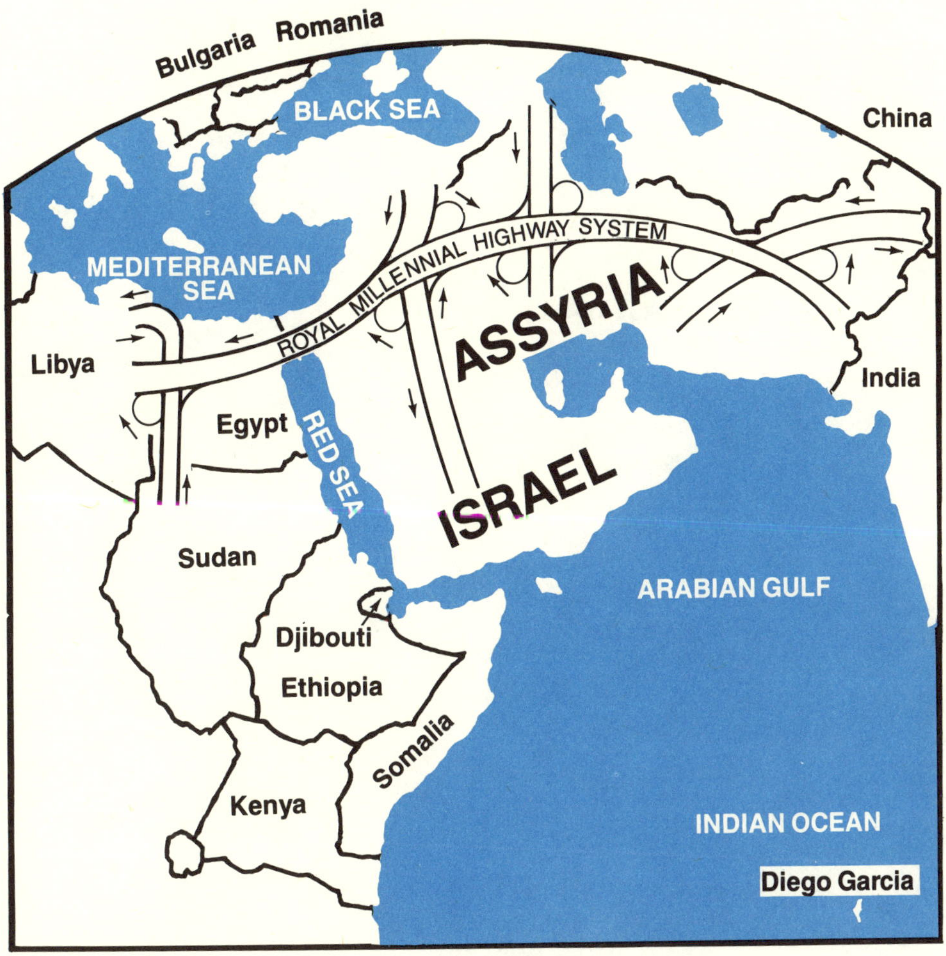

FIRST CONQUESTS OF ANTICHRIST AND THE MIDDLE EAST CONFEDERACY
West Germany
France
Switz
Italy
Southern Confeder

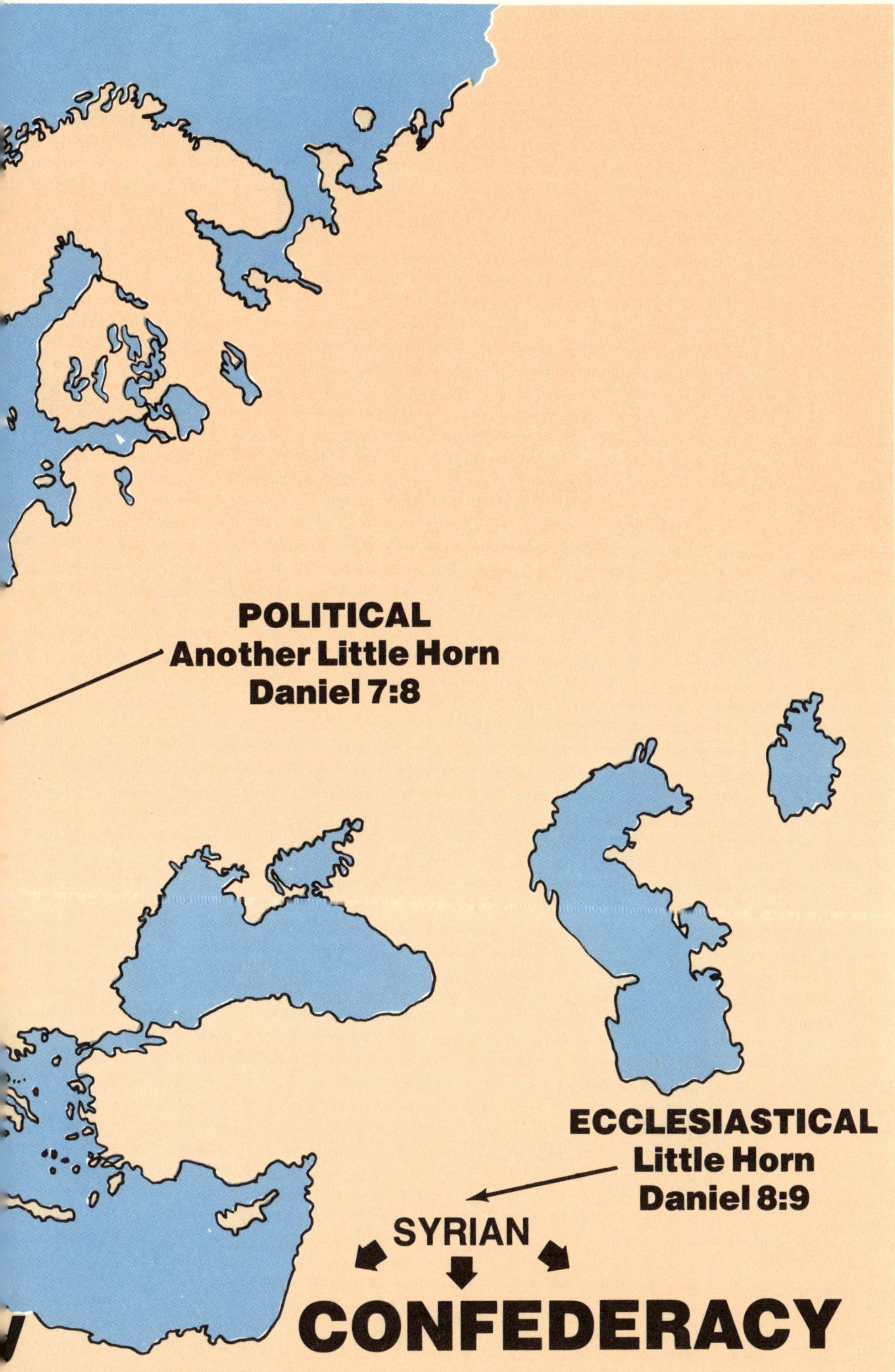
POLITICAL
Another Little Horn
Daniel 7:8
ECCLESIASTICAL
Little Horn
Daniel 8:9
SYRIAN
CONFEDERACY

Last Days
BABYLON
POLITICAL
ROME

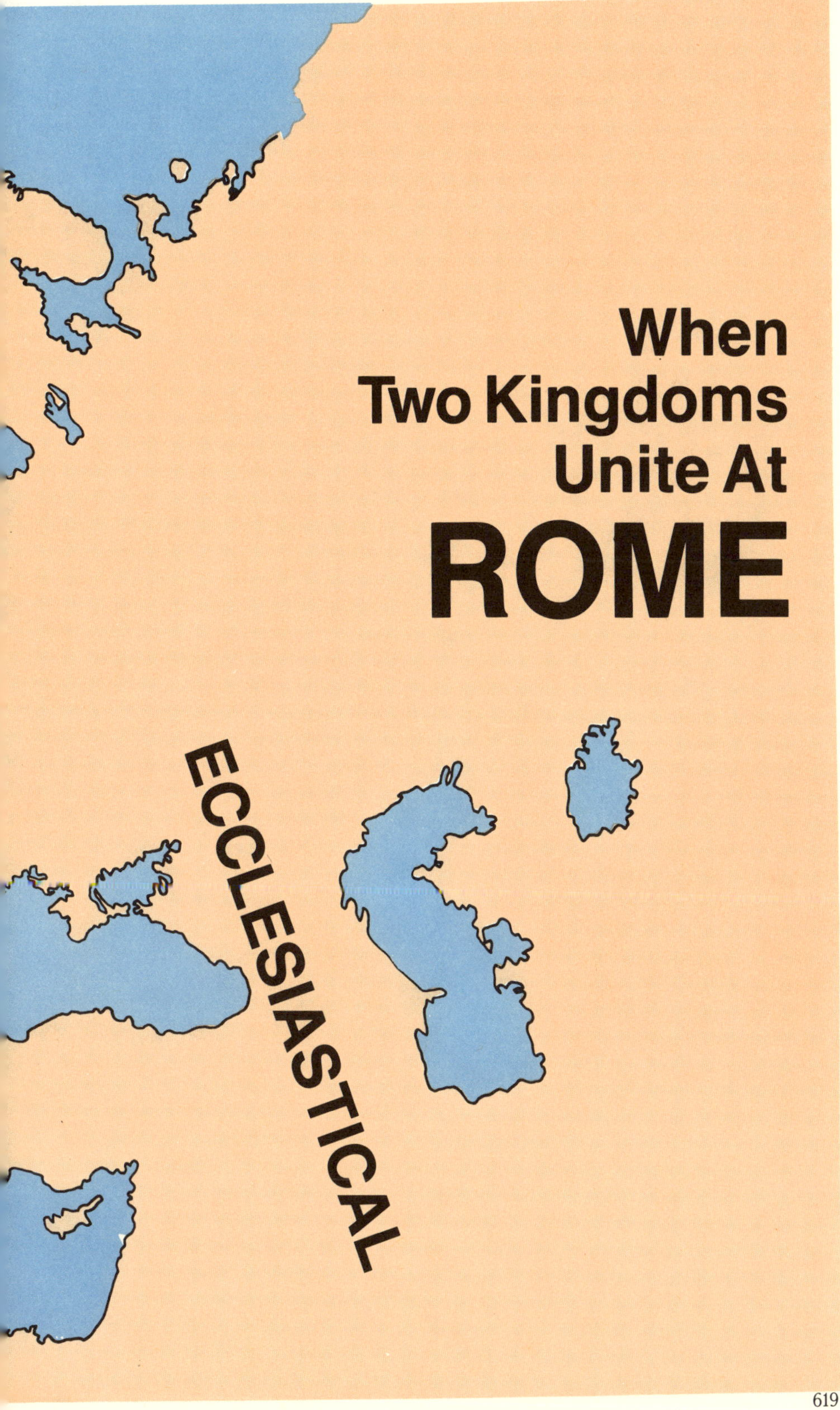
When
Two Kingdoms
Unite At
ROME
ECCLESIASTICAL

TREATY OF ROME
AND THE ANCIENT
ROMAN EMPIRE
IRELAND
ENGLAND
DENMARK
ROMAN
EMPIRE

OUTSIDE THE EMPIRE

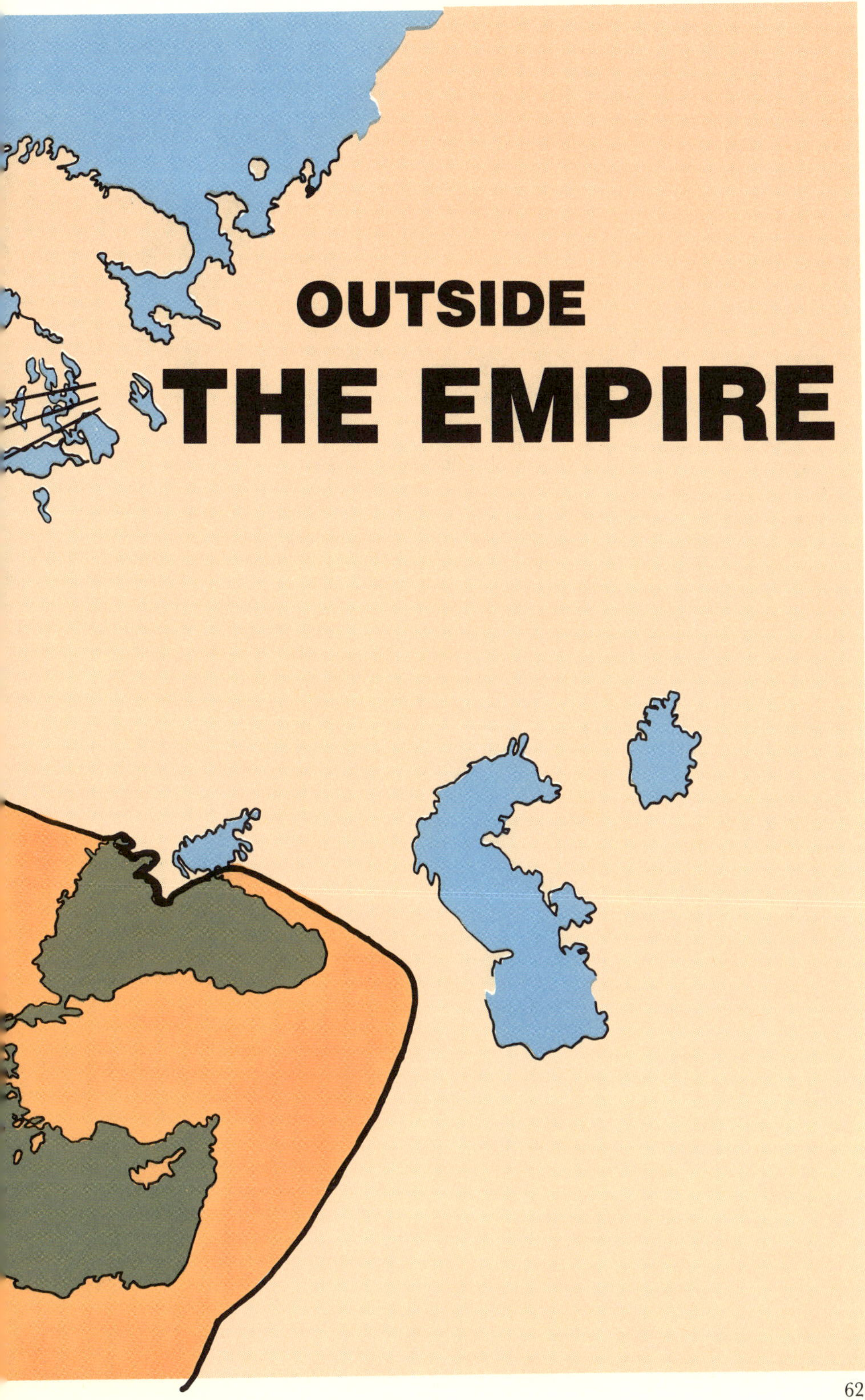

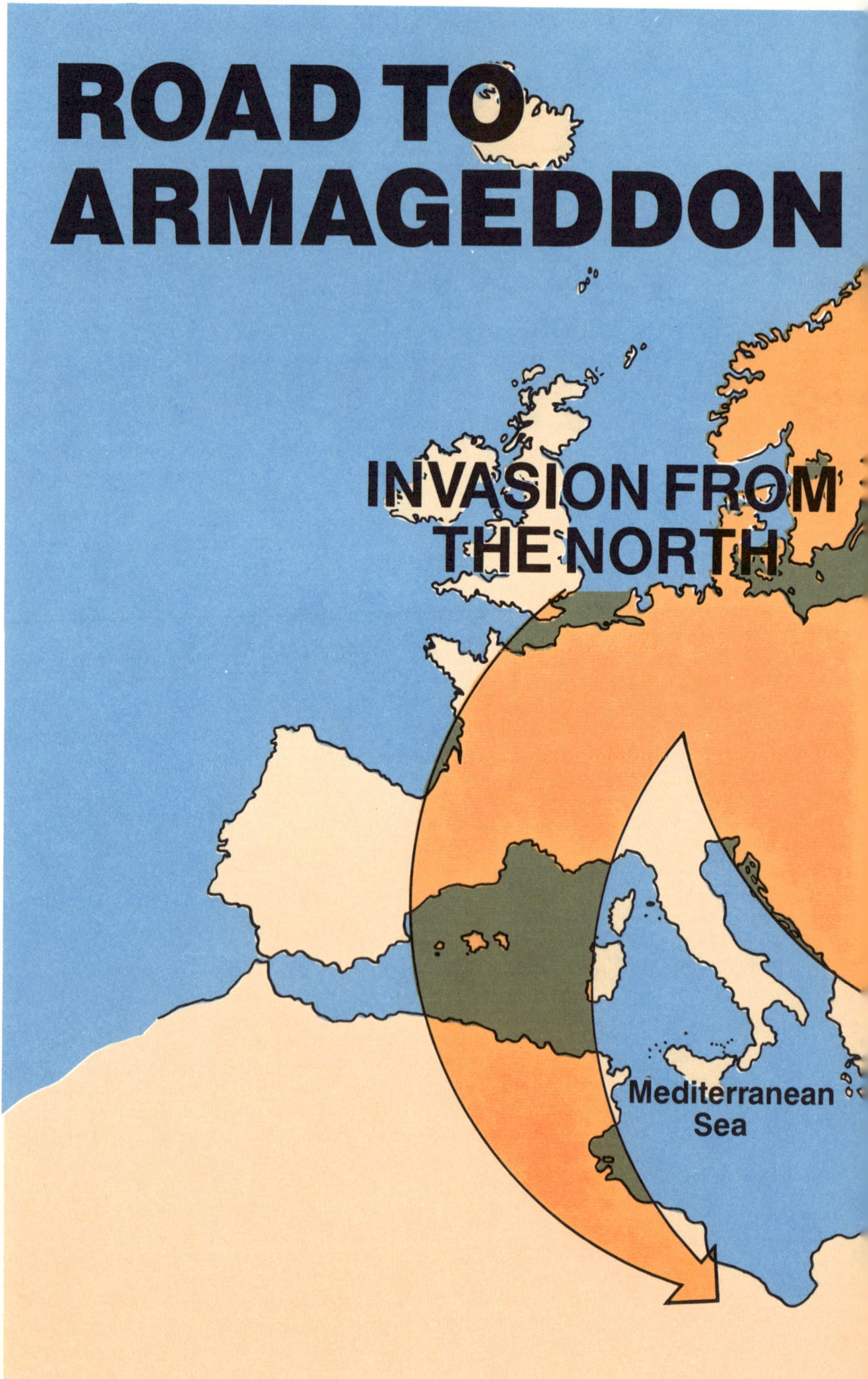
ROAD TO ARMAGEDDON
INVASION FROM THE NORTH
Mediterranean Sea

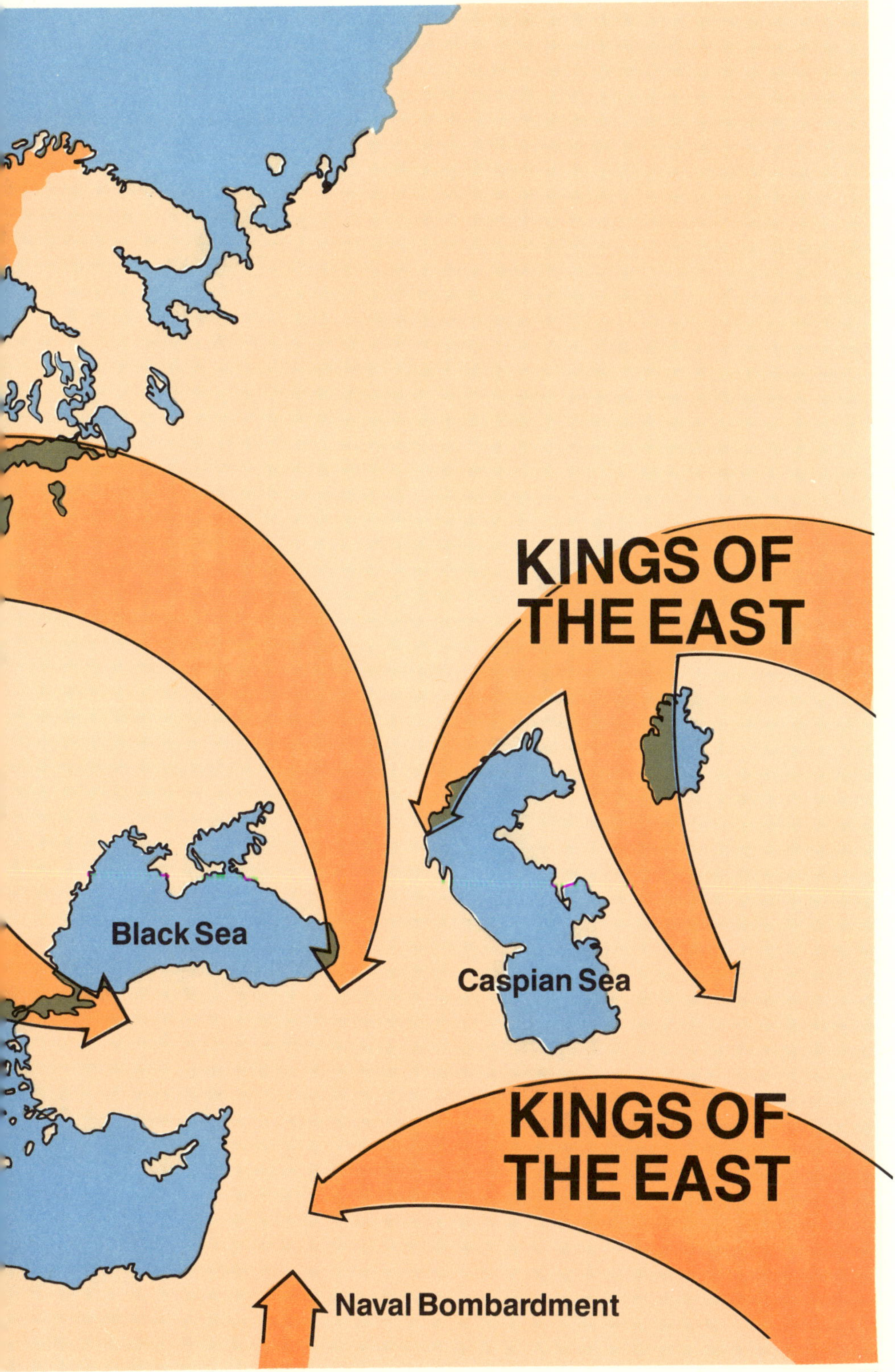
KINGS OF
THE EAST
Black Sea
Caspian Sea
KINGS OF
THE EAST
Naval Bombardment

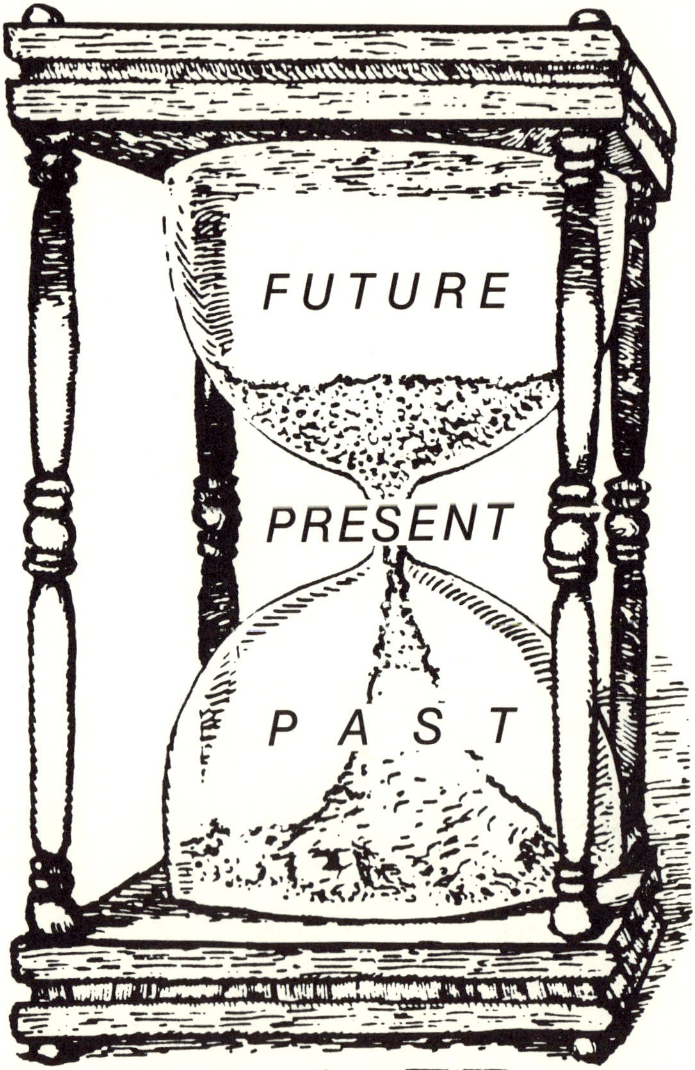
FUTURE
PRESENT
PAST

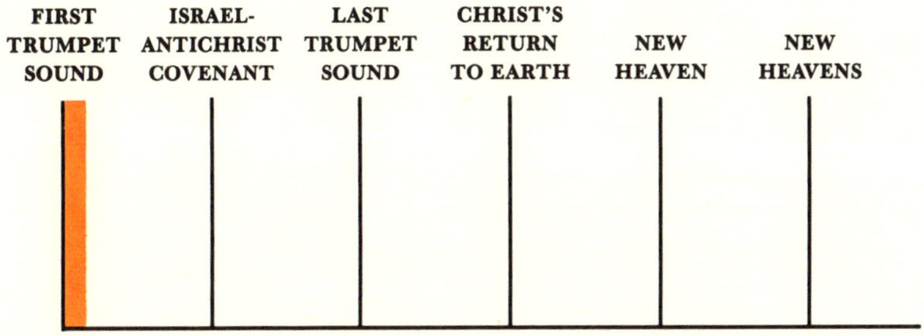

A. DEFEAT OF THE SOVIET UNION'S FORCES

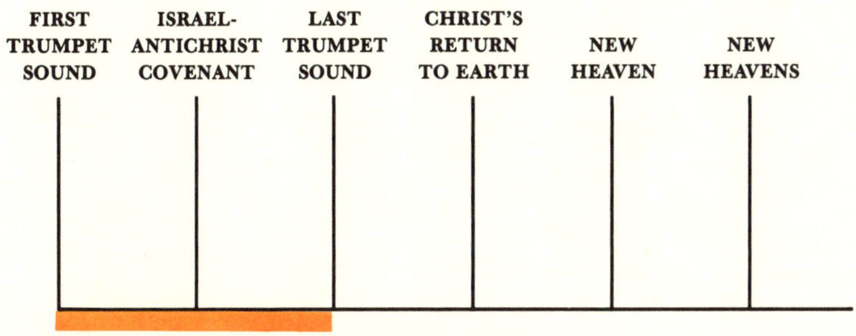

B. ISRAEL'S SEVEN YEARS OF PEACE

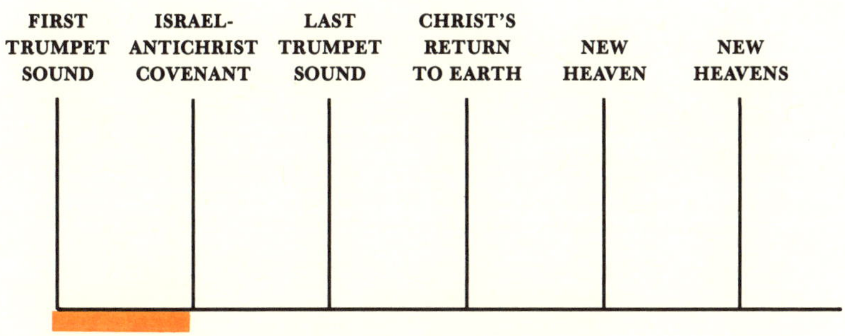

C. RISE OF THE EUROPEAN CONQUEROR

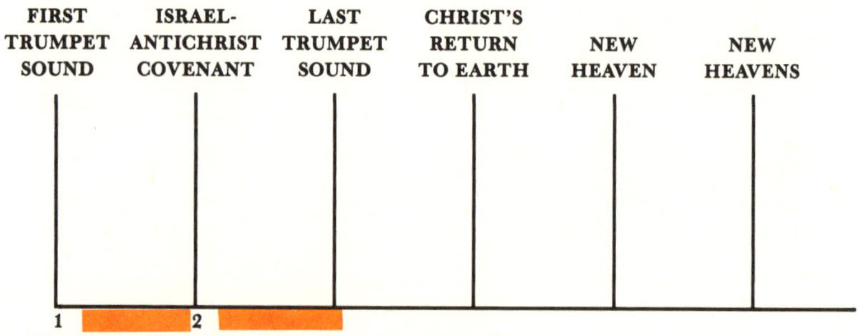

D. TRIBULATION OF THE SAINTS

1. Lesser Tribulation - Revelation 6:9-11
2. Great Tribulation - Revelation 7:9-17

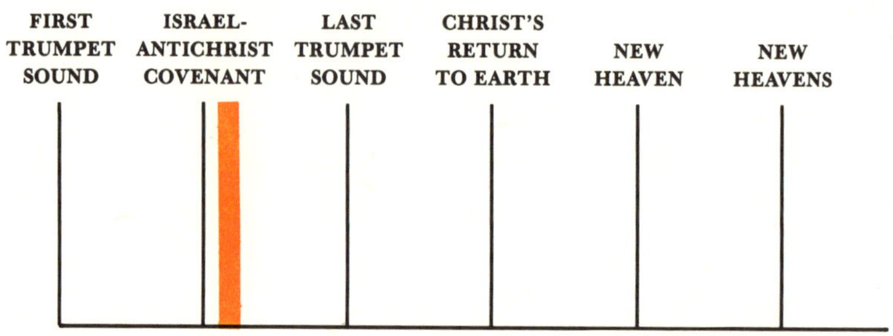

E. DEDICATION OF ISRAEL'S TRIBULATION TEMPLE

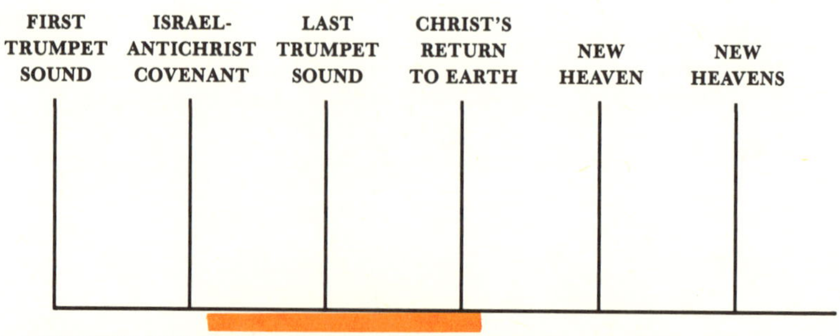

F. THE REIGN OF ANTICHRIST AS THE BEAST

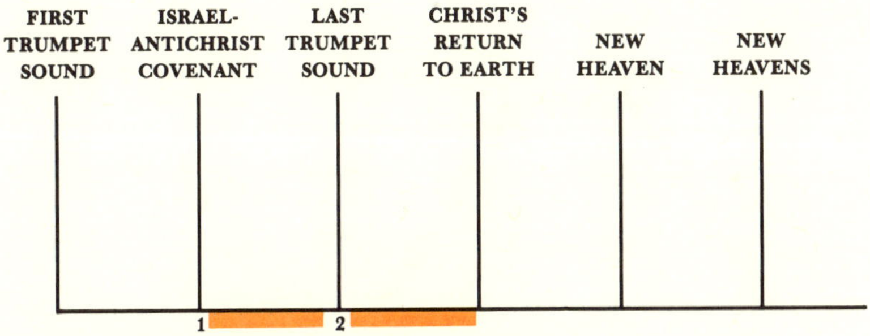

G. THE TWO GREAT TRIBULATIONS

1. Great Tribulation against the Christians
2. Great Tribulation against Israel

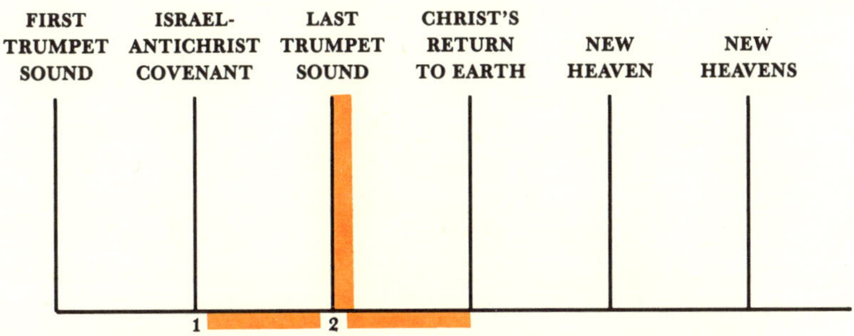

H. THE SEVEN YEAR PEACE COVENANT IS BROKEN

1. When the peace covenant is signed between the Zionists and Antichrist
2. When the peace covenant is broken by the Antichrist

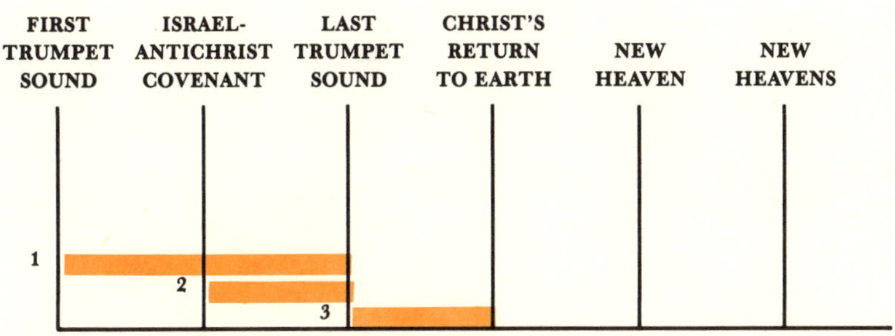

I.

1. Seven seals
2. Seven trumpets
3. Seven vials

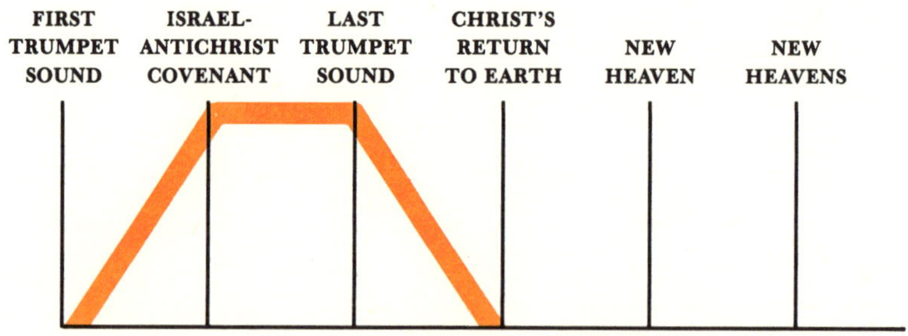

J. RISE AND FALL OF ANTICHRIST AND THE FALSE PROPHET

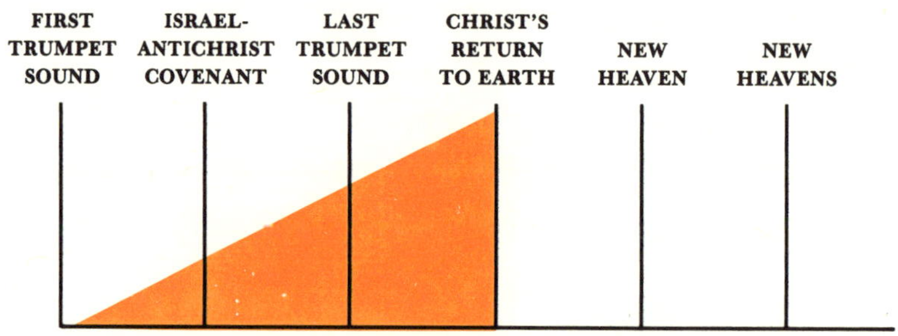

K. HOW THE JUDGMENTS OF THE LORD INCREASE

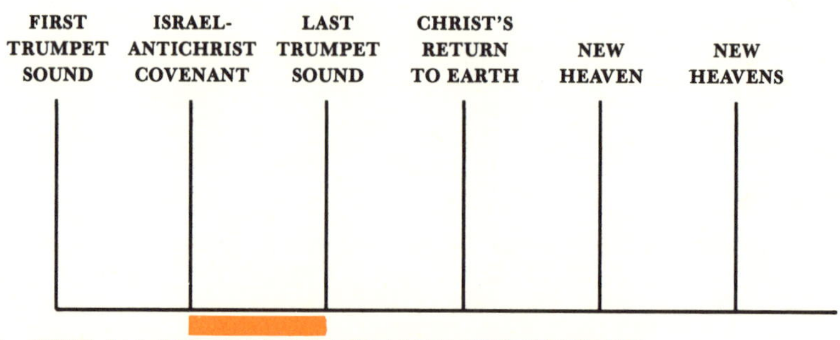

L. THE 144,000 SEALED ISRAELI CHRISTIANS

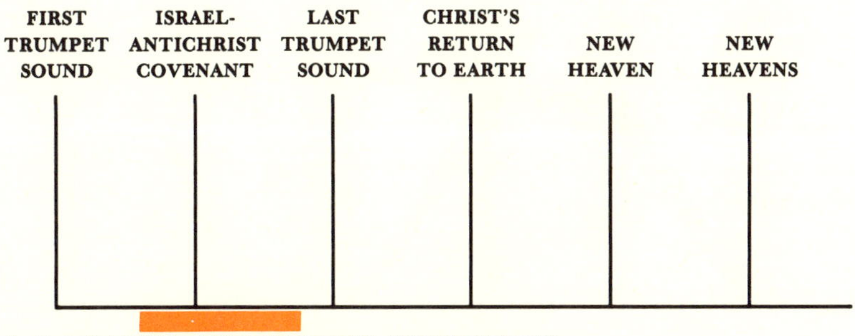

M. LAST TWO WITNESSES (PROPHETS)

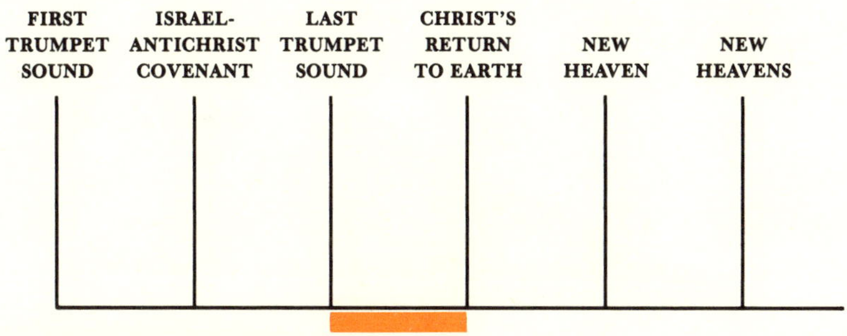

N. THE DAY OF THE LORD

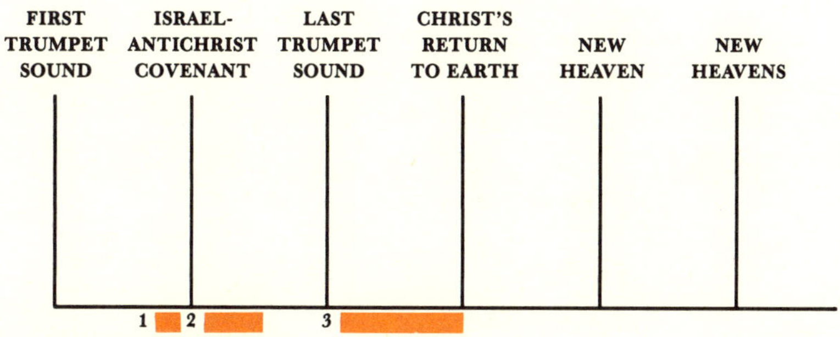

O. THE THREE WOES

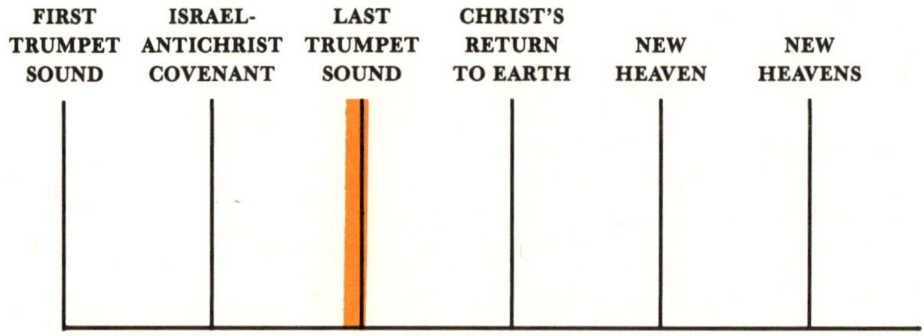

P. THE SEVEN THUNDERS

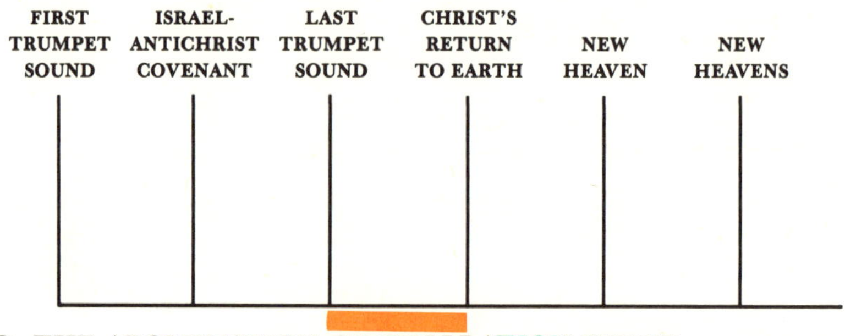

Q. THE ABOMINATION OF DESOLATION SETUP

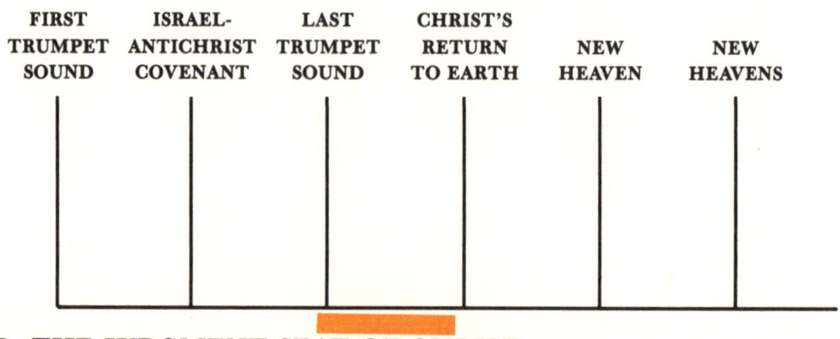

**R. THE JUDGMENT SEAT OF CHRIST —
MARRIAGE OF THE LAMB**

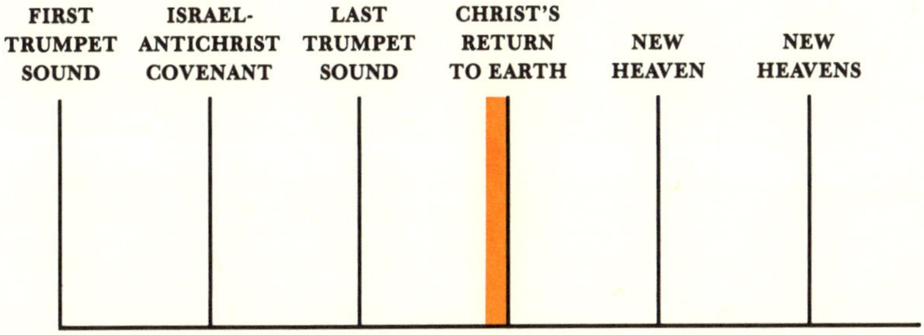

S. THE BATTLE OF ARMAGEDDON

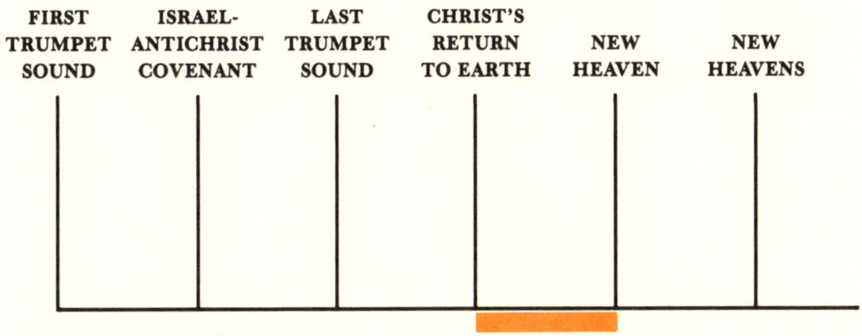

T. THE MILLENNIUM

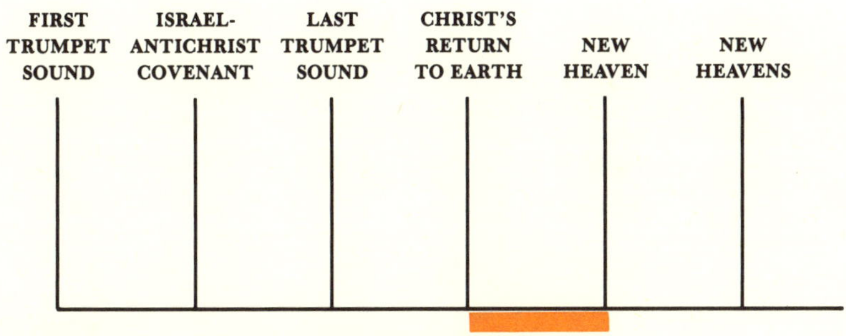

U. PERIOD OF THE OUTER DARKNESS

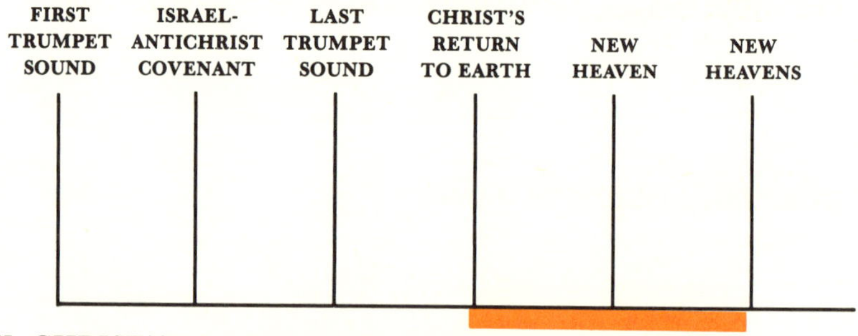

V. CHRIST'S MANIFESTED KINGDOM ON EARTH

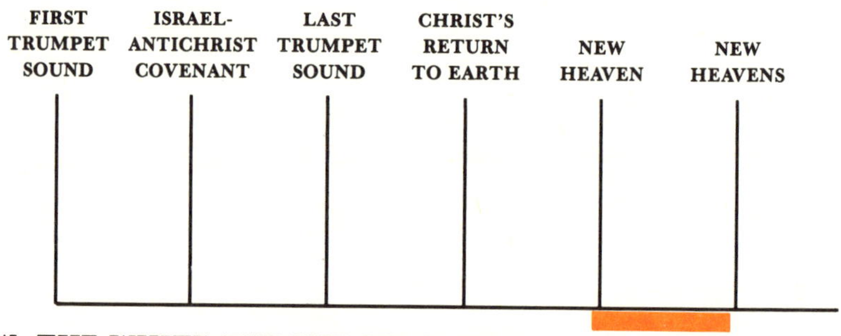

W. THE WHITE THRONE JUDGMENT

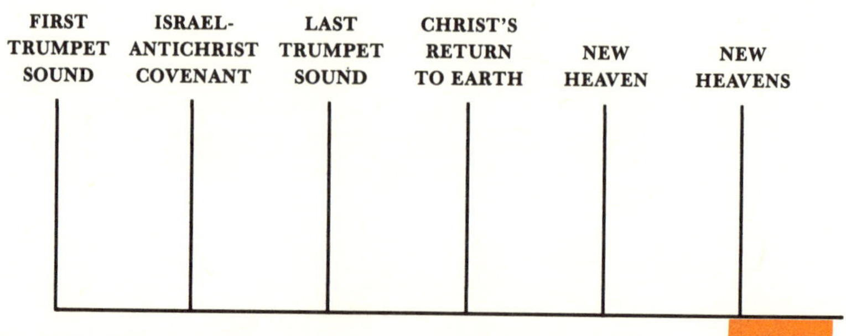

X. MANIFESTED KINGDOM OF THE HEAVENLY FATHER

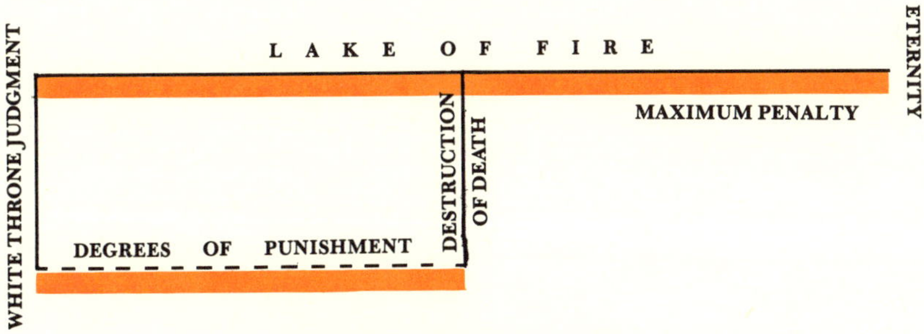

Y. VERDICTS FROM THE HIGHEST ROYAL SUPREME COURT

- **First death** can only destroy the **body**.
- **Second death** can destroy **both body and soul**. Matthew 10:28
- When those being punished have served their penalties under degrees of punishment then second death will end the existence of both body and soul. The "spirit" returns to God at first death (Ecclesiastes 12:7).
- All those still existing under punishment after the last death has been destroyed will never have an egress or redress.
- Fallen angels, as "spirit" beings, will all suffer maximum (eternal) sentences because a "spirit" can never be terminated or destroyed.

"The **last** (second) **enemy** that shall be destroyed is (the **second**) **death**"
I Corinthians 15:26

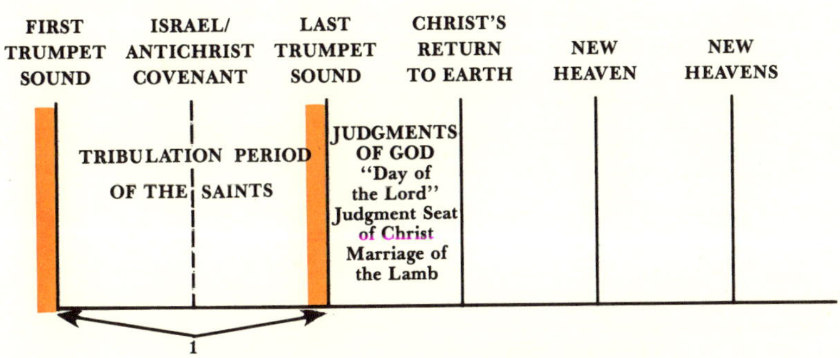

Z. THE TWO RESTRAINING FORCES OF 2 THESSALONIANS 2:6,7 (See page 673)

1. "WHAT" - An **Event**: the Rapture
   "HE" - The **Subject**: the redeemed Church

   NO CONVERSIONS TO CHRIST CAN RESULT WITHOUT THE PRESENCE OF THE HOLY SPIRIT:

   DANIEL 11:32-35; ZECHARIAH 4:6; JOHN 16:7-11; ACTS 2:16-21; ROMANS 8:16,23; I CORINTHIANS 3:16; I JOHN 5:6

   A) THE WRATH OF GOD ("THE DAY OF THE LORD") WILL NOT BE POURED UPON THE EARTH UNTIL **ALL** THE REDEEMED TO BE IN THE "FIRST RESURRECTION" ARE REMOVED FROM THE EARTH, INCLUDING THE LAST TWO PROPHETS AND THE 144,000 (LUKE 21:28-36; I THESSALONIANS 5:9).

   B) **ALL** THESE REDEEMED MUST APPEAR AT THE JUDGMENT SEAT OF CHRIST: 2 CORINTHIANS 5:10, ROMANS 14:10, DANIEL 7:10.

AA. THE UNIFICATION OF TWO GREAT KINGDOMS

REVELATION 17:13,17 — I CORINTHIANS 15:24

ANTICHRIST (POLITICAL KINGDOM) FALSE PROPHET (ECCLESIASTICAL KINGDOM)	BABYLON (UNREDEEMED) - MORTALITY
CHRIST'S KINGDOM (COMPLETION OF REDEMPTIVE PROCESS) HEAVENLY FATHER'S KINGDOM (ETERNAL INHERITANCE)	NEW HEAVENS (REDEEMED) - IMMORTALITY

Index

Alliances
COMECON, 336, 348; European Economic Community (EEC -Common Market), 273, 336, Treaty of Rome 348-351, 357, 365, 371; Kings of the East, 331, 346, 351, 397, 455-458; League of Nations, 365; NATO, 318, 336; United Nations, 365

Bodies, Human
Immortal, 183, 189-190, 208, 405; Paradisiacal, 183-184, 189, 207, 212, 405; Physical, 183-186, 189, 218, 220-221, 242-243, DNA 224

Books
Book of Life, 126, 141-147, 186, 187, 206, 218, 226, 228, 232, 240, 246, 249, 257-258, 329-330, 332, 359, 555, 577-589; Book, The, 231, 541-589; Book of Remembrance 257-258

Children
313-314, 221, 246
Deceased, 189, 233, 246, 258, 281; Unborn, 217-233, 259

Cities (Also see Israel)
Constantinople, 346; Ebro, 455; Hamonah, 445; Jerusalem, 426, 458; Medinah, 339; Rome, 343-344, 350, 357, 448, 451, 455, 458, Treaty of Rome 348-349; Sodom/ Gomorrah, 361; Tarshish, 460; Tyrus, 451-452

Climate
Storms, 494, 499; Vortex Winds, 488, 493-494

Continents
Africa, 347, 377, 419, 422-423, 449, 505; Americas, 347, 377-378, 386, 419, 433; Asia, 319, 368, 422-423, 433, 435, 449, Seven Asian Churches 263-264; Europe, 346-347, 370-

371; Middle East, 319, 327, 355, 363-371

Creations
Doctrine of, 74-75, 89-101, 109-114, 155-159, 161-162, 167-172, 235; Primordial, 90-95, 96-99, 103-108, 155, Dispensational Calendar 155, 167-172, 211, DNA 89, Evolution 79, 84, 89, "Let" 93, 155, 157

Death
84-87, 126, 168-170, 236
First, 13, 34, 185-186, 241-243, 296; Second 85-87, 242-243, 248-249, 589

Dimensions
Earth (Third), 71, 77-78, 179, 236; Earth (Fourth), 10-24, 36-37, 90, 478, 569; Eternity, 25-26, 31, 72-75; Infinities, 235-243, 280-281; Time, 235-236, 282, 286-287, 514-515, 548

Doctrines
Atonement, 165, 265-266, 268, 269, 309, 312; Baptism in the Holy Spirit, 276; Conscience, 185, 218, 226; Dimensions, Seven, 235-243; Faith, 81, 146-147; Foreknowledge, God's, 141-147, 228, 232-233, 253-254, 257-258, 261; Gnosticism, xxxii; Infinities, 235-243; Innocence, 231; Jealousy, 106-108; Perenial Grace, 284, 541-542; Prayers and Baptism for the Dead, 200; Predestination, 258, 260-261; Pride, 106-108; Process Theology, 10-11; Sanctification, 544-545; Spiritism, 109; Temptations, 129; Transubstantiation, 481; Water Baptism, 276

Empires
Assyrian, 345, 469, 470-471; Babylonian (Ancient), 345, 448, 457-458; Babylonian (Prophetic), 341-344, 368; Byzantium, 346; Egyptian, 337, 345; Germany, Imperial, 332-333, 336, 346-348; Grecian, 337, 345; Hapsburg, 346, Hitler, Adolf 342, 346, Caesars 346, Czars 346; Medo-Persian, 336-337, 345; Ottoman, 364; Roman (Ancient), 27, 319, 322, 335-336, 345-349, 357, 377, 409-410, 505; Roman (Reconstituted), 273, 319, 340-351, 355-356

Heavens
Heaven of Heavens (Third), 27, 35-36, Giant Oysters 36, 41, 73, 90, 180, 193, 301, 556; Heavens of Heavens (Angels, habitations), 23-26, 41-42, 73, 125, 547; New Heaven (Second), 4-5, 8, 18-21, 25-29, 74, 156, 228, 249, 314, 502-503, 507, 519, 529, 531, 535, 555-556, 563-564, 568, 595-596, Mansions 26, Unicorn 99; New Heavens (Future), 531, 591-595; New Jerusalem (Holy City), 25-27, 30, 202-208, 478, 535, 552-564, 591; Planetary (First), 4-5, 17-18, 30, 35-55, 57-59, 61-62, 72, 77-84, 135, 275-279, 388, 424, 523, 534, 536, 539, 541-543; Alpha Centauri, 285; Andromeda, 46, 93, 285; Asteroids, 57-58; Black Holes, 45-46, 48, 281-282, 550-551; Comets, 242, 274; Cygnus Constellation, 286; Dry Places, 164, 542-543; Graffiti, 543; Gravity, 40; Helium, 47; Herculean/Globular Galaxies, 285; Hydrogen, 40; Hyperspace, 39, 281; Jupiter, 57; Mars, 57; Meteors, 57; Neutron Stars, 48; Novas, 39, 45; Parsec, 282; Pulsars, 550; Quasars, 72-73, 241; UFOs, 301, 543-544; White Holes, 51-52, 281-282

Incarcerated Places
Bottomless Pit, 100, 123, 135, 137, 175, 330-333, Tartarus 537, 537-540; Hell (in the Earth), 17-18, Hades 17, 37, 192-196, 242; Lake of Fire, 242, Gehenna 538, 538-540; Outer Darkness, 117, 394-395, 536-537, 551, 560

Incarnations
Antediluvians, 152, 196, 522; Antichrist, The (as "The Beast"), 58-59, 74-116, 134, 270, 292, 298, 308, 312, 322, 324, 327, 329, 335-338, 340, 347, 350, 353-357, 359-361, 374-378, 379-381, 388-396, 412, 416-417, 421, 430, 440, 442-443, 446-452, 453, 455, 461-462; Antichrist, The (as "The Conqueror"), 319, 321-322, 327, 341-342, 348-351, 353, 370-371; Christ, Jesus (Birth), 80-81, 110, 219, 264, 271-274, 277, 354, 384, Christ's Portrait 266; False Prophet, The, 340, 389

Incorporeals
Abaddon/Apollyon, 117, 177; Angels, Fallen, 94, 100-

108, 111, 115-120, 121-138, 163-165, 177, 197-200, 331, 388, 462, 542-545; Angels, Holy, 119, 124-125, 131, 163-165, 170-172, 173-178, 197-200, 233, 236, 237, 259, 301, 305, 388-389, 462, 464, 467, 517, 566, 568; Angel of the Lord, The, 67-69; Christ, Jesus (Pre-incarnation), 23-24, 31, 74, 109-111, 143, 165, 173-177, 202, 277; Demons, 93-94, 98-99, 135-136; Fire, 79-80; Gabriel/Michael, 173; God, 41-42, 47-48, 77, 79-83, 106-107, 110, 143, 218-219, 235-236, 238-240, 241, 486-487, Godhead 109-110, 143, 250, 277; Holy Spirit, 47, 49-50, 61-69, 78-83, 91-92, 97, 137, 138, 140, 145, 157, 167, 183, 200, 208, 224-225, 226, 236, 240, 247-248, 255, 276-278, 281, 297, 299, 357, 390, 465, 507-508, 544,; Soul (Animal), 236, 241; Soul (Human), 86, 95-96, 184, 187-188, 191-194, 218, 236-237, 240-242, 284, 296, 544; Spirit (Entity), 96-97, 110-111, 116, 134, 187-188, 218-219, 238-241, 281, 284, 286, 296, 545; Star of the East, 67-69, 272; Visions and Dreams, xxxii-xxxiii, 113-114

Inventions
263, 264, 426-427, 432, 459-460, 522-525, 531, 567, 600-601
Airplanes, 426-427, 459-460; Steamships, 427

Israel
17, 204, 264, 267-270, 272-273, 363-366, 370, 374, 380, 381, 392, 413, 415-417, 420, 436, 446-447, 500-501, 565-566, 594
Ark of the Covenant, 17, 267-269, 376, 389, 416-417, 422, 458, 565-566; Feasts, 511, 597-601; Judah, 375; Peace Covenant (with Antichrist), 366-367, 374, 449-450; Seven Years of Peace, 75, 366-367, 442-443; Tribes: British-Israel Movement, 456; Celts, 455; Dan, 455-458; Hidden (Lost) Tribes, 380; Zionists, 379, 413, 416, 442, 447, 458, Camp David 365, 422, Israel Declares Independence 365

Judgments
Flood, The, 3, 99, 151-154, 283, 332, 361, 484-494, 522, 534, Polar Ice, 153-154; Incorrigibles, 240, 241, 247; Jerusalem, 4, 123-124, 267-270, 410; Nations, 258, 389-393, 397, 417, 434, 458, 466, 472-

473, 501, 517; Nephilims, 196; Seat of Christ, 195, 221, 241, 248, 258-259, 290, 299, 305, 310, 379, 387, 394-395, 401-402, 413, 502-503, Hypoglycemia 185, Tears 537; Universe, 53-55; White Throne Judgment (see Resurrections)

Justice of God
220-221, 222, 242, 258-259, 577-589
Injustice, 7-8, 217-222; Justification, 7-8, 223; Unredeemed, 9, 242, 314

Kingdoms
Antichrist's, The (Political), 337-343, 345, 346, 350, 355-356, 370-371, 374-375, 379-381, 387-396, 398-399, 416-417, 422, 447-452, 454-455, 534; False Prophet's, The (Ecclesiastical), 325, 328, 337-339, 350, 363, 369-370; Heavenly Father's, 13, 19-20, 33-34, 148-149, 246, 411, 530, 552, 561-566, 568-569, 591-595; Millennial, 4-5, 18-21, 31-32, 33, 35, 113-114, 127-128, 147, 148-149, 153-154, 155-156, 159, 175, 206, 208, 212, 247-249, 290, 297-298, 312, 314-315, 325, 329, 332, 338, 360, 370, 381, 386, 392-393, 397, 411, 413, 416, 427, 434, 436-437, 445, 454, 458 (Also see Pangaea); Mystery, 148, 221-222, 232, 242-243, 245-250, 254, 264, 296, 298, 381, 383-386, 387-391, 402, 406, 411, 413, 437, 467, 551-552

Maps
324, 341, 357, 458

Music
170-172, 194, 206, 564-565

Nations (Also see Alliances)
Albania, 336, 346; Algeria, 346, Arabia, 64, 377; Armenia, 365; Assyria, 269, 370, 454, 468; Austria, 335-336, 341, 350-351; Belgium, 345, 456; Bulgaria, 336, 346, 455; Chaldea, 324; China, 346; Cuba, 336; Czechoslovakia, 336; Denmark, 349, 357, 456; Edom, 337-338; Egypt, 260, 269-270, 319, 369, 376, 380, 389, 417, 421-422, 454, 468, 470-471, 512, Five Cities of Refuge for Israel 270, 421-422; England, 377, 426, 455-456; Ethiopia, 377, 422, 425, 427; France, 345, 456; Germany, 336, 345, 371, 456; Greece, 346; Greenland, 488;

Hungary, 336, 346, 456; Ireland, 357, 456-457; Italy, 345, 456; Jordan, 368-369, 455; Lebanon, 368-370; Libya, 346, 365, 377, 451; Luxembourg, 345; Macedonia, 433; Media, 336-337; Mongolia, 336; Morocco, 346, 422; Netherlands, 345; Persia (Iran), 451; Poland, 336, 456; Portugal, 345; Rumania, 336, 346, 456; Scotland, 455-456; Soviet Union, 304, 318-319, 327-329, 339, 346-347, 354-355, 365-367, 369, 451; Spain, 345, 348, 456; Sweden, 456; Switzerland, 345, 349-351, 456; Syria, 319; Turkey, 346, 455; United States, 304, 318, 327, 329, 346-347, 355, 365, 386, 419-424, 429-430, 434-436, 439-443, American Indians 430, Metes and Bounds Act 431-432; Vietnam, 336; Yugoslavia, 346, 456

Nuclear Energy
Atom's Composition, 36-39, 42, 45-47, 52-55, 58-59, 78-79, 87, 90, 92-93, 98, 138, 156-157, 170-172, 192, 235-236, 239, 241, 275, 277, 304, 343; Alamogordo, 53; ARM, 87; Antimatter, 54; Colors, 84; Dust Bombs, Radioactive, 321-323; Electromagnetism, 83, 278; Electromagnetic Pulse (EMP), 326-327; Electrons, 88, 279; Fissioning, 54; Gluons, 36; Gravitons, 52; Isotope, 87; Lightwave lengths, 84; Mesons, 87; Molecules, 58-59, 78, 209; Nuclei, 551; Nuclear Forces, 267-268; Photons, 78-79, 86, 238, 281; Positrons, 50; Quantum Mechanics, 279-280; Quarks, 36, 38; Rays, 77; Rest Particles, 78-79; Ultra-matter, 84; Uranium, 53; Wastes, 491; Z-Zero Particle, 38

Oil
346-347

Pangaea
Ancient, 419, 575; Millennium, 20-21, 460, 482, 505, 528-529; New Earth, 20-21, 471, 473, 564, 569-570, Salt Pits, 20

Paradises
Primordial, 179-181, Occupants 180-181, God's Throne 180, A Planet 181; On the Earth, 183-190, Paradisiacal Bodies 183-184, Second Death 187, DNA 185, Cursed

Ground 185, Leaves From the Tree of Life 185; In the Earth, 191-196, Awakening of Souls 195-198, Place of Rest 186, Earth's Wobble 191, State of Souls 192-194, Preached To 189; In Transition, 197-200, Little Children 188; In Heaven, 201-215, 237, 246, Beauty of Occupants 214, Chariots 188, Epidermis 208, Frame of Mind 202, Garden of God 201-202, Horses 207, Incorruption 184, Landscape 209, Language 206-208, Lighting 206, Sex 210-211, Tangibility, 209, Tranquility 212-213, Tree of Knowledge 189, Wardrobes 202-204

Personalities

Biblical: Adam/Eve, 21, 50, 134, 157, 161-162, 185, 189, 209, 210, 214, 479, 522; Abraham/ Sarah, 14, 147-148, 176, 380, 384; Daniel, 3, 254; Elijah, 185, 199, 534; Enoch, 181, 199, 534; Esther, 433; Herod, 272-273; Isaac, 384; Iscariot, Judas, 265; Job, 3; John the Baptist, 219, 268, 409; John the Revelator, 309; Joshua, 269; Lot, 3; Mary, 198; Moses, 15, 79, 197, 269, 415, 424, 878; Nebuchadnezzar, 348; Nehemia, 265; Noah, 3, 152; Peleg, 419; Pharaoh, 259-260; Philip, 534, 558

Non-Biblical: Alexander the Great, 337; Allenby, Lord, 364; Bah'u'llah, 366; Balfour, L., 364; Bethe, H., 322; Blomfield, A., 254; Bohr, Niel, xxx; Brown, H., 322; Eusebius, 268; Carlos, King Juan, 348; Czar, 346; Dalton, xxix; Darwin, Charles, 98, 133-134; Edison, Thomas, 522; Einstein, Albert, xxix-xxx; xxxiii, xxxv, 8, 38, 46, 47, 72, 255, 278-280, 516, 534; Eusebus, 410; Hipparchus, 424; Josephus, 4, 268, 409; Laurent, xxix; Lavoister, xxi; Louth, Bishop, 421; Luther, Martin, xxix; Otto the Great, 336; Parker, D., 322; Planck, xxix; Rutherford, xxxix; Seitz, 322; Sziland, L., 322; Titus, Emperor, 4, 268; Truman, Harry, 421; Wagner, A., xxix;, Wesley, J., xxix; Wilhelm II, Kaiser, 347, 371

Prophets

Pseudo, xxxvi-xxxvii, 207-208, 290-291, 318-320, 325, 373, 406-414, 450-451, 506;

True, 412-413, 415-417, 469, 486

Millennium (Also see Kingdoms)
Acid Rains, 489; Alcohol, 479; Arbor Days, 488-489; Biochips, 529; Cancer, 495; Computers, 525; Desserts, 482; Drugs, 478-479; Dew, 491; Earthquakes, 482-483; Electronics, 525; Gun Control, 480-481; Harvests, 491; Herbs, 481; Insects, 490; Inventions, 521, 523-524; Microchips, 523-525; Polar Regions, 485-486; Sabbatical Years, 482-484; Scicntists, 522-523; Technology, 522; Tobacco, 479; Water Canopy, 482, 484-486; Weather, 484-495; Volcanoes, 486, 493,; Wildlife, 480; Wine, 481-482, 485

Redemption
126, 147-149, 165, 169-170, 239, 241-242, 248-249, 263, 266, 309-310, 555-556, 561-571, 573-576
Body, Of the, 13-14, 82-83, 185, 189-190, 237-238, 242-243, 276, 278, 301-305, 313, 317, 397, 507; Earth, Of the, 4, 7-8, 13-21, 48, 65, 74, 98, 137-140, Pangaea 460, 482, 505; Jerusalem, Of, 14-15, 273; National Israel, Of, 15-16, Old Testament Sacrifices 266; Soul, Of the, 7-11, 63, 119-120, 143-144, 199-200, 221-222, 232, 544-545

Redeemed Persons
Immortality, 146, 190, 198, 237-238, 260, 276, 280-281, 284, 291, 296; Religious Sects, xxx, 314-315

Resurrections
Christ's, 63, 75, 110-111, 307; First, 10, 283, 290/ 313, 385, 394, 397-398, 569, DNA 224; Second, 249, 258, 577-589, Infanticide/Euthanasia 233

Return of Christ
Pre-Tribulation Rapture (To Midair), 174-175, 185, 192, 237, 264, 268, 279, 290, 291-292, 295-300, 301-305, 307-315, 325, 332, 339-340, 355, 357, 373, 391, 403, 439, 442, 445-446, 449-450, 565, Royal Crowns 295, Twenty-Four Elders 298, 312, Four Living Creatures 298, 312; Tribulation Rapture (To Midair), 174-175, 332, 385, 387-396; Revelation of Christ

(To the Earth), 268, 270, 273, 308, 398-399, 461-475; Day of God, 534; Day of the Lord, 534; Sword of the Lord, 461, 468, 472, 474

River of Life
198

Sites
Aachen, 348; Babel, 420, 522; Bering Sea, 355; Caspian Sea, 455; Constantinople, 346; Danube River, 455; Dardenelles, 455; Dedan, 377; Ebro River, 455; Gaza Strip, 364,; Geneva, 350; Hebrides, 455; Hiroshima, 326, Istanbul, 346; Kidron, 473; Lucerne, 350; Medina, 339; Megiddon, 445, 457; Rome, 342, 343, 357, 455; Sheba, 380; Tyrus, 343; Vienna, 335-336; 341; Zurich, 650

Wars
Armageddon, 122, 128, 130, 177, 270, 304, 331, 343, 346, 351, 356, 376, 393, 397, 414, 445, 456, 458-459, 462-463, 472, 504, 519, 521; Civil War (USA), 431; European: World War I, 371, 414, 420-421, 427; Middle East: Israel's Six Day War 435, 457-458, Soviet Union (Prophetic) 304, 355, 374, 376-377, 414, 439-442, 445-446, 447, 451, 457; Nuclear, 121, 304-305, 321-323, 326-337, 343, 439-440, 465, 523, 548-551, Dust Bomb Weapons 321-322, Particle Beam Weapon 464, Super-Radiation Bombs 323; Post-Millennial, 414, 471-472, 480-481; Revolutionary (USA), 431

Word of God
Bible Study, 253, 271, 544,; Prophecies, 253-266; Nescience, xxxii-xxxiii; Written, xxxii, 64, 112-113, 117-118, 207-208, 253-254, 274, 544; Greek Words: Logos xxxiii, Rhema xxxiii, Hades 17, Kosmos 146, Brephos 219, Ekklestia 246, Tachy 275, Keleuo 301, Atomos 301, Ekpheugo 303, Apoluttrosis 305, Parousia 473, Pharmakeia 479; Hebrew Words: Sheol 17, Nephilim 196, 283, Howy 426, Gama 427, Halah 427, Baza 429, Mashak 430, Mbulucah 431, Nephesh 196, 241, Eber 426; Latin: Rapere 303